Where Mathematics Comes From

Where Mathematics Comes From

HOW THE EMBODIED MIND BRINGS MATHEMATICS INTO BEING

George Lakoff
Rafael E. Núñez

BASIC
BOOKS

A Member of the Perseus Books Group

All figures were drawn by Rafael E. Núñez, with the exception of the following.

Figure 1.1 on page 17 is reprinted by permission from *Nature* (K. Wynn, "Addition and Subtraction by Human Infants," Vol. 358: 749–750), copyright 1992, Macmillan Magazines Ltd.

The graphic in Figure 1.2 on page 20 is adapted from G. Mandler and B. J. Shebo, "Subitizing: An Analysis of Its Component Processes," Journal of Experimental Psychology: General, 111 (1): 1–22.

Figure 1.3 on page 25 is modified from *Atlas of Human Anatomy* by Frank Netter (plate 99, "Meninges and Brain"), copyright 1989 by Havas MediMedia.

Drawings in Figure 2.1a on page 32 and Figure 2.3 on page 40 are credited to Ginger Beringer.

Figure 2.4 on page 45, drawn by Rafael E. Núñez, is adapted from *Taschenbuch der Mathematik* by I. N. Bronstein and K. A. Semendjajew, 25th edition, copyright 1991 by B. G. Teubner Verlagsgesellschaft, Stuttgart.

Figures 12.2a and 12.2b on page 269 are from *Elements of the Topology of Plane Sets of Points* by M. H. A. Newman, copyright 1992 by Dover Publications Inc. Reprinted by permission.

Figure 12.2c on page 269 is reprinted by permission of Open Court Publishing Company, a division of Carus Publishing, Peru, IL, from *Foundations of Geometry* by David Hilbert, copyright 1971 by Open Court Publishing Company.

Figure 12.2d page 269 is from *What Is Mathematics? An Elementary Approach to Ideas and Methods*, second edition by Richard Courant and Herbert Robbins, edited by Ian Stewart, copyright 1941 by the late Richard Courant, Herbert Robbins and revised by Ian Stewart. Used by permission of Oxford University Press, Inc.

Figure 12.2e on page 269 is from *Taschenbuch der Mathematik* by I. N. Bronstein and K. A. Semendjajew, 25th edition, copyright 1991 by B. G. Teubner Verlagsgesellschaft, Stuttgart. Reprinted with permission.

Figure 12.4 on page 285 is modified from *To Infinity and Beyond: A Cultural History of the Infinite* by Eli Maor, copyright 1987 by Birkhäuser Boston, Inc., with permission.

Figure CS 4.5 on page 443 is a picture of Rafael E. Núñez's father's slide rule. (Thanks, Dad!)

Copyright © 2000 by George Lakoff and Rafael E. Núñez

Published by Basic Books,
A Member of the Perseus Books Group

Designed by Rachel Hegarty

Library of Congress cataloging-in-publication data is available.
ISBN 0-465-03771-2

01 02 03 / 10 9 8 7 6 5 4 3 2 1

To Rafael's parents,
César Núñez and Eliana Errázuriz,

to George's wife,
Kathleen Frumkin,

and to the memory of two dear friends
James D. McCawley
and
Francisco J. Varela

Contents

Acknowledgments   ix
Preface   xi

Introduction: Why Cognitive Science Matters to Mathematics   1

Part I
THE EMBODIMENT OF BASIC ARITHMETIC

1 The Brain's Innate Arithmetic   15
2 A Brief Introduction to the Cognitive Science of the Embodied Mind 27
3 Embodied Arithmetic: The Grounding Metaphors   50
4 Where Do the Laws of Arithmetic Come From?   77

Part II
ALGEBRA, LOGIC, AND SETS

5 Essence and Algebra   107
6 Boole's Metaphor: Classes and Symbolic Logic   121
7 Sets and Hypersets   140

Part III
THE EMBODIMENT OF INFINITY

8 The Basic Metaphor of Infinity   155
9 Real Numbers and Limits   181

10 Transfinite Numbers   208
11 Infinitesimals   223

Part IV

BANNING SPACE AND MOTION: THE DISCRETIZATION PROGRAM THAT SHAPED MODERN MATHEMATICS

12 Points and the Continuum   259
13 Continuity for Numbers: The Triumph of Dedekind's Metaphors   292
14 Calculus Without Space or Motion: Weierstrass's
    Metaphorical Masterpiece   306

Le trou normand:
A CLASSIC PARADOX OF INFINITY

325

Part V

IMPLICATIONS FOR THE PHILOSOPHY OF MATHEMATICS

15 The Theory of Embodied Mathematics   337
16 The Philosophy of Embodied Mathematics   364

Part VI

$e^{\pi i} + 1 = 0$

A CASE STUDY OF THE COGNITIVE STRUCTURE OF CLASSICAL MATHEMATICS

Case Study 1. Analytic Geometry and Trigonometry   383
Case Study 2. What Is e?   399
Case Study 3. What Is i?   420
Case Study 4. $e^{\pi i} + 1 = 0$—How the Fundamental Ideas of
    Classical Mathematics Fit Together   433

References   453
Index   473

Acknowledgments

T HIS BOOK WOULD NOT HAVE BEEN POSSIBLE without a lot of help, especially from mathematicians and mathematics students, mostly at the University of California at Berkeley.

Our most immediate debt is to Reuben Hersh, who has long been one of our heroes and who has supported this endeavor since its inception.

We have been helped enormously by conversations with mathematicians, especially Ferdinando Arzarello, Elwyn Berlekamp, George Bergman, Felix Browder, Martin Davis, Joseph Goguen, Herbert Jaeger, William Lawvere, Leslie Lamport, Lisa Lippincott, Giuseppe Longo, Robert Osserman, Jean Petitot, David Steinsalz, William Thurston, and Alan Weinstein. Other scholars concerned with mathematical cognition and education have made crucial contributions to our understanding, especially Mariolina Bartolini Bussi, Paolo Boero, Janete Bolite Frant, Stanislas Dehaene, Jean-Louis Dessalles, Laurie Edwards, Lyn English, Gilles Fauconnier, Jerry Feldman, Deborah Forster, Cathy Kessel, Jean Lassègue, Paolo Mancosu, Maria Alessandra Mariotti, João Filipe Matos, Srini Narayanan, Bernard Plancherel, Jean Retschitzki, Adrian Robert, and Mark Turner. In spring 1997 we gave a graduate seminar at Berkeley, called "The Metaphorical Structure of Mathematics: Toward a Cognitive Science of Mathematical Ideas," as a way of approaching the task of writing this book. We are grateful to the students and faculty in that seminar, especially Mariya Brodsky, Mireille Broucke, Karen Edwards, Ben Hansen, Ilana Horn, Michael Kleber, Manya Raman, and Lisa Webber. We would also like to thank the students in George Lakoff's course on The Mind and Mathematics in spring 2000, expecially Lydia Chen, Ralph Crowder, Anton Dochterman, Markku Hannula, Jeffrey Heer, Colin Holbrook, Amit Khetan, Richard Mapplebeck-Palmer, Matthew Rodriguez, Eric Scheel, Aaron Siegel, Rune Skoe, Chris Wilson, Grace Wang, and Jonathan Wong. Brian Denny, Scott Nailor, Shweta Narayan, and

Katherine Talbert helped greatly with their detailed comments on a preliminary version of the manuscript.

A number of readers have pointed out errors in the first printing, some typographical, some affecting content. We extend our thanks for this assistance to Joseph Auslander, Bonnie Gold, Alan Jennings, Hector Lomeli, Michael Reeken, and Norton Starr.

We are not under the illusion that we can thank our partners, Kathleen Frumkin and Elizabeth Beringer, sufficiently for their immense support in this enterprise, but we intend to work at it.

The great linguist and lover of mathematics James D. McCawley, of the University of Chicago, died unexpectedly before we could get him the promised draft of this book. As someone who gloried in seeing dogma overturned and who came to intellectual maturity having been taught that mathematics provided foundations for linguistics, he would have delighted in the irony of seeing arguments for the reverse.

Just before the publication of this paperback edition, our friend the brilliant neuroscientist Francisco J. Varela, of the Laboratory of Cognitive Neurosciences and Brain Imaging in Paris, died after a long illness. His support had been fundamental to our project, and we had looked forward to many more years of fruitful dialog, which his untimely death has brought to an abrupt end. His deep and pioneering understanding of the embodied mind will continue to inspire us, as it has for the past two decades.

It has been our great good fortune to be able to do this research in the magnificent, open intellectual community of the University of California at Berkeley. We would especially like to thank the Institute of Cognitive Studies and the International Computer Science Institute for their support and hospitality over many years.

Rafael Núñez would like to thank the Swiss National Science Foundation for fellowship support that made the initial years of this research possible. George Lakoff offers thanks to the university's Committee on Research for grants that helped in the preparation and editing of the manuscript.

Finally, we can think of few better places in the world to brainstorm and find sustenance than O Chame, Café Fanny, Bistro Odyssia, Café Nefeli, The Musical Offering, Café Strada, and Le Bateau Ivre.

Preface

W̲E̲ ̲A̲R̲E̲ ̲C̲O̲G̲N̲I̲T̲I̲V̲E̲ ̲S̲C̲I̲E̲N̲T̲I̲S̲T̲S̲—a linguist and a psychologist—each with a long-standing passion for the beautiful ideas of mathematics. As specialists within a field that studies the nature and structure of ideas, we realized that despite the remarkable advances in cognitive science and a long tradition in philosophy and history, there was still no discipline of *mathematical idea analysis* from a cognitive perspective—no cognitive science of mathematics.

With this book, we hope to launch such a discipline.

A discipline of this sort is needed for a simple reason: Mathematics is deep, fundamental, and essential to the human experience. As such, it is crying out to be understood.

It has not been.

Mathematics is seen as the epitome of precision, manifested in the use of symbols in calculation and in formal proofs. Symbols are, of course, just symbols, not ideas. The intellectual content of mathematics lies in its ideas, not in the symbols themselves. In short, the intellectual content of mathematics does not lie where the mathematical rigor can be most easily seen—namely, in the symbols. Rather, it lies in human ideas.

But mathematics by itself does not and cannot *empirically* study human ideas; human cognition is simply not its subject matter. It is up to cognitive science and the neurosciences to do what mathematics itself cannot do—namely, apply the science of mind to human mathematical ideas. That is the purpose of this book.

One might think that the nature of mathematical ideas is a simple and obvious matter, that such ideas are just what mathematicians have consciously taken them to be. From that perspective, the commonplace formal symbols do as good a job as any at characterizing the nature and structure of those ideas. If that were true, nothing more would need to be said.

But those of us who study the nature of concepts within cognitive science know, from research in that field, that the study of human ideas is not so simple. Human ideas are, to a large extent, grounded in sensory-motor experience. Abstract human ideas make use of precisely formulatable cognitive mechanisms such as conceptual metaphors that import modes of reasoning from sensory-motor experience. It is *always* an empirical question just what human ideas are like, mathematical or not.

The central question we ask is this: How can cognitive science bring systematic *scientific rigor* to the realm of human mathematical ideas, which lies outside the rigor of mathematics itself? Our job is to help make precise what mathematics itself cannot—the nature of mathematical ideas.

Rafael Núñez brings to this effort a background in mathematics education, the development of mathematical ideas in children, the study of mathematics in indigenous cultures around the world, and the investigation of the foundations of embodied cognition. George Lakoff is a major researcher in human conceptual systems, known for his research in natural-language semantics, his work on the embodiment of mind, and his discovery of the basic mechanisms of everyday metaphorical thought.

The general enterprise began in the early 1990s with the detailed analysis by one of Lakoff's students, Ming Ming Chiu (now a professor at the Chinese University in Hong Kong), of the basic system of metaphors used by children to comprehend and reason about arithmetic. In Switzerland, at about the same time, Núñez had begun an intellectual quest to answer these questions: How can human beings understand the idea of actual infinity—infinity conceptualized as a thing, not merely as an unending process? What is the concept of actual infinity in its mathematical manifestations—points at infinity, infinite sets, infinite decimals, infinite intersections, transfinite numbers, infinitesimals? He reasoned that since we do not encounter actual infinity directly in the world, since our conceptual systems are finite, and since we have no cognitive mechanisms to perceive infinity, there is a good possibility that metaphorical thought may be necessary for human beings to conceptualize infinity. If so, new results about the structure of metaphorical concepts might make it possible to precisely characterize the metaphors used in mathematical concepts of infinity. With a grant from the Swiss NSF, he came to Berkeley in 1993 to take up this idea with Lakoff.

We soon realized that such a question could not be answered in isolation. We would need to develop enough of the foundations of mathematical idea analysis so that the question could be asked and answered in a precise way. We would need to understand the cognitive structure not only of basic arithmetic but also of symbolic logic, the Boolean logic of classes, set theory, parts of algebra, and a fair amount of classical mathematics: analytic geometry, trigonometry, calculus, and complex numbers. That would be a task of many lifetimes. Because of other commitments, we had only a few years to work on the project—and only part-time.

So we adopted an alternative strategy. We asked, What would be the minimum background needed

- to answer Núñez's questions about infinity,
- to provide a serious beginning for a discipline of mathematical idea analysis, and
- to write a book that would engage the imaginations of the large number of people who share our passion for mathematics and want to understand what mathematical ideas are?

As a consequence, our discussion of arithmetic, set theory, logic, and algebra are just enough to set the stage for our subsequent discussions of infinity and classical mathematics. Just enough for that job, but not trivial. We seek, *from a cognitive perspective,* to provide answers to such questions as, Where do the laws of arithmetic come from? Why is there a unique empty class and why is it a subclass of all classes? Indeed, why is the empty class a class at all, if it cannot be a class *of* anything? And why, in formal logic, does every proposition follow from a contradiction? Why should anything at all follow from a contradiction?

From a cognitive perspective, these questions cannot be answered merely by giving definitions, axioms, and formal proofs. That just pushes the question one step further back: How are those definitions and axioms understood? To answer questions at this level requires an account of ideas and cognitive mechanisms. Formal definitions and axioms are *not* basic cognitive mechanisms; indeed, they themselves require an account in cognitive terms.

One might think that the best way to understand mathematical ideas would be simply to ask mathematicians what they are thinking. Indeed, many famous mathematicians, such as Descartes, Boole, Dedekind, Poincaré, Cantor, and Weyl, applied this method to themselves, introspecting about their own thoughts. Contemporary research on the mind shows that as valuable a method as this can be, it can at best tell a partial and not fully accurate story. Most of our thought and our systems of concepts are part of the cognitive unconscious (see Chapter 2). We human beings have no direct access to our deepest forms of understanding. The analytic techniques of cognitive science are necessary if we are to understand how we understand.

One of the great findings of cognitive science is that our ideas are shaped by our bodily experiences—not in any simpleminded one-to-one way but indirectly, through the grounding of our entire conceptual system in everyday life. The cognitive perspective forces us to ask, Is the system of mathematical ideas also grounded indirectly in bodily experiences? And if so, exactly how?

The answer to questions as deep as these requires an understanding of the cognitive superstructure of a whole nexus of mathematical ideas. This book is concerned with how such cognitive superstructures are built up, starting for the most part with the commonest of physical experiences.

To make our discussion of classical mathematics tractable while still showing its depth and richness, we have limited ourselves to one profound and central question: What does Euler's classic equation, $e^{\pi i} + 1 = 0$, mean? This equation links all the major branches of classical mathematics. It is proved in introductory calculus courses. The equation itself mentions only numbers and mathematical operations on them. What is lacking, from a cognitive perspective, is an analysis of the *ideas* implicit in the equation, the *ideas* that characterize those branches of classical mathematics, the way those *ideas* are linked in the equation, and why the truth of the equation follows from those *ideas*. To demonstrate the utility of mathematical idea analysis for classical mathematics, we set out to provide an initial idea analysis for that equation that would answer all these questions. This is done in the case-study chapters at the end of the book.

To show that mathematical idea analysis has some importance for the philosophy of mathematics, we decided to apply our techniques of analysis to a pivotal moment in the history of mathematics—the arithmetization of real numbers and calculus by Dedekind and Weierstrass in 1872. These dramatic developments set the stage for the age of mathematical rigor and the Foundations of Mathematics movement. We wanted to understand exactly what ideas were involved in those developments. We found the answer to be far from obvious: The modern notion of mathematical rigor and the Foundations of Mathematics movement both rest on a sizable collection of crucial conceptual metaphors.

In addition, we wanted to see if mathematical idea analysis made any difference at all in how mathematics is understood. We discovered that it did: What is called the *real-number line* is not a line as most people understand it. What is called the *continuum* is not continuous in the ordinary sense of the term. And what are called *space-filling curves* do not fill space as we normally conceive of it. These are not mathematical discoveries but discoveries about how mathematics is conceptualized—that is, discoveries in the cognitive science of mathematics.

Though we are not primarily concerned here with mathematics education, it *is* a secondary concern. Mathematical idea analysis, as we seek to develop it, asks what theorems *mean* and *why* they are true *on the basis of what they mean*. We believe it is important to reorient mathematics teaching more toward understanding mathematical ideas and understanding *why* theorems are true.

In addition, we see our job as helping to make mathematical ideas precise in an area that has previously been left to "intuition." Intuitions are not necessarily vague. A cognitive science of mathematics should study the precise nature of clear mathematical intuitions.

The Romance of Mathematics

In the course of our research, we ran up against a mythology that stood in the way of developing an adequate cognitive science of mathematics. It is a kind of "romance" of mathematics, a mythology that goes something like this.

- Mathematics is abstract and disembodied—yet it is real.
- Mathematics has an objective existence, providing structure to this universe and any possible universe, independent of and transcending the existence of human beings or any beings at all.
- Human mathematics is just a part of abstract, transcendent mathematics.
- Hence, mathematical proof allows us to discover transcendent truths of the universe.
- Mathematics is part of the physical universe and provides rational structure to it. There are Fibonacci series in flowers, logarithmic spirals in snails, fractals in mountain ranges, parabolas in home runs, and π in the spherical shape of stars and planets and bubbles.
- Mathematics even characterizes logic, and hence structures reason itself—any form of reason by any possible being.
- To learn mathematics is therefore to learn the language of nature, a mode of thought that would have to be shared by any highly intelligent beings anywhere in the universe.
- Because mathematics is disembodied and reason is a form of mathematical logic, reason itself is disembodied. Hence, machines can, in principle, think.

It is a beautiful romance—the stuff of movies like *2001*, *Contact*, and *Sphere*. It initially attracted us to mathematics.

But the more we have applied what we know about cognitive science to understand the cognitive structure of mathematics, the more it has become clear that this romance cannot be true. Human mathematics, the only kind of mathematics that human beings know, cannot be a subspecies of an abstract, transcendent mathematics. Instead, it appears that mathematics as we know it arises from the nature of our brains and our embodied experience. As a consequence, *every* part of the romance appears to be false, for reasons that we will be discussing.

Perhaps most surprising of all, we have discovered that a great many of the most fundamental mathematical ideas are inherently metaphorical in nature:

- The *number line*, where numbers are conceptualized metaphorically as points on a line.
- Boole's *algebra of classes*, where the formation of classes of objects is conceptualized metaphorically in terms of algebraic operations and elements: plus, times, zero, one, and so on.
- *Symbolic logic*, where reasoning is conceptualized metaphorically as mathematical calculation using symbols.
- *Trigonometric functions*, where angles are conceptualized metaphorically as numbers.
- The *complex plane*, where multiplication is conceptualized metaphorically in terms of rotation.

And as we shall see, Núñez was right about the centrality of conceptual metaphor to a full understanding of infinity in mathematics. There are two infinity concepts in mathematics—one literal and one metaphorical. The literal concept ("in-finity"—lack of an end) is called "potential infinity." It is simply a process that goes on without end, like counting without stopping, extending a line segment indefinitely, or creating polygons with more and more sides. No metaphorical ideas are needed in this case. Potential infinity is a useful notion in mathematics, but the main event is elsewhere. The idea of "actual infinity," where infinity becomes a *thing*—an infinite set, a point at infinity, a transfinite number, the sum of an infinite series—is what is really important. Actual infinity is fundamentally a metaphorical idea, just as Núñez had suspected. The surprise for us was that *all* forms of actual infinity—points at infinity, infinite intersections, transfinite numbers, and so on—appear to be special cases of just one Basic Metaphor of Infinity. This is anything but obvious and will be discussed at length in the course of the book.

As we have learned more and more about the nature of human mathematical cognition, the Romance of Mathematics has dissolved before our eyes. What has emerged in its place is an even more beautiful picture—a picture of what mathematics really is. One of our main tasks in this book is to sketch that picture for you.

None of what we have discovered is obvious. Moreover, it requires a prior understanding of a fair amount of basic cognitive semantics and of the overall cognitive structure of mathematics. That is why we have taken the trouble to write a book of this breadth and depth. We hope you enjoy reading it as much as we have enjoyed writing it.

Introduction:
Why Cognitive Science
Matters to Mathematics

MATHEMATICS AS WE KNOW IT HAS BEEN CREATED and used by human beings: mathematicians, physicists, computer scientists, and economists—all members of the species *Homo sapiens*. This may be an obvious fact, but it has an important consequence. Mathematics as we know it is limited and structured by the human brain and human mental capacities. The only mathematics we know or can know is a brain-and-mind-based mathematics.

As cognitive science and neuroscience have learned more about the human brain and mind, it has become clear that the brain is not a general-purpose device. The brain and body co-evolved so that the brain could make the body function optimally. Most of the brain is devoted to vision, motion, spatial understanding, interpersonal interaction, coordination, emotions, language, and everyday reasoning. Human concepts and human language are not random or arbitrary; they are highly structured and limited, because of the limits and structure of the brain, the body, and the world.

This observation immediately raises two questions:

1. Exactly what mechanisms of the human brain and mind allow human beings to formulate mathematical ideas and reason mathematically?
2. Is brain-and-mind-based mathematics all that mathematics *is*? Or is there, as Platonists have suggested, a disembodied mathematics transcending all bodies and minds and structuring the universe—this universe and every possible universe?

I

Question 1 asks where mathematical ideas come from and how mathematical ideas are to be analyzed from a cognitive perspective. Question 1 is a scientific question, a question to be answered by cognitive science, the interdisciplinary science of the mind. As an empirical question about the human mind and brain, it cannot be studied purely within mathematics. And as a question for empirical science, it cannot be answered by an a priori philosophy or by mathematics itself. It requires an understanding of human cognitive processes and the human brain. Cognitive science matters to mathematics because only cognitive science can answer this question.

Question 1 is what this book is mostly about. We will be asking how normal human cognitive mechanisms are employed in the creation and understanding of mathematical ideas. Accordingly, we will be developing techniques of mathematical idea analysis.

But it is Question 2 that is at the heart of the philosophy of mathematics. It is the question that most people want answered. Our answer is straightforward:

- Theorems that human beings prove are within a human mathematical conceptual system.
- All the mathematical knowledge that we have or can have is knowledge within human mathematics.
- There is no way to know whether theorems proved by human mathematicians have any objective truth, external to human beings or any other beings.

The basic form of the argument is this:

1. The question of the existence of a Platonic mathematics cannot be addressed *scientifically*. At best, it can only be a matter of faith, much like faith in a God. That is, Platonic mathematics, like God, cannot in itself be perceived or comprehended via the human body, brain, and mind. Science alone can neither prove nor disprove the existence of a Platonic mathematics, just as it cannot prove or disprove the existence of a God.
2. As with the conceptualization of God, all that is possible for human beings is an understanding of mathematics in terms of what the human brain and mind afford. The only conceptualization that we can have of mathematics is a human conceptualization. Therefore, mathematics as we know it and teach it can only be humanly created and humanly conceptualized mathematics.

3. What human mathematics is, is an empirical scientific question, not a mathematical or a priori philosophical question.

4. Therefore, it is only through cognitive science—the interdisciplinary study of mind, brain, and their relation—that we can answer the question: What is the nature of the only mathematics that human beings know or can know?

5. Therefore, if you view the nature of mathematics as a scientific question, then mathematics *is* mathematics as conceptualized by human beings using the brain's cognitive mechanisms.

6. However, you may view the nature of mathematics itself not as a scientific question but as a philosophical or religious one. The burden of scientific proof is on those who claim that an external Platonic mathematics does exist, and that theorems proved in human mathematics are objectively true, external to the existence of any beings or any conceptual systems, human or otherwise. At present there is no known way to carry out such a scientific proof in principle.

This book aspires to tell you what human mathematics, conceptualized via human brains and minds, is like. Given the present and foreseeable state of our scientific knowledge, human mathematics *is* mathematics. What human mathematical concepts are is what mathematical concepts are.

We hope that this will be of interest to you whatever your philosophical or religious beliefs about the existence of a transcendent mathematics.

There is an important part of this argument that needs further elucidation. What accounts for what the physicist Eugene Wigner has referred to as "the unreasonable effectiveness of mathematics in the natural sciences" (Wigner, 1960)? How can we make sense of the fact that scientists have been able to find or fashion forms of mathematics that accurately characterize many aspects of the physical world and even make correct predictions? It is sometimes assumed that the effectiveness of mathematics as a scientific tool shows that mathematics itself exists *in the structure of the physical universe*. This, of course, is not a scientific argument with any empirical scientific basis.

We will take this issue up in detail in Part V of the book. Our argument, in brief, will be that whatever "fit" there is between mathematics and the world occurs in the minds of scientists who have observed the world closely, learned the appropriate mathematics well (or invented it), and fit them together (often effectively) using their all-too-human minds and brains.

Finally, there is the issue of whether human mathematics is an instance of, or an approximation to, a transcendent Platonic mathematics. This position presupposes a nonscientific faith in the existence of Platonic mathematics. We will argue that even this position cannot be true. The argument rests on analyses we will give throughout this book to the effect that human mathematics makes fundamental use of conceptual metaphor in characterizing mathematical concepts. Conceptual metaphor is limited to the minds of living beings. Therefore, human mathematics (which is constituted in significant part by conceptual metaphor) cannot be a part of Platonic mathematics, which—if it existed—would be purely literal.

Our conclusions will be:

1. Human beings can have no access to a transcendent Platonic mathematics, if it exists. A belief in Platonic mathematics is therefore a matter of faith, much like religious faith. There can be no scientific evidence for or against the existence of a Platonic mathematics.
2. The only mathematics that human beings know or can know is, therefore, a *mind-based mathematics*, limited and structured by human brains and minds. The only scientific account of the nature of mathematics is therefore an account, via cognitive science, of human mind-based mathematics. Mathematical idea analysis provides such an account.
3. Mathematical idea analysis shows that human mind-based mathematics uses conceptual metaphors as part of the mathematics itself.
4. Therefore human mathematics cannot be a part of a transcendent Platonic mathematics, if such exists.

These arguments will have more weight when we have discussed in detail what human mathematical concepts are. That, as we shall see, depends upon what the human body, brain, and mind are like. A crucial point is the argument in (3)—that conceptual metaphor structures mathematics as human beings conceptualize it. Bear that in mind as you read our discussions of conceptual metaphors in mathematics.

Recent Discoveries about the Nature of Mind

In recent years, there have been revolutionary advances in cognitive science—advances that have an important bearing on our understanding of mathematics. Perhaps the most profound of these new insights are the following:

1. *The embodiment of mind.* The detailed nature of our bodies, our brains, and our everyday functioning in the world structures human concepts and human reason. This includes mathematical concepts and mathematical reason.

2. *The cognitive unconscious.* Most thought is unconscious—not repressed in the Freudian sense but simply inaccessible to direct conscious introspection. We cannot look directly at our conceptual systems and at our low-level thought processes. This includes most mathematical thought.

3. *Metaphorical thought.* For the most part, human beings conceptualize abstract concepts in concrete terms, using ideas and modes of reasoning grounded in the sensory-motor system. The mechanism by which the abstract is comprehended in terms of the concrete is called *conceptual metaphor.* Mathematical thought also makes use of conceptual metaphor, as when we conceptualize numbers as points on a line.

This book attempts to apply these insights to the realm of mathematical ideas. That is, we will be taking mathematics as a subject matter for cognitive science and asking how mathematics is created and conceptualized, especially how it is conceptualized metaphorically.

As will become clear, it is only with these recent advances in cognitive science that a deep and grounded mathematical idea analysis becomes possible. Insights of the sort we will be giving throughout this book were not even imaginable in the days of the old cognitive science of the disembodied mind, developed in the 1960s and early 1970s. In those days, thought was taken to be the manipulation of purely abstract symbols and all concepts were seen as literal—free of all biological constraints and of discoveries about the brain. Thought, then, was taken by many to be a form of symbolic logic. As we shall see in Chapter 6, symbolic logic is itself a mathematical enterprise that requires a cognitive analysis. For a discussion of the differences between the old cognitive science and the new, see *Philosophy in the Flesh* (Lakoff & Johnson, 1999) and *Reclaiming Cognition* (Núñez & Freeman, eds., 1999).

Mathematics is one of the most profound and beautiful endeavors of the imagination that human beings have ever engaged in. Yet many of its beauties and profundities have been inaccessible to nonmathematicians, because most of the cognitive structure of mathematics has gone undescribed. Up to now, even the basic ideas of college mathematics have appeared impenetrable, mysterious, and paradoxical to many well-educated people who have approached them. We believe that cognitive science can, in many cases, dispel the paradoxes and clear away the shrouds of mystery to reveal in full clarity the magnificence of those ideas. To do so, it must reveal how mathematics is grounded in embodied experience and how conceptual metaphors structure mathematical ideas.

Many of the confusions, enigmas, and seeming paradoxes of mathematics arise because conceptual metaphors that are part of mathematics are not recognized as metaphors but are taken as literal. When the full metaphorical character of mathematical concepts is revealed, such confusions and apparent paradoxes disappear.

But the conceptual metaphors themselves do not disappear. They cannot be analyzed away. Metaphors are an essential part of mathematical thought, not just auxiliary mechanisms used for visualization or ease of understanding. Consider the metaphor that Numbers Are Points on a Line. Numbers don't have to be conceptualized as points on a line; there are conceptions of number that are not geometric. But the number line is one of the most central concepts in all of mathematics. Analytic geometry would not exist without it, nor would trigonometry.

Or take the metaphor that Numbers Are Sets, which was central to the Foundations movement of early-twentieth-century mathematics. We don't have to conceptualize numbers as sets. Arithmetic existed for over two millennia without this metaphor—that is, without zero conceptualized as being the empty set, 1 as the set containing the empty set, 2 as the set containing 0 and 1, and so on. But if we do use this metaphor, then forms of reasoning about sets can also apply to numbers. It is only by virtue of this metaphor that the classical Foundations of Mathematics program can exist.

Conceptual metaphor is a cognitive mechanism for allowing us to reason about one kind of thing as if it were another. This means that metaphor is not simply a linguistic phenomenon, a mere figure of speech. Rather, it is a cognitive mechanism that belongs to the realm of thought. As we will see later in the book, "conceptual metaphor" has a technical meaning: It is a *grounded, inference-preserving cross-domain mapping*—a neural mechanism that allows us to use the inferential structure of one conceptual domain (say, geometry) to reason about another (say, arithmetic). Such conceptual metaphors allow us to apply what we know about one branch of mathematics in order to reason about another branch.

Conceptual metaphor makes mathematics enormously rich. But it also brings confusion and apparent paradox if the metaphors are not made clear or are taken to be literal truth. Is zero a point on a line? Or is it the empty set? Or both? Or is it just a number and neither a point nor a set? There is no one answer. Each answer constitutes a choice of metaphor, and each choice of metaphor provides different inferences and determines a different subject matter.

Mathematics, as we shall see, layers metaphor upon metaphor. When a single mathematical idea incorporates a dozen or so metaphors, it is the job of the cognitive scientist to tease them apart so as to reveal their underlying cognitive structure.

This is a task of inherent scientific interest. But it also can have an important application in the teaching of mathematics. We believe that revealing the cognitive structure of mathematics makes mathematics much more accessible and comprehensible. Because the metaphors are based on common experiences, the mathematical ideas that use them can be understood for the most part in everyday terms.

The cognitive science of mathematics asks questions that mathematics does not, and cannot, ask about itself. How do we understand such basic concepts as infinity, zero, lines, points, and sets using our everyday conceptual apparatus? How are we to make sense of mathematical ideas that, to the novice, are paradoxical—ideas like space-filling curves, infinitesimal numbers, the point at infinity, and non-well-founded sets (i.e., sets that "contain themselves" as members)?

Consider, for example, one of the deepest equations in all of mathematics, the Euler equation, $e^{\pi i} + 1 = 0$, e being the infinite decimal 2.718281828459045. . . , a far-from-obvious number that is the base for natural logarithms. This equation is regularly taught in elementary college courses. But what exactly does it mean? We are usually told that an exponential of the form q^n is just the number q multiplied by itself n times; that is, $q \cdot q \cdot \ldots \cdot q$. This makes perfect sense for 2^5, which would be $2 \cdot 2 \cdot 2 \cdot 2 \cdot 2$, which multiplies out to 32. But this definition of an exponential makes no sense for $e^{\pi i}$. There are at least three mysteries here.

1. What does it mean to multiply an infinite decimal like e by itself? If you think of multiplication as an algorithmic operation, where do you start? Usually you start the process of multiplication with the last digit on the right, but there is no last digit in an infinite decimal.
2. What does it mean to multiply any number by itself π times? π is another infinite nonrepeating decimal. What could "π times" for performing an operation mean?
3. And even worse, what does it mean to multiply a number by itself an imaginary $(\sqrt{-1})$ number of times?

And yet we are told that the answer is −1. The typical proof is of no help here. It proves that $e^{\pi i} + 1 = 0$ is true, but it does not tell you what $e^{\pi i}$ means! In the course of this book, we will.

In this book, unlike most other books about mathematics, we will be concerned not just with *what* is true but with what mathematical ideas *mean*, how they can be understood, and *why* they are true. We will also be concerned with the nature of mathematical truth from the perspective of a mind-based mathematics.

One of our main concerns will be the concept of infinity in its various manifestations: infinite sets, transfinite numbers, infinite series, the point at infinity, infinitesimals, and objects created by taking values of sequences "at infinity," such as space-filling curves. We will show that there is a single Basic Metaphor of Infinity that all of these are special cases of. This metaphor originates outside mathematics, but it appears to be the basis of our understanding of infinity in virtually all mathematical domains. When we understand the Basic Metaphor of Infinity, many classic mysteries disappear and the apparently incomprehensible becomes relatively easy to understand.

The results of our inquiry are, for the most part, not mathematical results but results in the cognitive science of mathematics. They are results about the human conceptual system that makes mathematical ideas possible and in which mathematics makes sense. But to a large extent they are not results reflecting the conscious thoughts of mathematicians; rather, they describe the *unconscious* conceptual system used by people who do mathematics. The results of our inquiry should not change mathematics in any way, but they may radically change the way mathematics is understood and what mathematical results are taken to mean.

Some of our findings may be startling to many readers. Here are some examples:

- Symbolic logic is not the basis of all rationality, and it is not absolutely true. It is a beautiful metaphorical system, which has some rather bizarre metaphors. It is useful for certain purposes but quite inadequate for characterizing anything like the full range of the mechanisms of human reason.
- The real numbers do not "fill" the number line. There is a mathematical subject matter, the hyperreal numbers, in which the real numbers are rather sparse on the line.
- The modern definition of *continuity* for functions, as well as the so-called *continuum*, do not use the idea of continuity as it is normally understood.

- So-called *space-filling curves* do not fill space.
- There is no absolute yes-or-no answer to whether 0.99999. . . . = 1. It will depend on the conceptual system one chooses. There is a mathematical subject matter in which 0.99999. . . . = 1, and another in which 0.99999. . . . ≠ 1.

These are not new mathematical findings but new ways of understanding well-known results. They are findings in the cognitive science of mathematics—results about the conceptual structure of mathematics and about the role of the mind in creating mathematical subject matters.

Though our research does not affect mathematical results in themselves, it does have a bearing on the understanding of mathematical results and on the claims made by many mathematicians. Our research also matters for the philosophy of mathematics. *Mind-based mathematics*, as we describe it in this book, is not consistent with any of the existing philosophies of mathematics: Platonism, intuitionism, and formalism. Nor is it consistent with recent postmodernist accounts of mathematics as a purely social construction. Based on our findings, we will be suggesting a very different approach to the philosophy of mathematics. We believe that the philosophy of mathematics should be consistent with scientific findings about the only mathematics that human beings know or can know. We will argue in Part V that the *theory of embodied mathematics*—the body of results we present in this book—determines an empirically based philosophy of mathematics, one that is coherent with the "embodied realism" discussed in Lakoff and Johnson (1999) and with "ecological naturalism" as a foundation for embodiment (Núñez, 1995, 1997).

Mathematics as we know it is human mathematics, a product of the human mind. Where does mathematics come from? It comes from us! We create it, but it is not arbitrary—not a mere historically contingent social construction. What makes mathematics nonarbitrary is that it uses the basic conceptual mechanisms of the embodied human mind as it has evolved in the real world. Mathematics is a product of the neural capacities of our brains, the nature of our bodies, our evolution, our environment, and our long social and cultural history.

By the time you finish this book, our reasons for saying this should be clear.

The Structure of the Book

Part I is introductory. We begin in Chapter 1 with the brain's innate arithmetic—the ability to subitize (i.e., to instantly determine how many objects are in a very small collection) and do very basic addition and subtraction. We move on in Chapter 2 to some of the basic results in cognitive science on which the remainder of the book rests. We then take up basic metaphors grounding our understanding of arithmetic (Chapter 3) and the question of where the laws of arithmetic come from (Chapter 4).

In Part II, we turn to the grounding and conceptualization of sets, logic, and forms of abstract algebra such as groups (Chapters 5, 6, and 7).

Part III deals with the concept of infinity—as fundamental a concept as there is in sophisticated mathematics. The question we ask is how finite human cognitive capacities and everyday conceptual mechanisms can give rise to the full range of mathematical notions of infinity: points at infinity, infinite sets, mathematical induction, infinite decimals, limits, transfinite numbers, infinitesimals, and so on. We argue that the concept of actual infinity is metaphorical in nature and that there is a single conceptual metaphor—the Basic Metaphor of Infinity (Chapter 8)—underlying most if not all infinite notions in mathematics (Chapters 8 through 11). We will then, in Part IV, point out the implications of this type of analysis for an understanding of the continuum (Chapter 12) and for continuity and the real numbers (Chapters 13 and 14).

At this point in the book, we take a break from our line of argumentation to address a commonly noticed apparent contradiction, which we name the Length Paradox. We call this interlude *le trou normand*, after the course in a rich French meal where a sorbet with calvados is served to refresh the palate.

We now have enough results for Part V, a discussion of an overall *theory of embodied mathematics* (Chapter 15) and a new philosophy of mathematics (Chapter 16).

To demonstrate the real power of the approach, we end the book with Part VI, a detailed case study of the equation that brings together the ideas at the heart of classical mathematics: $e^{\pi i} + 1 = 0$. To show exactly what this equation means, we have to look at the cognitive structure—especially the conceptual metaphors—underlying analytic geometry and trigonometry (Case Study 1), exponentials and logarithms (Case Study 2), imaginary numbers (Case Study 3), and the cognitive mechanisms combining them (Case Study 4).

We chose to place this case study at the end for three reasons. First, it is a detailed illustration of how the cognitive mechanisms described in the book can shed light on the structure of classical mathematics. We have placed it after our discussion of the philosophy of mathematics to provide an example to the reader of how a change in the nature of what mathematics is can lead to a new understanding of familiar mathematical results.

Second, it is in the case study that mathematical idea analysis comes to the fore. Though we will be analyzing mathematical ideas from a cognitive perspective throughout the book, the study of Euler's equation demonstrates the power of the analysis of ideas in mathematics, by showing how a single equation can bring an enormously rich range of ideas together—even though the equation itself contains nothing but numbers: e, π, $\sqrt{-1}$, 1, and 0. We will be asking throughout how mere *numbers* can express *ideas*. It is in the case study that the power of the answer to this question becomes clear.

Finally, there is an educational motive. We believe that classical mathematics can best be taught with a cognitive perspective. We believe that it is important to teach mathematical ideas and to explain why mathematical truths follow from those ideas. This case study is intended to illustrate to teachers of mathematics how this can be done.

We see our book as an early step in the development of a cognitive science of mathematics—a discipline that studies the cognitive mechanisms used in the human creation and conceptualization of mathematics. We hope you will find this discipline stimulating, challenging, and worthwhile.

Part I

The Embodiment of Basic Arithmetic

1

The Brain's Innate Arithmetic

THIS BOOK ASKS A CENTRAL QUESTION: What is the cognitive structure of sophisticated mathematical ideas? What are the simplest mathematical ideas, and how do we build on them and extend them to develop complex mathematical ideas: the laws of arithmetic, set theory, logic, trigonometry, calculus, complex numbers, and various forms of infinity—transfinite numbers, infinitesimals, limits, and so on? Let us begin with the most fundamental aspects of number and arithmetic, the part we are all born with.

Number Discrimination by Babies

The very idea that babies have mathematical capacities is startling. Mathematics is usually thought of as something inherently difficult that has to be taught with homework and exercises. Yet we come into life prepared to do at least some rudimentary form of arithmetic. Recent research has shown that babies have the following numerical abilities:

1. At three or four days, a baby can discriminate between collections of two and three items (Antell & Keating, 1983). Under certain conditions, infants can even distinguish three items from four (Strauss & Curtis, 1981; van Loosbroek & Smitsman, 1990).
2. By four and a half months, a baby "can tell" that one plus one is two and that two minus one is one (Wynn, 1992a).
3. A little later, infants "can tell" that two plus one is three and that three minus one is two (Wynn, 1995).

4. These abilities are not restricted to visual arrays. Babies can also discriminate numbers of sounds. At three or four days, a baby can discriminate between sounds of two or three syllables (Bijeljac-Babic, Bertoncini, & Mehler, 1991).

5. And at about seven months, babies can recognize the numerical equivalence between arrays of objects and drumbeats of the same number (Starkey, Spelke, & Gelman, 1990).

How do we know that babies can make these numerical distinctions? Here is one of the now-classic experimental procedures (Starkey & Cooper, 1980): Slides were projected on a screen in front of babies sitting on their mother's lap. The time a baby spent looking at each slide before turning away was carefully monitored. When the baby started looking elsewhere, a new slide appeared on the screen. At first, the slides contained two large black dots. During the trials, the baby was shown the same numbers of dots, though separated horizontally by different distances. After a while, the baby would start looking at the slides for shorter and shorter periods of time. This is technically called *habituation*; nontechnically, the baby got bored.

The slides were then changed without warning to three black dots. Immediately the baby started to stare longer, exhibiting what psychologists call a longer *fixation time*. The consistent difference of fixation times informs psychologists that the baby could tell the difference between two and three dots. The experiment was repeated with the three dots first, then the two dots. The results were the same. These experiments were first tried with babies between four and five months of age, but later it was shown that newborn babies at three or four days showed the same results (Antell & Keating, 1983). These findings have been replicated not just with dots but with slides showing objects of different shapes, sizes, and alignments (Strauss & Curtis, 1981). Such experiments suggest that the ability to distinguish small numbers is present in newborns, and thus that there is at least some innate numerical capacity.

The ability to do the simplest arithmetic was established using similar habituation techniques. Babies were tested using what, in the language of developmental psychology, is called the *violation-of-expectation paradigm*. The question asked was this: Would a baby at four and a half months expect, given the presence of one object, that the addition of one other object would result in the presence of two objects? In the experiment, one puppet is placed on a stage. The stage is then covered by a screen that pops up in front of it. Then the baby sees someone placing a second identical puppet behind the screen. Then the screen is lowered. If there are two puppets there, the baby shows no surprise;

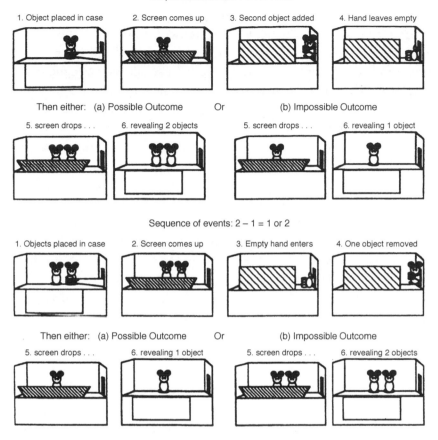

FIGURE 1.1 A usual design for studying the arithmetic capabilities of infants (Wynn, 1992a). These studies indicate that infants as young as four-and-a-half months old have a primitive form of arithmetic. They react normally to events in which 1 puppet + 1 puppet = 2 puppets or in which 2 puppets − 1 puppet = 1 puppet. But they exhibit startled reactions (e.g., systematically longer staring) for impossible outcomes in which 1 puppet + 1 puppet = 1 puppet or in which 2 puppets − 1 puppet = 2 puppets.

that is, it doesn't look at the stage any longer than otherwise. If there is only one puppet, the baby looks at the stage for a longer time. Presumably, the reason is that the baby expected two puppets, not one, to be there. Similarly, the baby stares longer at the stage if three puppets are there when the screen is lowered. The conclusion is that the baby can tell that one plus one is supposed to be two, not one or three (see Figure 1.1).

Similar experiments started with two puppets being placed on-stage, the screen popping up to cover them, and then one puppet being visibly removed from behind the screen. The screen was then lowered. If there was only one puppet there, the babies showed no surprise; that is, they didn't look at the screen for any longer time. But if there were still two puppets on the stage after one had apparently been removed, the babies stared at the stage for a longer time. They presumably knew that two minus one is supposed to leave one, and they were surprised when it left two. Similarly, babies at six months expected that two plus one would be three and that three minus one would be two. In order to show that this was not an expectation based merely on the location of the puppets, the same experiment was replicated with puppets moving on turntables, with the same results (Koechlin, Dehaene, & Mehler, 1997). These findings suggest that babies use mechanisms more abstract than object location.

Finally, to show that this result had to do with abstract number and not particular objects, other experimenters had the puppets change to balls behind the screen. When two balls appeared instead of two puppets, four- and five-month-olds (unlike older infants) manifested no surprise, no additional staring at the stage. But when one ball or three balls appeared where two were expected, the babies did stare longer, indicating surprise (Simon, Hespos, & Rochat, 1995). The conclusion was that only number, not object identity, mattered.

In sum, newborn babies have the ability to discern the number of discrete, separate arrays of objects in space and the number of sounds produced sequentially (up to three or four). And at about five months they can distinguish correct from incorrect addition and subtraction of objects in space, for very small numbers.

The evidence that babies have these abilities is robust, but many questions remain open. What exactly are the mechanisms—neurophysiological, psychological, and others—underlying these abilities? What are the exact situational conditions under which these abilities can be confirmed experimentally? When an infant's expectations are violated in such experiments, exactly what expectation is being violated? How do these abilities relate to other developmental processes? And so on.

The experimental results to date do not give a complete picture. For example, there is no clear-cut evidence that infants have a notion of order before the age of fifteen months. If they indeed lack the concept of order before this age, this would suggest that infants can do what they do without realizing that, say, three is larger than two or that two is larger than one (Dehaene, 1997). In other words, it is conceivable that babies make the distinctions they make, but without a rudimentary concept of order. If so, when, exactly, does order emerge from the rudiments of baby arithmetic—and how?

Despite the evidence discussed, experimenters do not necessarily agree on how to answer these questions and how to interpret many of the findings. The new field of baby arithmetic is going through the usual growing pains. (For a brief summary, see Bideaud, 1996.) But as far as the present book is concerned, what matters is that such abilities do exist at a very early age. We will refer to these abilities as *innate arithmetic*. (This term has also been used in Butterworth, 1999, p. 108.)

Subitizing

All human beings, regardless of culture or education, can instantly tell at a glance whether there are one, two, or three objects before them. This ability is called *subitizing*, from the Latin word for "sudden." It is this ability that allows newborn babies to make the distinctions discussed above. We can subitize—that is, accurately and quickly discern the number of—up to about four objects. We cannot as quickly tell whether there arc thirteen as opposed to fourteen objects, or even whether there are seven as opposed to eight. To do that takes extra time and extra cognitive operations—grouping the objects into smaller, subitizable groups and counting them. In addition to being able to subitize objects in arrays, we can subitize sequences. For example, given a sequence of knocks or beeps or flashes of light, we can accurately and quickly tell how many there are, up to five or six (Davis & Pérusse, 1988). These results are well established in experimental studies of human perception, and have been for half a century (Kaufmann, Lord, Reese, & Volkmann, 1949). Kaufmann et al. observed that subitizing was a different process from counting or estimating. Today there is a fair amount of robust evidence suggesting that the ability to subitize is inborn. A survey of the range of subitizing experiments can be found in Mandler and Shebo (1982).

The classic subitizing experiment involves reaction time and accuracy. A number of items are flashed before subjects for a fraction of a second and they have to report as fast as they can how many there are. As you vary the number of items presented (the independent variable), the reaction time (the dependent variable) is roughly about half a second (actually about 600 milliseconds) for arrays of three items. After that, with arrays of four or five items, the reaction time begins increasing linearly with the number of items presented (see Figure 1.2). Accuracy varies according to the same pattern: For arrays of three or four items, there are virtually no errors. Starting with four items, the error rate rises linearly with the number of items presented. These results hold when the objects presented are in different spatial locations. When they overlap spatially, as with concentric circles, the results no longer hold (Trick & Pylyshyn, 1993, 1994).

There is now a clear consensus that subitizing is not merely a pattern-recognition process. However, the neural mechanism by which subitizing works is

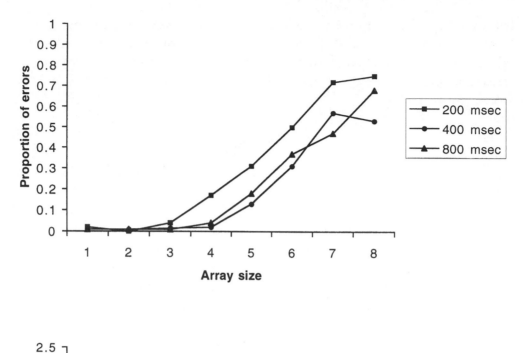

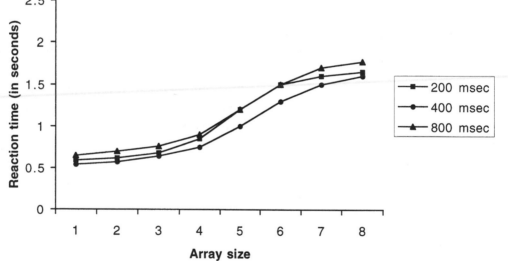

FIGURE 1.2   Fifty years ago, experimental studies established the human capacity for making quick, error-free, and precise judgments of the numerosity of small collections of items.
The capacity was called *subitizing*. The figure shows results for these judgments under three
experimental conditions. The levels of accuracy (top graphic) and reaction time (bottom
graphic) stay stable and low for arrays of sizes of up to four items. The numbers increase dra-
matically for larger arrays (adapted from Mandler & Shebo, 1982).

still in dispute. Randy Gallistel and Rochel Gelman have claimed that subitizing is just very fast counting—serial processing, with visual attention placed on each item (Gelman & Gallistel, 1978). Stanislas Dehaene has hypothesized instead that subitizing is all-at-once; that is, it is accomplished via "parallel preattentive processing," which does not involve attending to each item one at a time (Dehaene, 1997). Dehaene's evidence for his position comes from patients with brain damage that prevents them from attending to things in their environment serially and therefore from counting them. They can nonetheless subitize accurately and quickly up to three items (Dehaene & Cohen, 1994).

The Numerical Abilities of Animals

Animals have numerical abilities—not just primates but also raccoons, rats, and even parrots and pigeons. They can subitize, estimate numbers, and do the simplest addition and subtraction, just as four-and-a-half-month-old babies can. How do we know? Since we can't ask animals directly, indirect evidence must be gathered experimentally.

Experimental methods designed to explore these questions have been conceived for more than four decades (Mechner, 1958). Here is a demonstration for showing that rats can learn to perform an activity a given number of times. The task involves learning to estimate the number of times required. The rats are first deprived of food for a while. Then they are placed in a cage with two levers, which we will call A and B. Lever B will deliver food, but only after lever A has been pressed a certain fixed number of times—say, four. If the rat presses A the wrong number of times or not at all and then presses B, it is punished. The results show that rats learn to press A about the right number of times. If the number of times is eight, the rats learn to press a number close to that—say, seven to nine times.

To show that the relevant parameter is number and not just duration of time, experimenters conceived further manipulations: They varied the degree of food deprivation. As a result, some of the rats were very hungry and pressed the lever much faster, in order to get food quickly. But despite this, they still learned to press the lever close to the right number of times (Mechner & Guevrekian, 1962). In other series of experiments, scientists showed that rats have an ability to learn and generalize when dealing with numbers or with duration of time (Church & Meck, 1984).

Rats show even more sophisticated abilities, extending across different action and sensory modalities. They can learn to estimate numbers in association not just with motor actions, like pressing a bar, but also with the perception of tones or light flashes. This shows that the numerical estimation capacity of rats is not limited to a specific sensory modality: It applies to number independent of modality. Indeed, modalities can be combined: Following the presentation of a sequence of, say, two tones synchronized with two light flashes (for a total of four events), the rats will systematically respond to four (Church & Meck, 1984).

Nonhuman primates display abilities that are even more sophisticated. Rhesus monkeys in the wild have arithmetic abilities similar to those of infants, as revealed by studies with the violation-of-expectation paradigm. For example, a monkey was first presented with one eggplant placed in an open box. Then a partition was placed in front of the eggplant, blocking the monkey's view. Then a second eggplant was placed in the box, in such a way that the monkey could see it being put there. The partition was then removed to reveal either one or two eggplants in the box. The monkey looked significantly longer at the "impossible" one-eggplant case, reacting even more strongly than the babies. According to the primatologists' interpretation, this was an indication that the monkey expected to see two eggplants in the box (Hauser, MacNeilage, & Ware, 1996).

In another line of research, primatologists have found that a chimpanzee can do arithmetic, combining simple physical fractions: one-quarter, one-half, and three-quarters. When one-quarter of an apple and one-half glass of a colored liquid were presented as a stimulus, the chimpanzee would choose as a response a three-quarter disc over a full disc (Woodruff & Premack, 1981).

Chimpanzees have even been taught to use numerical symbols, although the training required years of painstaking work. About twenty years ago at Kyoto University, Japanese primatologists trained chimpanzees to use arbitrary visual signs to characterize collections of objects (and digits to characterize numbers). One of their best "students," a chimpanzee named Ai, learned to report the kind, color, and numerosity of collections of objects. For instance, she would appropriately select sequences of signs, like "pencil-red-three" for a group of three red pencils and "toothbrush-blue-five" for a collection of five blue toothbrushes (Matsuzawa, 1985). Reaction-time data indicate that beyond the numbers three or four, Ai used a very humanlike serial form of counting. Recently Ai has made improvements, mastering the labeling of collections of objects up to nine, as well as ordering digits according to their numerical size (Matsuzawa, 1997).

Other researchers have shown that chimpanzees are also able to calculate using numerical symbols. For example, Sarah Boysen succeeded, through years of training, in teaching her chimpanzee, Sheba, to perform simple comparisons and additions. Sheba was progressively better able to match a collection of objects with the corresponding Arabic numeral, from 0 to 9. Sheba also learned to choose, from among several collections of objects, the one correctly matching a given numeral. Later, using an ingenious experimental design, Boysen demonstrated that Sheba was able to mentally perform additions using symbols alone. For example, given symbols "2" and "4," Sheba would pick out the symbol "6" as a result (Boysen & Capaldi, 1993; Boysen & Berntson, 1996). Although these impressive capacities require years of training, they show that our closest relative, the chimpanzee, shares with us a nontrivial capacity for at least some innate arithmetic along with abilities that can be learned through long-term, explicit, guided training.

The Inferior Parietal Cortex

Up to now, we have mainly discussed human numerical capacities that are non-symbolic, where numbers of objects and events were involved, but not symbols for those numbers. Numbers are, of course, distinct from numerals, the symbols for numbers. The capacity for using numerals is more complex than that for number alone, as we shall discuss below in detail. Moreover, there are two aspects of the symbolization of number: written symbols (say, Arabic numerals like "6") and words, both spoken and written, for those symbols (say, "six"). The words and the numerals have different grammatical structure. The grammatical structure of the words is highly language-dependent, as can be seen from the English "eighty-one" versus the French "quatre-vingt-un" ("four-twenty-one"). Thus, the capacity for naming numbers involves a capacity for number plus two symbolic capacities—one for written numerals and one for characterizing the structure of the (typically complex) words for those numerals.

There is a small amount of evidence suggesting that the inferior parietal cortex is involved in symbolic numerical abilities. One bit of evidence comes from patients with *Epilepsia arithmetices,* a rare form of epileptic seizure that occurs when doing arithmetic calculations. About ten cases in the world have been studied. In each case, the electroencephalogram (EEG) showed abnormalities in the inferior parietal cortex. From the moment the patients started doing even very simple arithmetic calculations, their brain waves showed abnormal rhythmic discharges and triggered epileptic fits. Other intellectual activities, such as reading, had no ill effects. (For discussion, see Dehaene, 1997, p. 191.)

A second piece of evidence comes from Mr. M, a patient of Laurent Cohen and Stanislas Dehaene, who has a lesion in the inferior parietal cortex. Mr. M cannot tell what number comes between 3 and 5, but he can tell perfectly well what letter comes between A and C and what day comes between Tuesday and Thursday. Knowledge of number sequence has been lost, but other sequential information is unaffected.

Mr. M can correctly give names to numerals. Shown the symbol 5, he can respond "five." But he cannot do simple arithmetic. He says that two minus one makes two, that nine minus eight is seven, that three minus one makes four. He has lost the sense of the structure of integers. He also fails "bisection" tasks—deciding which number falls in a given interval. Between three and five, he places three. Between ten and twenty, he places thirty, then corrects his answer to twenty-five. Yet his rote arithmetic memory is intact. He knows the multiplication table by heart; he knows that three times nine is twenty-seven, but fails when the result of addition goes beyond ten. Asked to add 8 + 5, he cannot break 5 down into 2 + 3, to give (8 + 2) + 3, or 10 + 3. He has lost every intuition about arithmetic, but he preserves rote memory. He can perform simple rote calculations, but he does not understand them.

The inferior parietal cortex is a highly associative area, located anatomically where neural connections from vision, audition, and touch come together—a location appropriate for numerical abilities, since they are common to all sensory modalities. Lesions in this area have been shown to affect not only arithmetic but also writing, representing the fingers of the hand, and distinguishing right from left. Mr. M has all these disabilities. However, some patients have only one or another, which suggests that the inferior parietal cortex is divided into microregions associated with each.

Dehaene (1997) asks why these capacities come together in a single region. "What," he asks, "is the relationship between numbers, writing, fingers, and space?" He speculates as follows: Numbers are connected to fingers because children learn to count on their fingers. Numbers are related to writing because they are symbolized by written numerals. Numbers are related to space in various ways; subitizing, for example, requires objects to be distributed over space, and integers are conceptualized as being spread in space over a number line. And mathematical talent often correlates with spatial abilities. Thus, Dehaene reasons that, despite limited evidence at present, it makes sense to conclude that basic arithmetic abilities make major use of the inferior parietal cortex.

Other mathematical abilities appear to involve other areas of the brain. For example, the prefrontal cortex, which is involved in complex structuring—complex motor routines, plans, and so on—seems to be used in complex arithmetic calculation, though not in rote memory (say, for multiplication tables). Patients with frontal lesions have difficulty using the multiplication algorithm, adding when they should multiply, forgetting to carry over, not processing digits in the right order; in short, they are unable to carry out complex sequential operations correctly.

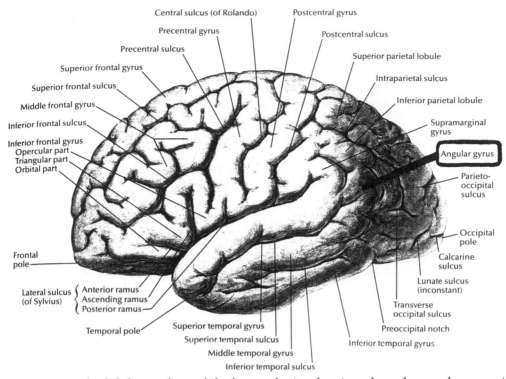

FIGURE 1.3   The left hemisphere of the human brain, showing where the angular gyrus is in the inferior parietal cortex. This is an area in which brain lesions appear to severely affect arithmetic capacities.

The capacity for basic arithmetic is separate from the capacity for rote memorization of addition and multiplication tables. Such rote abilities seem to be subcortical, associated with the basal ganglia. One patient of Cohen and Dehaene has a lesion of the basal ganglia and has lost many rote abilities: A former teacher and devout Christian, she can no longer recite the alphabet, familiar nursery rhymes, or the most common prayers. She has also lost the use of memorized addition and multiplication tables. Yet, with her inferior parietal cortex intact, she retains other nonrote arithmetic abilities. She can compare two numbers and find which number falls in between them. And although she does not remember what two times three is, she can calculate it by mentally counting three groups of two objects. She has no trouble with subtraction. This suggests that rote mathematical abilities involve the subcortical basal ganglia and cortico-subcortical loops.

Not only is rote calculation localized separately from basic arithmetic abilities but algebraic abilities are localized separately from the capacity for basic arithmetic. Dehaene (1997) cites a patient with a Ph.D. in chemistry who has *acalculia*, the inability to do basic arithmetic. For example, he cannot solve $2 \cdot 3$, $7 - 3$, $9 \div 3$, or $5 \cdot 4$. Yet he can do abstract algebraic calculations. He can simplify $(a \cdot b) / (b \cdot a)$ into 1 and $a \cdot a \cdot a$ into a^3, and could recognize that $(d/c) + a$ is not generally equal to $(d + a) / (c + a)$. Dehaene concludes that algebraic calculation and arithmetic calculation are processed in different brain regions.

From Brains to Minds and from Basic Arithmetic to Mathematics

Very basic arithmetic uses at least the following capacities: subitizing, perception of simple arithmetic relationships, the ability to estimate numerosity with close approximation (for bigger arrays), and the ability to use symbols, calculate, and memorize short tables. At present, we have some idea of what areas of the brain are active when we use such capacities, and we have some idea of which of these capacities are innate (for a general discussion, see Butterworth, 1999).

But that is not very much. First, it is not much of mathematics. In fact, when compared to the huge edifice of mathematics it is almost nothing. Second, knowing *where* is far from knowing *how*. To know what parts of the brain "light up" when certain tasks are performed is far from knowing the neural mechanism by which those tasks are performed. Identifying the parts of the brain involved is only a small, albeit crucial, part of the story.

For us, the hard question is how we go from such simple capacities to sophisticated forms of mathematics and how we employ ordinary cognitive mechanisms to do so. In the next chapter, we will introduce the reader to the basic cognitive mechanisms that are needed to begin answering these questions and that we will refer to throughout the book.

2

A Brief Introduction to the Cognitive Science of the Embodied Mind

The Cognitive Unconscious

Perhaps the most fundamental, and initially the most startling, result in cognitive science is that most of our thought is unconscious—that is, fundamentally inaccessible to our direct, conscious introspection. Most everyday thinking occurs too fast and at too low a level in the mind to be thus accessible. Most cognition happens backstage. That includes mathematical cognition.

We all have systems of concepts that we use in thinking, but we cannot consciously inspect our conceptual inventory. We all draw conclusions instantly in conversation, but we cannot consciously look at each inference and our own inference-drawing mechanisms while we are in the act of inferring on a massive scale second by second. We all speak in a language that has a grammar, but we do not consciously put sentences together word by word, checking consciously that we are following the grammatical rules of our language. To us, it seems easy: We just talk, and listen, and draw inferences without effort. But what goes on in our minds behind the scenes is enormously complex and largely unavailable to us.

Perhaps the most startling realization of all is that we have unconscious memory. The very idea of an unconscious memory seems like a contradiction in terms, since we usually think of remembering as a conscious process. Yet hundreds of experimental studies have confirmed that we remember without being aware that we are remembering—that experiences we don't recall do in fact have a detectable and sometimes measurable effect on our behavior. (For an excellent overview, see Schacter, 1996.)

What has not been done so far is to extend the study of the cognitive unconscious to mathematical cognition—that is, the way we implicitly understand mathematics as we do it or talk about it. A large part of unconscious thought involves automatic, immediate, implicit rather than explicit understanding—making sense of things without having conscious access to the cognitive mechanisms by which you make sense of things. Ordinary everyday mathematical sense-making is not in the form of conscious proofs from axioms, nor is it always the result of explicit, conscious, goal-oriented instruction. Most of our everyday mathematical understanding takes place without our being able to explain exactly what we understood and how we understood it. Indeed, when we use the term "understanding" throughout this book, this automatic unconscious understanding is the kind of understanding we will be referring to, unless we say otherwise.

Therefore, this book is not about those areas of cognitive science concerned with conscious, goal-oriented mathematical cognition, like conscious approaches to problem solving or to constructing proofs. Though this book may have implications for those important fields, we will not discuss them here.

Our enterprise here is to study everyday mathematical understanding of this automatic unconscious sort and to ask a crucial question: How much of mathematical understanding makes use of the same kinds of conceptual mechanisms that are used in the understanding of ordinary, nonmathematical domains? Are the same cognitive mechanisms that we use to characterize ordinary ideas also used to characterize mathematical ideas?

We will argue that a great many cognitive mechanisms that are not specifically mathematical are used to characterize mathematical ideas. These include such ordinary cognitive mechanisms as those used for the following ordinary ideas: basic spatial relations, groupings, small quantities, motion, distributions of things in space, changes, bodily orientations, basic manipulations of objects (e.g., rotating and stretching), iterated actions, and so on.

To be more specific, we will suggest that:

- Conceptualizing the technical mathematical concept of a class makes use of the everyday concept of a collection of objects in a bounded region of space.
- Conceptualizing the technical mathematical concept of recursion makes use of the everyday concept of a repeated action.

- Conceptualizing the technical mathematical concept of complex arithmetic makes use of the everyday concept of rotation.
- Conceptualizing derivatives in calculus requires making use of such everyday concepts as motion, approaching a boundary, and so on.

From a nontechnical perspective, this should be obvious. But from the technical perspective of cognitive science, one must ask:

Exactly what everyday concepts and cognitive mechanisms are used in exactly what ways in the unconscious conceptualization of technical ideas in mathematics?

Mathematical idea analysis, as we will be developing it, depends crucially on the answers to this question. Mathematical ideas, as we shall see, are often grounded in everyday experience. Many mathematical ideas are ways of mathematicizing ordinary ideas, as when the idea of a derivative mathematicizes the ordinary idea of instantaneous change.

Since the cognitive science of mathematics is a new discipline, not much is known for sure right now about just how mathematical cognition works. Our job in this book is to explore how the general cognitive mechanisms used in everyday nonmathematical thought can create mathematical understanding and structure mathematical ideas.

Ordinary Cognition and Mathematical Cognition

As we saw in the previous chapter, it appears that all human beings are born with a capacity for subitizing very small numbers of objects and events and doing the simplest arithmetic—the arithmetic of very small numbers. Moreover, if Dehaene (1997) is right, the inferior parietal cortex, especially the angular gyrus, "plays a crucial role in the mental representation of numbers as quantities" (p. 189). In other words, there appears to be a part of the brain innately specialized for a sense of quantity—what Dehaene, following Tobias Dantzig, refers to as "the number sense."

But there is a lot more to mathematics than the arithmetic of very small numbers. Trigonometry and calculus are very far from "three minus one equals two." Even realizing that zero is a number and that negative numbers are numbers took centuries of sophisticated development. Extending numbers to the rationals, the reals, the imaginaries, and the hyperreals requires an enormous cognitive apparatus and goes well beyond what babies and animals, and even a normal adult without instruction, can do. The remainder of this book will be concerned with the embodied cognitive capacities that allow one to go from innate basic numerical abilities to a deep and rich understanding of, say, college-level mathematics.

From the work we have done to date, it appears that such advanced mathematical abilities are not independent of the cognitive apparatus used outside mathematics. Rather, it appears that the cognitive structure of advanced mathematics makes use of the kind of conceptual apparatus that is the stuff of ordinary everyday thought. This chapter presents prominent examples of the kinds of everyday conceptual mechanisms that are central to mathematics—especially advanced mathematics—as it is embodied in human beings. The mechanisms we will be discussing are (a) image schemas, (b) aspectual schemas, (c) conceptual metaphor, and (d) conceptual blends.

Spatial Relations Concepts and Image Schemas

Every language has a system of spatial relations, though they differ radically from language to language. In English we have prepositions like *in, on, through, above,* and so on. Other languages have substantially different systems. However, research in cognitive linguistics has shown that spatial relations in a given language decompose into conceptual primitives called *image schemas,* and these conceptual primitives appear to be universal.

For example, the English word *on,* in the sense used in "The book is *on* the desk," is a composite of three primitive image schemas:

- The *Above schema* (the book is *above* the desk)
- The *Contact schema* (the book is *in contact with* the desk)
- The *Support schema* (the book is *supported by* the desk)

The Above schema is orientational; it specifies an orientation in space relative to the gravitational pull one feels on one's body. The Contact schema is one of a number of topological schemas; it indicates the absence of a gap. The Support schema is force-dynamic in nature; it indicates the direction and nature of a force. In general, static image schemas fall into one of these categories: orientational, topological, and force-dynamic. In other languages, the primitives combine in different ways. Not all languages have a single concept like the English *on.* Even in a language as close as German, the *on* in *on the table* is rendered as *auf,* while the *on* in *on the wall* (which does not contain the Above schema) is translated as *an.*

A common image schema of great importance in mathematics is the *Container schema,* which occurs as the central part of the meaning of words like *in* and *out.* The Container schema has three parts: an Interior, a Boundary, and an

Exterior. This structure forms a gestalt, in the sense that the parts make no sense without the whole. There is no Interior without a Boundary and an Exterior, no Exterior without a Boundary and an Interior, and no Boundary without sides, in this case an Inside and an Outside. This structure is topological in the sense that the boundary can be made larger, smaller, or distorted and still remain the boundary of a Container schema.

To get schemas for the concepts In and Out, more must be added to the Container schema. The concept In requires that the Interior of the Container schema be "profiled"—that is, highlighted or activated in some way over the Exterior and Boundary. In addition, a figure/ground distinction must be added. For example, in a sentence like "The car is in the garage," the garage is the ground; that is, it is the landmark relative to which the car (the figure) is located. In cognitive linguistics, the ground in an image schema is called the *Landmark*, and the figure is called the *Trajector*. Thus, the In schema has the structure:

- Container schema, with Interior, Boundary, and Exterior
- Profiled: the Interior
- Landmark: the Interior

Image schemas have a special cognitive function: They are both *per*ceptual and *con*ceptual in nature. As such, they provide a bridge between language and reasoning on the one hand and vision on the other. Image schemas can fit visual perception, as when we see the milk as being *in* the glass. They can also be imposed on visual scenes, as when we see the bees swarming *in* the garden, where there is no physical container that the bees are in. Because spatial-relations terms in a given language name complex image schemas, image schemas are the link between language and spatial perception.

In addition, complex image schemas like *In* have built-in spatial "logics" by virtue of their image-schematic structures. Figure 2.1 illustrates the spatial logic built into the Container schema. In connection with this figure, consider the following two statements:

1. Given two Container schemas A and B and an object X, if A is *in* B and X is *in* A, then X is *in* B.
2. Given two Container schemas A and B and an object Y, if A is *in* B and Y is *outside of* B, then Y is *outside of* A.

We don't have to perform deductive operations to draw these conclusions. They are self-evident simply from the images in Figure 2.1. Because image

(a)

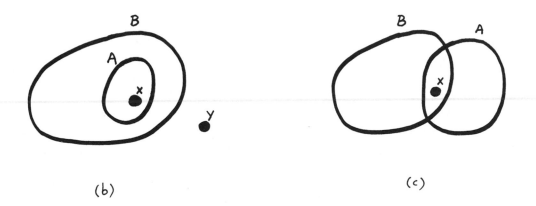

(b)   (c)

FIGURE 2.1   The logic of cognitive Container schemas. In (a), one cognitive Container schema, *A*, occurs inside another, *B*. By inspection, one can see that if *X* is in *A*, then *X* is in *B*. Similarly, if *Y* is outside *B*, then *Y* is outside *A*. We conceptualize physical containers in terms of cognitive containers, as shown in (b), which has the same logic as (a). However, conceptual containers, being part of the mind, can do what physical containers usually cannot— namely, form intersections, as in (c). In that case, an imagined entity *X* can be in two Container schemas *A* and *B* at once. Cognitive Container schemas are used not only in perception and imagination but also in conceptualization, as when we conceptualize bees as swarming *in* the garden. Container schemas are the cognitive structures that allow us to make sense of familiar Venn diagrams (see Figure 2.4).

schemas have spatial logics built into their imagistic structure, they can function as spatial concepts and be used directly in spatial reasoning. Reasoning about space seems to be done directly in spatial terms, using image schemas rather than symbols, as in mathematical proofs and deductions in symbolic logic.

Ideas do not float abstractly in the world. Ideas can be created only by, and instantiated only in, brains. Particular ideas have to be generated by neural structures in brains, and in order for that to happen, exactly the right kind of neural processes must take place in the brain's neural circuitry. Given that image schemas are conceptual in nature—that is, they constitute ideas with a structure of a very special kind—they must arise through neural circuitry of a very special kind.

Terry Regier (1996) has used the techniques of structured connectionism to build a computational neural model of a number of image schemas, as part of a neural simulation of the learning of spatial-relations terms in various languages. The research involved in Regier's simulation makes certain things clear. First, topographic maps of the visual field are needed in order to link cognition to vision. Second, a visual "filling-in" mechanism (Ramachandran & Gregory, 1991), in which activation spreads from outside to inside in a map of the visual field, will, in combination with other neural structures required, yield the topological properties of the Container schema. Third, orientation-sensitive cell assemblies found in the visual cortex are employed by orientational schemas. Fourth, map comparisons, requiring neural connections across maps, are needed. Such map-comparison structures are the locus of the relationship between the Trajector and the Landmark. Whatever changes are made in future models of spatial-relations concepts, it appears that at least these features will be needed.

Here is the importance of this for embodied mathematics: The concept of containment is central to much of mathematics. Closed sets of points are conceptualized as containers, as are bounded intervals, geometric figures, and so on. The concept of orientation is equally central. It is used in notions like angles, direction of change (tangents to a curve), rotations, and so on. The concepts of containment and orientation are not special to mathematics but are used in thought and language generally. Like any other concepts, these arise only via neural mechanisms in the right kind of neural circuitry. It is of special interest that the neural circuitry we have evolved for other purposes is an inherent part of mathematics, which suggests that embodied mathematics does not exist independently of other embodied concepts used in everyday life. Instead, mathematics makes use of our adaptive capacities—our ability to adapt other cognitive mechanisms for mathematical purposes.

Incidentally, the visual system of the brain, where such neural mechanisms as orientational cell assemblies reside, is not restricted to vision. It is also the locus of mental imagery. Mental imagery experiments, using fMRI techniques, have shown that much of the visual system, down to the primary visual cortex, is active when we create mental imagery without visual input. The brain's visual system is also active when we dream (Hobson, 1988, 1994). Moreover, congenitally blind people, most of whom have the visual system of the brain intact, can perform visual imagery experiments perfectly well, with basically the same results as sighted subjects, though a bit slower (Marmor & Zaback, 1976; Carpenter & Eisenberg, 1978; Zimler & Keenan, 1983; Kerr, 1983). In short, one should not think of the visual system as operating purely on visual input. Thus, it makes neurological sense that structures in the visual system can be used for conceptual purposes, even by the congenitally blind.

Moreover, the visual system is linked to the motor system, via the prefrontal cortex (Rizzolatti, Fadiga, Gallese, & Fogassi, 1996; Gallese, Fadiga, Fogassi, & Rizzolatti, 1996). Via this connection, motor schemas can be used to trace out image schemas with the hands and other parts of the body. For example, you can use your hands to trace out a seen or imagined container, and correspondingly you can visualize the structure of something whose shape you trace out with your hands in the dark. Thus, congenitally blind people can get "visual" image-schematic information from touch. Image schemas are kinesthetic, going beyond mere seeing alone, even though they use neural structures in the visual system. They can serve general conceptual purposes and are especially well suited for a role in mathematical thought.

There are many image schemas that characterize concepts important for mathematics: centrality, contact, closeness, balance, straightness, and many, many more. Image schemas and their logics are essential to mathematical reasoning.

Motor Control and Mathematical Ideas

One might think that nothing could be further from mathematical ideas than motor control, the neural system that governs how we move our bodies. But certain recent discoveries about the relation between motor control and the human conceptual system suggest that our neural motor-control systems may be centrally involved in mathematical thought. Those discoveries have been made in the field of structured connectionist neural modeling.

Building on work by David Bailey (1997), Srini Narayanan (1997) has observed that neural motor-control programs all have the same superstructure:

- *Readiness:* Before you can perform a bodily action, certain conditions of readiness have to be met (e.g., you may have to reorient your body, stop doing something else, rest for a moment, and so on).

- *Starting up:* You have to do whatever is involved in beginning the process (e.g., to lift a cup, you first have to reach for it and grasp it).
- *The main process:* Then you begin the main process.
- *Possible interruption and resumption:* While you engage in the main process, you have an option to stop, and if you do stop, you may or may not resume.
- *Iteration or continuing:* When you have done the main process, you can repeat or continue it.
- *Purpose:* If the action was done to achieve some purpose, you check to see if you have succeeded.
- *Completion:* You then do what is needed to complete the action.
- *Final state:* At this point, you are in the final state, where there are results and consequences of the action.

This might look superficially like a flow diagram used in classical computer science. But Narayanan's model of motor-control systems differs in many significant respects: It operates in real time, is highly resource- and context-dependent, has no central controller or clock, and can operate concurrently with other processes, accepting information from them and providing information to them. According to the model, these are all necessary properties for the smooth function of a neural motor-control system.

One might think the motor-control system would have nothing whatever to do with concepts, especially abstract concepts of the sort expressed in the grammars of languages around the world. But Narayanan has observed that this general motor-control schema has the same structure as what linguists have called *aspect*—the general structuring of events. Everything that we perceive or think of as an action or event is conceptualized as having that structure. We reason about events and actions in general using such a structure. And languages throughout the world all have means of encoding such a structure in their grammars. What Narayanan's work tells us is that *the same neural structure used in the control of complex motor schemas can also be used to reason about events and actions* (Narayanan, 1997).

We will call such a structure an *Aspect schema*.

One of the most remarkable of Narayanan's results is that exactly the same general neural control system modeled in his work can carry out a complex bodily movement when providing input to muscles, or carry out a rational inference when input to the muscles is inhibited. What this means is that neural control systems for bodily motions have the same characteristics needed for rational inference in the domain of aspect—that is, the structure of events.

Among the logical entailments of the aspectual system are two inferential patterns important for mathematics:

- The stage characterizing the completion of a process is further along relative to the process than any stage within the process itself.
- There is no point in a process further along than the completion stage of that process.

These fairly obvious inferences, as we shall see in Chapter 8 on infinity, take on considerable importance for mathematics.

Verbs in the languages of the world have inherent aspectual structure, which can be modified by various syntactic and morphological means. What is called *imperfective aspect* focuses on the internal structure of the main process. *Perfective aspect* conceptualizes the event as a whole, not looking at the internal structure of the process, and typically focusing on the completion of the action. Some verbs are inherently imperfective, like *breathe* or *live*. The iterative activity of breathing and the continuous activity of living—as we conceptualize them—do not have completions that are part of the concept. Just as the neural motor-control mechanism governing breathing does not have a completion (stopping, as in holding one's breath, is quite different from completion), so the concept is without a notion of completion. Death follows living but is *not* the *completion* of living, at least in our culture. Death is conceptualized, rather, as the cutting-off of life, as when a child is killed in an auto accident: Death follows life, but life is not completed. And you can say, "I have lived" without meaning that your life has been completed. Thus, an inherently imperfective concept is one that is conceptualized as being open-ended—as not having a completion.

There are two ways in which processes that have completions can be conceptualized: The completion may be either (1) internal to the process or (2) external to the process. This is not a matter of how the natural world really works but of how we conceptualize it and structure it through language. Take an example of case 1: If you jump, there are stages of jumping—namely, taking off, moving through the air, and landing. Landing completes the process of jumping. The completion, landing, is conceptualized as part of the jumping, as *internal* to what "jump" means. There is a minimally contrasting case that exemplifies case 2: flying. In the everyday concept of flying, as with birds and planes, landing is part of the conceptual frame. Landing follows flying and is a completion of flying. But *landing* is not conceptualized as part of *flying*. *Landing* is a completion of *flying* but it is *external* to *flying*. The distinction between an internal completion, as in *jump*, and an external completion, as in *fly*, is crucial in aspect.

Aspectual ideas occur throughout mathematics. A rotation through a certain number of degrees, for example, is conceptualized as a process with a starting point and an ending point. The original notion of continuity for a function was conceptualized in terms of a continuous process of motion—one without intermediate ending points. The very idea of an algorithmic process of calculation involves a starting point, a process that may or may not be iterative, and a well-defined completion. As we shall see in Chapters 8 through 11, all notions of infinity and infinitesimals use aspectual concepts.

The Source-Path-Goal Schema

Every language includes ways of expressing spatial sources (e.g., "from") and goals (e.g., "to," "toward") and paths intermediate between them (e.g., "along," "through," "across"). These notions do not occur isolated from one another but, rather, are part of a larger whole, the *Source-Path-Goal schema*. It is the principal image schema concerned with motion, and it has the following elements (or *roles*):

- A trajector that moves
- A source location (the starting point)
- A goal—that is, an intended destination of the trajector
- A route from the source to the goal
- The actual trajectory of motion
- The position of the trajector at a given time
- The direction of the trajector at that time
- The actual final location of the trajector, which may or may not be the intended destination.

Extensions of this schema are possible: the speed of motion, the trail left by the thing moving, obstacles to motion, forces that move one along a trajectory, additional trajectors, and so on.

This schema is topological in the sense that a path can be expanded or shrunk or deformed and still remain a path. As in the case of the Container schema, we can form spatial relations from this schema by the addition of profiling and a Trajector-Landmark relation. The concept expressed by *to* profiles the goal and identifies it as the landmark relative to which the motion takes place. The concept expressed by *from* profiles the source, taking the source as the landmark relative to which the motion takes place.

The Source-Path-Goal schema also has an internal spatial logic and built-in inferences (see Figure 2.2):

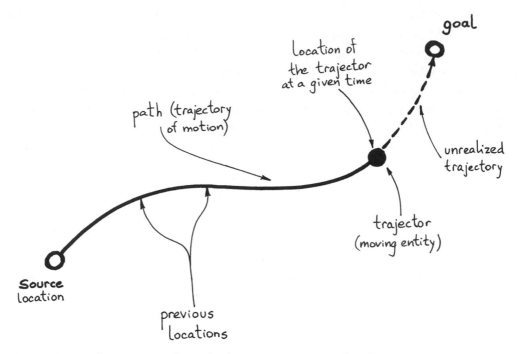

FIGURE 2.2   The Source-Path-Goal schema. We conceptualize linear motion using a conceptual schema in which there is a moving entity (called a trajector), a source of motion, a trajectory of motion (called a path), and a goal with an unrealized trajectory approaching that goal. There is a logic inherent in the structure of the schema. For example, if you are at a given location on a path, you have been at all previous locations on that path.

- If you have traversed a route to a current location, you have been at all previous locations on that route.
- If you travel from A to B and from B to C, then you have traveled from A to C.
- If there is a direct route from A to B and you are moving along that route toward B, then you will keep getting closer to B.
- If X and Y are traveling along a direct route from A to B and X passes Y, then X is further from A and closer to B than Y is.

The Source-Path-Goal schema is ubiquitous in mathematical thought. The very notion of a directed graph (see Chapter 7), for example, is an instance of the Source-Path-Goal schema. Functions in the Cartesian plane are often conceptualized in terms of motion along a path—as when a function is described as "going up," "reaching" a maximum, and "going down" again.

One of the most important manifestations of the Source-Path-Goal schema in natural language is what Len Talmy (1996, 2000) has called *fictive motion*. In one form of fictive motion, a line is thought of in terms of motion tracing that line, as in sentences like "The road *runs* through the woods" or "The fence *goes* up the hill." In mathematics, this occurs when we think of two lines "*meeting* at a point" or the graph of a function as "*reaching* a minimum at zero."

Conceptual Composition

Since image schemas are conceptual in nature, they can form complex composites. For example, the word "into" has a meaning—the *Into schema*—that is the composite of an *In schema* and a *To schema*. The meaning of "out of" is the composite of an *Out schema* and a *From schema*. These are illustrated in Figure 2.3. Formally, they can be represented in terms of correspondences between elements of the schemas that are part of the composite.

The following notations indicate composite structures.

The Into schema

- The In schema: A Container schema, with the Interior profiled and taken as Landmark
- The To schema: A Source-Path-Goal schema, with the Goal profiled and taken as Landmark
- Correspondences: (Interior; Goal) and (Exterior; Source)

The Out-of schema

- The Out schema: A Container schema, with the Exterior profiled and taken as Landmark
- The From schema: A Source-Path-Goal schema, with the Source profiled and taken as Landmark
- Correspondences: (Interior; Source) and (Exterior; Goal)

Conceptual Metaphor

Metaphor, long thought to be just a figure of speech, has recently been shown to be a central process in everyday thought. Metaphor is not a mere embellishment; it is the basic means by which abstract thought is made possible. One of the principal results in cognitive science is that abstract concepts are typically understood, via metaphor, in terms of more concrete concepts. This phenome-

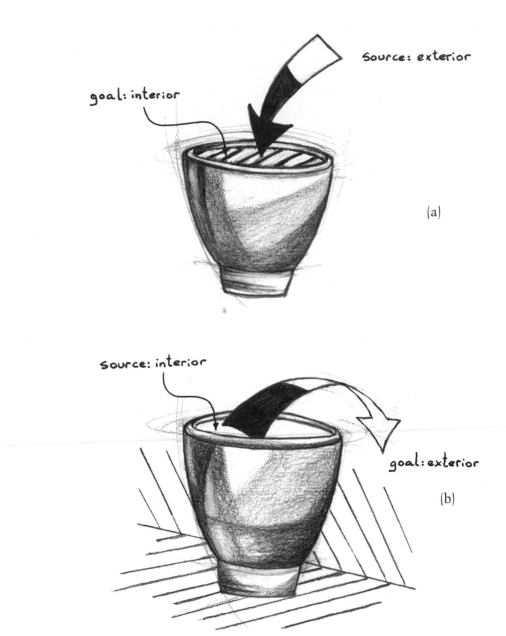

goal: interior

Source: exterior

(a)

Source: interior

goal: exterior

(b)

FIGURE 2.3 Conceptual composition of schemas. The English expressions "into" and "out of" have composite meanings. "In" profiles the interior of a Container schema, while "out" profiles the exterior. "To" profiles the goal of the Source-Path-Goal schema, while "from" profiles the source. With "into" (a), the interior is the goal and the exterior is the source. With "out of" (b), the interior is the source and the exterior is the goal.

non has been studied scientifically for more than two decades and is in general as well established as any result in cognitive science (though particular details of analyses are open to further investigation). One of the major results is that metaphorical mappings are systematic and not arbitrary.

Affection, for example, is understood in terms of physical warmth, as in sentences like "She *warmed* up to me," "You've been *cold* to me all day," "He gave me an *icy* stare," "They haven't yet *broken the ice.*" As can be seen by this example, the metaphor is not a matter of words, but of conceptual structure. The words are all different (*warm, cold, icy, ice*), but the conceptual relationship is the same in all cases: Affection is conceptualized in terms of warmth and disaffection in terms of cold.

This is hardly an isolated example:

- Importance is conceptualized in terms of size, as in "This is a big issue," "He's a giant in the meatpacking business," and "It's a small matter; we can ignore it."
- Similarity is conceptualized in terms of physical closeness, as in "These colors are very close," "Our opinions on politics are light-years apart," "We may not agree, but our views are in the same ballpark," and "Over the years, our tastes have diverged."
- Difficulties are conceptualized as burdens, as in "I'm weighed down by responsibilities," "I've got a light load this semester," and "He's overburdened."
- Organizational structure is conceptualized as physical structure, as in "The theory is full of holes," "The fabric of this society is unraveling," "His proposed plan is really tight; everything fits together very well."

Hundreds of such conceptual metaphors have been studied in detail. They are extremely common in everyday thought and language (see Lakoff & Johnson, 1980, 1999; Grady, 1998; Núñez, 1999). On the whole, they are used unconsciously, effortlessly, and automatically in everyday discourse; that is, they are part of the cognitive unconscious. Many arise naturally from correlations in our commonplace experience, especially our experience as children. Affection correlates with warmth in the experience of most children. The things that are important in their lives tend to be big, like their parents, their homes, and so on. Things that are similar tend to occur close together: trees, flowers, dishes, clouds. Carrying something heavy makes it difficult to move and to perform other activities. When we examine a complex physical object with an internal structure, we can perceive an organization in it. Not all conceptual metaphors arise in this way, but most of the basic ones do.

Such correlations in experience are special cases of the phenomenon of *conflation* (see C. Johnson, 1997). Conflation is part of embodied cognition. It is the simultaneous activation of two distinct areas of our brains, each concerned with distinct aspects of our experience, like the physical experience of warmth and the emotional experience of affection. In a conflation, the two kinds of experience occur inseparably. The coactivation of two or more parts of the brain generates a single complex experience—an experience of affection-with-warmth, say, or an experience of difficulty-with-a-physical-burden. It is via such conflations that neural links across domains are developed—links that often result in conceptual metaphor, in which one domain is conceptualized in terms of the other.

Each such conceptual metaphor has the same structure. Each is a unidirectional mapping from entities in one conceptual domain to corresponding entities in another conceptual domain. As such, conceptual metaphors are part of our system of thought. Their primary function is to allow us to reason about relatively abstract domains using the inferential structure of relatively concrete domains. The structure of image schemas is preserved by conceptual metaphorical mappings. In metaphor, conceptual cross-domain mapping is primary; metaphorical language is secondary, deriving from the conceptual mapping. Many words for source-domain concepts also apply to corresponding target-domain concepts. When words for source-domain concepts do apply to corresponding target concepts, they do so systematically, not haphazardly.

To see how the inferential structure of a concrete source domain gives structure to an abstract target domain, consider the common conceptual metaphor that States Are Locations, as in such expressions as "I'm in a depression," "He's close to hysteria; don't push him over the edge," and "I finally came out of my funk." The source domain concerns bounded regions in physical space. The target domain is about the subjective experience of being in a state.

STATES ARE LOCATIONS

Source Domain SPACE		Target Domain STATES
Bounded Regions in Space	→	States

Here is an example of how the patterns of inference of the source domain are carried over to the target domain.

If you're in a *bounded region*, you're not out of that *bounded region*.	→	If you're in a *state*, you're not out of that *state*.
If you're out of a *bounded region*, you're not in that *bounded region*.	→	If you're out of a *state*, you're not in that *state*.
If you're deep in a *bounded region*, you are far from being out of that *bounded region*.	→	If you're deep in a *state*, you are far from being out of that *state*.
If you are on the edge of a *bounded region*, you are close to being in that *bounded region*.	→	If you are on the edge of a *state*, you are close to being in that *state*.

Throughout this book we will use the common convention that *names* of metaphorical mappings are given in the form "A Is B," as in "States Are Bounded Regions in Space." It is important to distinguish between such names for metaphorical mappings and the metaphorical mappings themselves, which are given in the form "B → A," as in "Bounded Regions in Space → States." Here the source domain is to the left of the arrow and the target domain is to the right.

An enormous amount of our everyday abstract reasoning arises through such metaphorical cross-domain mappings. Indeed, much of what is often called logical inference is in fact spatial inference mapped onto an abstract logical domain. Consider the logic of the Container schema. There is a commonplace metaphor, Categories Are Containers, through which we understand a category as being a bounded region in space and members of the category as being objects inside that bounded region. The metaphorical mapping is stated as follows:

CATEGORIES ARE CONTAINERS

Source Domain CONTAINERS		Target Domain CATEGORIES
Bounded regions in space	→	Categories
Objects inside the bounded regions	→	Category members
One bounded region inside another	→	A subcategory of a larger category

Suppose we apply this mapping to the two inference patterns mentioned above that characterize the spatial logic of the Container schema, as follows:

Source Domain CONTAINER SCHEMA INFERENCES		*Target Domain* CATEGORY INFERENCES
Excluded Middle		*Excluded Middle*
Every object X is either in *Container schema A* or out of *Container schema A*.	→	Every entity X is either in *category A* or out of *category A*.
Modus Ponens		*Modus Ponens*
Given two *Container schemas* A and B and an object X, if A is in B and X is in A, then X is in B.	→	Given two *categories A* and B and an entity X, if A is in B and X is in A, then X is in B.
Hypothetical Syllogism		*Hypothetical Syllogism*
Given three *Container schemas* A, B, and C, if A is in B and B is in C, then A is in C.	→	Given three *categories A*, B and C, if A is in B and B is in C, then A is in C.
Modus Tollens		*Modus Tollens*
Given two *Container schemas* A and B and an object Y, if A is in B and Y is outside B, then Y is outside A.	→	Given two *categories A* and B and an entity Y, if A is in B and Y is outside B, then Y is outside A.

The point here is that the logic of Container schemas is an embodied spatial logic that arises from the neural characterization of Container schemas. The excluded middle, modus ponens, hypothetical syllogism, and modus tollens of classical categories are metaphorical applications of that spatial logic, since the Categories Are Containers metaphor, like conceptual metaphors in general, preserves the inferential structure of the source domain.

Moreover, there are important entailments of the Categories Are Containers metaphor:

The overlap of the interiors of two bounded regions	→	The conjunction of two categories
The totality of the interiors of two bounded regions	→	The disjunction of two categories

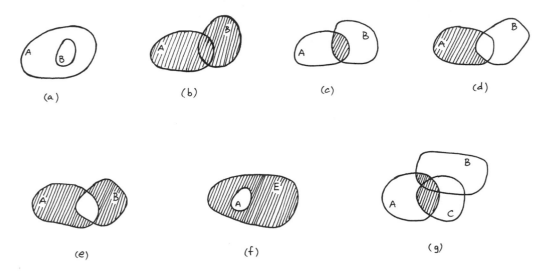

FIGURE 2.4   Venn diagrams. Here is a common set of Venn diagrams of the sort one finds in texts on classes and sets, which are typically conceptualized metaphorically as containers and derive their logics from the logic of conceptual Container schemas. When one "visualizes" classes and sets in this way, one is using cognitive Container schemas in the visualization. The diagrams depict various mathematical ideas: (a) the relation $B \subseteq A$; (b) $A \cup B$; (c) $A \cap B$; (d) the difference $A \backslash B$; (e) the symmetric difference $A \vartriangle B$; (f) the complement $C_E A$; and (g) $A \cap (B \cup C)$, which equals $(A \cap B) \cup (A \cap C)$.

In short, given the spatial logic of Container schemas, the Categories Are Containers metaphor yields an everyday version of what we might call folk Boolean logic, with intersections and unions. That is why the Venn diagrams of Boolean logic look so natural to us (see Figure 2.4), although there are differences between folk Boolean logic and technical Boolean logic, which will be discussed in Chapter 6. Folk Boolean logic, which is *conceptual*, arises from a *perceptual* mechanism—the capacity for perceiving the world in terms of contained structures.

From the perspective of the embodied mind, spatial logic is primary and the abstract logic of categories is secondarily derived from it via conceptual metaphor. This, of course, is the very opposite of what formal mathematical logic suggests. It should not be surprising, therefore, that embodied mathematics will look very different from disembodied formal mathematics.

Metaphors That Introduce Elements

Conceptual metaphors do not just map preexisting elements of the source domain onto preexisting elements of the target domain. They can also *introduce new elements* into the target domain. Consider, for example, the concept of love. There is a common metaphor in the contemporary Western world in which Love Is a Partnership. Here is the mapping.

LOVE IS A PARTNERSHIP

Source Domain BUSINESS		*Target Domain* LOVE
Partners	→	Lovers
Partnership	→	Love relationship
Wealth	→	Well-being
Profits from the business	→	"Profits" from the love relationship
Work for the business	→	"Work" put into the relationship
Sharing of work for the business	→	Sharing of "work" put into the relationship
Sharing of profits from the business	→	Sharing of "profits" from the relationship

Love need not always be conceptualized via this metaphor as a partnership. Romeo and Juliet's love was not a partnership, nor was Tristan and Isolde's. Similarly, love in many cultures around the world is not conceptualized in terms of business—and it need not be so conceptualized for individual cases in the Western world. But this is a common metaphorical way of understanding love—so common that it is sometimes taken as literal. For example, sentences like "I'm putting all the work into this relationship and you're getting everything out of it," "It was hard work, but worth it," and "The relationship was so unrewarding that it wasn't worth the effort" are so commonplace in discussions of love relationships that they are rarely noticed as metaphorical at all.

From the perspective of this book, there is an extremely important feature of this metaphor: It *introduces* elements into the target domain that are not inherent to the target domain. It is not inherent in love-in-itself that there be "work in the relationship," "profits (increases in well-being) from the relationship," and a "sharing of relationship work and profits." Romeo and Juliet would have been aghast at such ideas. These ideas are elements introduced into the target domain by the Love Is a Partnership metaphor, and they don't exist there without it.

The fact that metaphors can introduce elements into a target domain is extremely important for mathematics, as we shall see later when we take up various forms of actual infinity (Chapter 8).

Evidence

Over the past two decades, an enormous range of empirical evidence has been collected that supports this view of conceptual metaphor. The evidence comes from various sources:

- generalizations over polysemy (cases where the same word has multiple systematically related meanings)
- generalizations over inference patterns (cases where source and target domains have corresponding inference patterns)
- novel cases (new examples of conventional mappings, as in poetry, song, advertisements, and so on) (see Lakoff & Turner, 1989)
- psychological experiments (see Gibbs, 1994)
- historical semantic change (see Sweetser, 1990)
- spontaneous gesture (see McNeill, 1992)
- American Sign Language (see Taub, 1997)
- child language development (see C. Johnson, 1997)
- discourse coherence (see Narayanan, 1997)
- cross-linguistic studies.

For a thorough discussion of such evidence, see Lakoff and Johnson, 1999.

Sophisticated Mathematical Ideas

Sophisticated mathematics, as we have pointed out, is a lot more than just basic arithmetic. Mathematics extends the use of numbers to many other ideas, for example, the numerical study of angles (trigonometry), the numerical study of change (calculus), the numerical study of geometrical forms (analytic geometry), and so on. We will argue, in our discussion of all these topics and more, that conceptual metaphor is the central cognitive mechanism of extension from basic arithmetic to such sophisticated applications of number. Moreover, we will argue that a sophisticated understanding of arithmetic itself requires conceptual metaphors using nonnumerical mathematical source domains (e.g., geometry and set theory). We will argue further that conceptual metaphor is also the principal cognitive mechanism in the attempt to provide set-theoretical foundations for mathematics and in the understanding of set theory itself.

Finally, it should become clear in the course of this discussion that much of the "abstraction" of higher mathematics is a consequence of the systematic layering of metaphor upon metaphor, often over the course of centuries.

Each metaphorical layer, as we shall see, carries inferential structure systematically from source domains to target domains—systematic structure that gets lost in the layers unless they are revealed by detailed metaphorical analysis. A good part of this book is concerned with such metaphorical decomposition of sophisticated mathematical concepts. Because this kind of study has never been done before, we will not be able to offer the extensive forms of evidence that have been found in decades of studies of conceptual metaphor in everyday language and thought. For this reason, we will limit our study to cases that are relatively straightforward—cases where the distinctness of the source and target domains is clear, where the correspondences across the domains have been well established, and where the inferential structures are obvious.

Conceptual Blends

A *conceptual blend* is the conceptual combination of two distinct cognitive structures with fixed correspondences between them. In mathematics, a simple case is the unit circle, in which a circle is superimposed on the Cartesian plane with the following fixed correspondences: (a) The center of the circle is the origin (0,0), and (b) the radius of the circle is 1. This blend has entailments that follow from these correspondences, together with the inferential structure of *both domains*. For example, the unit circle crosses the x-axis at (1,0) and (–1,0), and it crosses the y-axis at (0,1) and (0,–1). The result is more than just a circle. It is a circle that has a fixed position in the plane and whose circumference is a length commensurate with the numbers on the x- and y-axes. A circle in the Euclidean plane, where there are no axes and no numbers, would not have these properties.

When the fixed correspondences in a conceptual blend are given by a metaphor, we call it a *metaphorical blend*. An example we will discuss extensively below is the Number-Line Blend, which uses the correspondences established by the metaphor Numbers Are Points on a Line. In the blend, new entities are created—namely, *number-points*, entities that are at once numbers and points on a line (see Fauconnier 1997; Turner & Fauconnier, 1995; Fauconnier & Turner, 1998). Blends, metaphorical and nonmetaphorical, occur throughout mathematics.

Many of the most important ideas in mathematics are metaphorical conceptual blends. As will become clear in the case-study chapters, understanding mathematics requires the mastering of extensive networks of metaphorical blends.

Symbolization

As we have noted, there is a critical distinction to be made among mathematical concepts, the written mathematical symbols for those concepts, and the words for the concepts. The words (e.g., "eighty-five" or "quatre-vingt-cinq") are part of some natural language, not mathematics proper.

In embodied mathematics, mathematical symbols, like 27, π, or $e^{\pi i}$, are meaningful by virtue of the mathematical concepts that they attach to. Those mathematical concepts are given in cognitive terms (e.g., image schemas; imagined geometrical shapes; metaphorical structures, like the number line; and so on), and those cognitive structures will ultimately require a neural account of how the brain creates them on the basis of neural structure and bodily and social experience. To understand a mathematical symbol is to associate it with a concept—something meaningful in human cognition that is ultimately grounded in experience and created via neural mechanisms.

As Stanislas Dehaene observed in the case of Mr. M—and as many of us experienced in grade school—numerical calculation may be performed with or without genuine understanding. Mr. M could remember his multiplication tables, but they were essentially meaningless to him.

The meaning of mathematical symbols is not in the symbols alone and how they can be manipulated by rule. Nor is the meaning of symbols in the interpretation of the symbols in terms of set-theoretical models that are themselves uninterpreted. Ultimately, mathematical meaning is like everyday meaning. It is part of embodied cognition.

This has important consequences for the teaching of mathematics. Rote learning and drill is not enough. It leaves out understanding. Similarly, deriving theorems from formal axioms via purely formal rules of proof is not enough. It, too, can leave out understanding. The point is not to be able to prove *that* $e^{\pi i} = -1$ but, rather, to be able to prove it knowing what $e^{\pi i}$ means, and knowing *why* $e^{\pi i} = -1$ on the basis of what $e^{\pi i}$ means, not just on the basis of the formal proof. In short, what is required is an adequate mathematical idea analysis to show *why* $e^{\pi i} = -1$ given our understanding of the ideas involved.

Euler's equation, $e^{\pi i} + 1 = 0$, ties together many of the most central ideas in classical mathematics. Yet on the surface it involves only numbers: e, π, i, 1, and 0. To show how this equation ties together *ideas*, we must have a theory of mathematical ideas and a theory of how they are mathematicized in terms of numbers.

Our interest, of course, goes beyond just $e^{\pi i}$ as such. Indeed, we are also interested in the all-too-common conception that mathematics is about calculation and about formal proofs from formal axioms and definitions and not about ideas and understanding. From the perspective of embodied mathematics, ideas and understanding are what mathematics is centrally about.

3

Embodied Arithmetic:
The Grounding Metaphors

ARITHMETIC IS A LOT MORE THAN SUBITIZING and the elementary numerical capacities of monkeys and newborn babies. To understand what arithmetic is from a cognitive perspective, we need to know much more. Why does arithmetic have the properties it has? Where do the laws of arithmetic come from? What cognitive mechanisms are needed to go from what we are born with to full-blown arithmetic? Arithmetic may seem easy once you've learned it, but there is an awful lot to it from the perspective of the embodied mind.

What Is Special About Mathematics?

As subsystems of the human conceptual system, arithmetic in particular and mathematics in general are special in several ways. They are:

- precise,
- consistent,
- stable across time and communities,
- understandable across cultures,
- symbolizable,
- calculable,
- generalizable, and
- effective as general tools for description, explanation, and prediction in a vast number of everyday activities, from business to building to sports to science and technology.

Any cognitive theory of mathematics must take these special properties into account, showing how they are possible given ordinary human cognitive capacities. That is the goal of this chapter.

The Cognitive Capacities Needed for Arithmetic

We are born with a minimal innate arithmetic, part of which we share with other animals. It is not much, but we do come equipped with it. Innate arithmetic includes at least two capacities: (1) a capacity for subitizing—instantly recognizing small numbers of items—and (2) a capacity for the simplest forms of adding and subtracting small numbers. (By "number" here, we mean a *cardinal* number, a number that specifies how many objects there are in a collection.) When we subitize, we have already limited ourselves to a grouping of objects in our visual field and we are distinguishing how many objects there are in that grouping.

In addition, we and many animals (pigeons, parrots, raccoons, rats, chimpanzees) have an innate capacity for "numerosity"—the ability to make consistent rough estimates of the number of objects in a group.

But arithmetic involves more than a capacity to subitize and estimate. Subitizing is certain and precise within its range. But we have additional capacities that allow us to extend this certainty and precision. To do this, we must count. Here are the cognitive capacities needed in order to count, say, on our fingers:

- *Grouping capacity:* To distinguish what we are counting, we have to be able to group discrete elements visually, mentally, or by touch.
- *Ordering capacity:* Fingers come in a natural order on our hands. But the objects to be counted typically do not come in any natural order in the world. They have to be ordered—that is, placed in a sequence, as if they corresponded to our fingers or were spread out along a path.
- *Pairing capacity:* We need a cognitive mechanism that enables us to sequentially pair individual fingers with individual objects, following the sequence of objects in order.
- *Memory capacity:* We need to keep track of which fingers have been used in counting and which objects have been counted.
- *Exhaustion-detection capacity:* We need to be able to tell when there are "no more" objects left to be counted.
- *Cardinal-number assignment:* The last number in the count is an ordinal number, a number in a sequence. We need to be able to assign that ordinal number as the size—the cardinal number—of the group counted. That cardinal number, the size of the group, has no notion of sequence in it.

- *Independent-order capacity:* We need to realize that the cardinal number assigned to the counted group is independent of the order in which the elements have been counted. This capacity allows us to see that the result is always the same.

When these capacities are used within the subitizing range between 1 and 4, we get stable results because cardinal-number assignment is done by subitizing, say, subitizing the fingers used for counting.

To count beyond four—the range of the subitizing capacity—we need not only the cognitive mechanisms listed above but the following additional capacities:

- *Combinatorial-grouping capacity:* You need a cognitive mechanism that allows you to put together perceived or imagined groups to form larger groups.
- *Symbolizing capacity:* You need to be able to associate physical symbols (or words) with numbers (which are conceptual entities).

But subitizing and counting are the bare beginnings of arithmetic. To go beyond them, to characterize arithmetic operations and their properties, you need much richer cognitive capacities:

- *Metaphorizing capacity:* You need to be able to conceptualize cardinal numbers and arithmetic operations in terms of your experiences of various kinds—experiences with groups of objects, with the part-whole structure of objects, with distances, with movement and locations, and so on.
- *Conceptual-blending capacity.* You need to be able to form correspondences across conceptual domains (e.g., combining *subitizing* with *counting*) and put together different conceptual metaphors to form complex metaphors.

Conceptual metaphor and conceptual blending are among the most basic cognitive mechanisms that take us beyond minimal innate arithmetic and simple counting to the elementary arithmetic of natural numbers. What we have found is that there are two types of conceptual metaphor used in projecting from subitizing, counting, and the simplest arithmetic of newborns to an arithmetic of natural numbers.

The first are what we call *grounding metaphors*—metaphors that allow you to project from everyday experiences (like putting things into piles) onto abstract concepts (like addition). The second are what we call *linking metaphors*, which link arithmetic to other branches of mathematics—for example, metaphors that allow you to conceptualize arithmetic in spatial terms, linking, say, geometry to arithmetic, as when you conceive of numbers as points on a line.

Two Kinds of Metaphorical Mathematical Ideas

Since conceptual metaphors play a major role in characterizing mathematical ideas, grounding and linking metaphors provide for two types of metaphorical mathematical ideas:

1. Grounding metaphors yield *basic, directly grounded ideas*. Examples: addition as adding objects to a collection, subtraction as taking objects away from a collection, sets as containers, members of a set as objects in a container. These usually require little instruction.
2. Linking metaphors yield *sophisticated ideas*, sometimes called *abstract ideas*. Examples: numbers as points on a line, geometrical figures as algebraic equations, operations on classes as algebraic operations. These require a significant amount of explicit instruction.

This chapter is devoted to *grounding* metaphors. The rest of the book is devoted primarily to *linking* metaphors.

Incidentally, there is another type of metaphor that this book is not about at all: what we will call *extraneous* metaphors, or metaphors that have nothing whatever to do with either the grounding of mathematics or the structure of mathematics itself. Unfortunately, the term "metaphor," when applied to mathematics, has mostly referred to such extraneous metaphors. A good example of an extraneous metaphor is the idea of a "step function," which can be drawn to look like a staircase. The staircase image, though helpful for visualization, has nothing whatever to do with either the inherent content or the grounding of the mathematics. Extraneous metaphors can be eliminated without any substantive change in the conceptual structure of mathematics, whereas eliminating grounding or linking metaphors would make much of the conceptual content of mathematics disappear.

Preserving Inferences About Everyday Activities

Since conceptual metaphors preserve inference structure, such metaphors allow us to ground our understanding of arithmetic in our prior understanding of extremely commonplace physical activities. Our understanding of elementary arithmetic is based on a correlation between (1) the most basic literal aspects of arithmetic, such as subitizing and counting, and (2) everyday activities, such as collecting objects into groups or piles, taking objects apart and putting them together, taking steps, and so on. Such correlations allow us to form metaphors by which we greatly extend our subitizing and counting capacities.

One of the major ways in which metaphor preserves inference is via the preservation of image-schema structure. For example, the formation of a collection or pile of objects requires conceptualizing that collection as a container—that is, a bounded region of space with an interior, an exterior, and a boundary—either physical or imagined. When we conceptualize numbers as collections, we project the logic of collections onto numbers. In this way, experiences like grouping that correlate with simple numbers give further logical structure to an expanded notion of number.

The Metaphorizing Capacity

The metaphorizing capacity is central to the extension of arithmetic beyond mere subitizing, counting, and the simplest adding and subtracting. Because of its centrality, we will look at it in considerable detail, starting with the Arithmetic Is Object Collection metaphor. This is a grounding metaphor, in that it grounds our conception of arithmetic directly in an everyday activity.

No metaphor is more basic to the extension of our concept of number from the innate cardinal numbers to the natural numbers (the positive integers). The reason is that the correlation of grouping with subitizing and counting the elements in a group is pervasive in our experience from earliest childhood.

Let us now begin an extensive guided tour of everything involved in this apparently simple metaphor. As we shall see, even the simplest and most intuitive of mathematical metaphors is incredibly rich, and so the tour will be extensive.

Arithmetic As Object Collection

If a child is given a group of three blocks, she will naturally subitize them automatically and unconsciously as being three in number. If one is taken away, she will subitize the resulting group as two in number. Such everyday experiences of subitizing, addition, and subtraction with small collections of objects involve correlations between addition and adding objects to a collection and between subtraction and taking objects away from a collection. Such regular correlations, we hypothesize, result in neural connections between sensory-motor physical operations like taking away objects from a collection and arithmetic operations like the subtraction of one number from another. Such neural connections, we believe, *constitute a conceptual metaphor* at the neural level—in this case, the metaphor that Arithmetic Is Object Collection. This metaphor, we hypothesize, is learned at an early age, prior to any formal arithmetic training. Indeed, arithmetic training assumes this unconscious conceptual (not linguistic!) metaphor: In teaching arithmetic, we all take it for granted that the adding and subtracting of numbers can be understood in terms of adding and taking away objects from collections. Of course, at this stage all of these are mental operations *with no symbols!* Calculating with symbols requires additional capacities.

The Arithmetic Is Object Collection metaphor is a precise mapping from the domain of physical objects to the domain of numbers. The metaphorical mapping consists of

1. the source domain of object collection (based on our commonest experiences with grouping objects);
2. the target domain of arithmetic (structured nonmetaphorically by subitizing and counting); and
3. a mapping across the domains (based on our experience subitizing and counting objects in groups). The metaphor can be stated as follows:

ARITHMETIC IS OBJECT COLLECTION

Source Domain OBJECT COLLECTION		*Target Domain* ARITHMETIC
Collections of objects of the same size	→	Numbers
The size of the collection	→	The size of the number
Bigger	→	Greater
Smaller	→	Less
The smallest collection	→	The unit (One)
Putting collections together	→	Addition
Taking a smaller collection from a larger collection	→	Subtraction

Linguistic Examples of the Metaphor

We can see evidence of this conceptual metaphor in our everyday language. The word *add* has the physical meaning of physically placing a substance or a number of objects into a container (or group of objects), as in "*Add* sugar to my coffee," "*Add* some logs to the fire," and "*Add* onions and carrots to the soup." Similarly, *take . . . from, take . . . out of,* and *take . . . away* have the physical meaning of removing a substance, an object, or a number of objects from some container or collection. Examples include "*Take* some books *out of* the box," "*Take* some water *from* this pot," "*Take away* some of these logs." By virtue of the Arithmetic Is Object Collection metaphor, these expressions are used for the corresponding arithmetic operations of addition and subtraction.

> If you *add* 4 apples *to* 5 apples, how many do you have? If you *take* 2 apples *from* 5 apples, how many apples are *left*? *Add* 2 *to* 3 and you have 5. *Take* 2 *from* 5 and you have 3 *left*.

It follows from the metaphor that adding yields something *bigger* (more) and subtracting yields something *smaller* (less). Accordingly, words like *big* and *small*, which indicate size for objects and collections of objects, are also used for numbers, as in "Which is *bigger*, 5 or 7?" and "Two is *smaller* than four." This metaphor is so deeply ingrained in our unconscious minds that we have to think twice to realize that numbers are not physical objects and so do not literally have a size.

Entailments of the Metaphor

The Arithmetic Is Object Collection metaphor has many entailments. Each arises in the following way: Take the basic truths about collections of physical objects. Map them onto statements about numbers, using the metaphorical mapping. The result is a set of "truths" about the natural numbers under the operations of addition and subtraction.

For example, suppose we have two collections, *A* and *B*, of physical objects, with *A* bigger than *B*. Now suppose we add the same collection *C* to each. Then *A* plus *C* will be a bigger collection of physical objects than *B* plus *C*. This is a fact about collections of physical objects of the same size. Using the mapping Numbers Are Collections of Objects, this physical truth that we experience in grouping objects becomes a mathematical truth about numbers: If *A* is greater than *B*, then *A* + *C* is greater than *B* + *C*. All of the following truths about numbers arise in this way, via the metaphor Arithmetic Is Object Collection.

The Laws of Arithmetic Are Metaphorical Entailments

In each of the following cases, the metaphor Arithmetic Is Object Collection maps a property of the source domain of *object collections* (stated on the left) to a unique corresponding property of the target domain of *numbers* (stated on the right). This metaphor extends properties of the innate subitized numbers 1 through 4 to an indefinitely large collection of natural numbers. In the cases below, you can see clearly how properties of object collections are mapped by the metaphor onto properties of natural numbers in general.

MAGNITUDE

Object collections have a magnitude	→	*Numbers* have a magnitude

STABILITY OF RESULTS FOR ADDITION

Whenever you add a fixed *object collection* to a second fixed *object collection,* you get the same result.	→	Whenever you add a fixed *number* to another fixed *number,* you get the same result.

STABILITY OF RESULTS FOR SUBTRACTION

Whenever you subtract a fixed *object collection* from a second fixed *object collection,* you get the same result.	→	Whenever you subtract a fixed *number* from another fixed *number,* you get the same result.

INVERSE OPERATIONS

For *collections:* Whenever you subtract what you added, or add what you subtracted, you get the original *collection.*	→	For *numbers:* Whenever you subtract what you added, or add what you subtracted, you get the original *number.*

UNIFORM ONTOLOGY

Object collections play three roles in addition. • what you add to something; • what you add something to; • the result of adding. Despite their differing roles, they all have the same nature with respect to the operation of the *addition* of *object collections.*	→	*Numbers* play three roles in addition. • what you add to something; • what you add something to; • the result of adding. Despite their differing roles, they all have the same nature with respect to the operation of the *addition* of *numbers.*

CLOSURE FOR ADDITION

The process of *adding* an *object collection* to *another object collection* yields a *third object collection*.	→	The process of *adding* a *number* to a *number* yields a third *number*.

UNLIMITED ITERATION FOR ADDITION

You can add *object collections* indefinitely.	→	You can add *numbers* indefinitely.

LIMITED ITERATION FOR SUBTRACTION

You can subtract *object collections* from other *object collections* until nothing is left.	→	You can subtract *numbers* from other *numbers* until nothing is left.

SEQUENTIAL OPERATIONS

You can do combinations of adding and subtracting *object collections*.	→	You can do combinations of adding and subtracting *numbers*.

Equational Properties

EQUALITY OF RESULT

You can obtain the same resulting *object collection* via different operations.	→	You can obtain the same resulting *number* via different operations.

PRESERVATION OF EQUALITY

For *object collections*, adding equals to equals yields equals.	→	For *numbers*, adding equals to equals yields equals.
For *object collections*, subtracting equals from equals yields equals.	→	For *numbers*, subtracting equals from equals yields equals.

COMMUTATIVITY

For *object collections*, adding *A* to *B* gives the same result as adding *B* to *A*.	→	For *numbers*, adding *A* to *B* gives the same result as adding *B* to *A*.

ASSOCIATIVITY

| For *object collections*, adding B to C and then adding A to the result is equivalent to adding A to B and adding C to that result. | → | For *numbers*, adding B to C and then adding A to the result is equivalent to adding A to B and adding C to that result. |

Relationship Properties

LINEAR CONSISTENCY

| For *object collections*, if A is bigger than B, then B is smaller than A. | → | For *numbers*, if A is greater than B, then B is less than A. |

LINEARITY

| If A and B are two *object collections*, then either A is bigger than B, or B is bigger than A, or A and B are the same size. | → | If A and B are two *numbers*, then either A is greater than B, or B is greater than A, or A and B are the same magnitude. |

SYMMETRY

| If *collection A* is the same size as *collection B*, then B is the same size as A. | → | If *number A* is the same magnitude as *number B*, then B is the same magnitude as A. |

TRANSITIVITY

| For *object collections*, if A is bigger than B and B is bigger than C, then A is bigger than C. | → | For *numbers*, if A is greater than B and B is greater than C, then A is greater than C. |

In order for there to be a metaphorical mapping from object collections to numbers, the entailments of such a mapping must be consistent with the properties of innate arithmetic and its basic extensions. This rudimentary form of arithmetic has some of these properties—for example, uniform ontology, linear consistency, linearity, symmetry, commutativity, and preservation of equality. The Arithmetic Is Object Collection metaphor will map the object-collection version of these properties onto the version of these properties in innate arithmetic (e.g., $2 + 1 = 1 + 2$).

However, this metaphor will also extend innate arithmetic, adding properties that the innate arithmetic of numbers 1 through 4 does not have, because of its limited range—namely, *closure* (e.g., under addition) and what follows from closure: unlimited iteration for addition, sequential operations, equality of result, and preservation of equality. The metaphor will map these properties from the domain of object collections to the expanded domain of number. The result is the elementary arithmetic of addition and subtraction for natural numbers, *which goes beyond innate arithmetic.*

Thus, the fact that there is an innate basis for arithmetic does not mean that all arithmetic is innate. Part of arithmetic arises from our experience in the world with object collections. The Arithmetic Is Object Collection metaphor arises naturally in our brains as a result of regularly using innate neural arithmetic while interacting with small collections of objects.

Extending Elementary Arithmetic

The version of the Arithmetic Is Object Collection metaphor just stated is limited to conceptualizing addition and subtraction of numbers in terms of addition and subtraction of collections. Operations in one domain (using only collections) are mapped onto operations in the other domain (using only numbers). There is no single operation characterized in terms of elements from both domains—that is, no single operation that uses both numbers *and* collections simultaneously.

But with multiplication, we *do* need to refer to numbers and collections simultaneously, since understanding multiplication in terms of collections requires performing operations on *collections* a certain *number* of times. This cannot be done in a domain with collections alone or numbers alone. In this respect, multiplication is cognitively more complex than addition or subtraction.

The cognitive mechanism that allows us to extend this metaphor from addition and subtraction to multiplication and division is *metaphoric blending*. This is not a new mechanism but simply a consequence of having metaphoric mappings.

Recall that each metaphoric mapping is characterized neurally by a fixed set of connections across conceptual domains. The results of inferences in the source domain are mapped to the target domain. If both domains, together with the mapping, are activated at once (as when one is doing arithmetic on object collections), the result is a metaphoric blend: the simultaneous activation of two domains with connections across the domains.

Two Versions of Multiplication and Division

Consider 3 times 5 in terms of collections of objects:

- Suppose we have 3 small collections of 5 objects each. Suppose we pool these collections. We get a single collection of 15 objects.
- Now suppose we have a big pile of objects. If we put 5 objects in a box 3 times, we get 15 objects in the box. This is repeated addition: We added 5 objects to the box repeatedly—3 times.

In the first case, we are doing multiplication by *pooling*, and in the second by *repeated addition*.

Division can also be characterized in two corresponding ways, *splitting up* and *repeated subtraction:*

- Suppose we have a single collection of 15 objects, then we can split it up into 3 collections of 5 objects each. That is, 15 divided by 3 is 5.
- Suppose again that we have a collection of 15 objects and that we repeatedly subtract 5 objects from it. Then, after 3 repeated subtractions, there will be no objects left. Again, 15 divided by 3 is 5.

In each of these cases we have used numbers with only addition and subtraction defined in order to characterize multiplication and division metaphorically in terms of object collection. From a cognitive perspective, we have used a metaphoric blend of object collections together with numbers to extend the Arithmetic Is Object Collection metaphor to multiplication and division.

We can state the pooling and iteration extensions of this metaphor precisely as follows:

THE POOLING/SPLITTING EXTENSION OF THE ARITHMETIC IS OBJECT COLLECTION METAPHOR

Source Domain THE OBJECT-COLLECTION/ ARITHMETIC BLEND		Target Domain ARITHMETIC
The pooling of A subcollections of size B to form an overall collection of size C.	$\rightarrow$	Multiplication $(A \cdot B = C)$
The splitting up of a collection of size C into A subcollections of size B.	$\rightarrow$	Division $(C \div B = A)$

THE ITERATION EXTENSION OF THE ARITHMETIC IS OBJECT COLLECTION METAPHOR

Source Domain THE OBJECT-COLLECTION/ ARITHMETIC BLEND		Target Domain ARITHMETIC
The repeated addition (A times) of a collection of size B to yield a collection of size C.	→	Multiplication ($A \cdot B = C$)
The repeated subtraction of collections of size B from an initial collection of size C until the initial collection is exhausted. A is the number of times the subtraction occurs.	→	Division ($C \div B = A$)

Note that in each case, the result of the operation is given in terms of the size of the collection as it is understood in the source domain of collections. Since the result of a multiplication or division is always a *collection* of a given size, multiplication and division (in this metaphor) can be combined with the addition and subtraction of collections to give further results in terms of collections.

What is interesting about these two equivalent metaphorical conceptions of multiplication and division is that they are both defined relative to the number-collection blend, but *they involve different ways of thinking about* operating on collections.

These metaphors for multiplication and division map the properties of the source domain onto the target domain, giving rise to the most basic properties of multiplication and division. Let us consider the commutative, associative, and distributive properties.

COMMUTATIVITY FOR MULTIPLICATION

Pooling A collections of size B gives a collection of the same resulting size as pooling B collections of size A.	→	Multiplying A times B gives the same resulting number as multiplying B times A.

ASSOCIATIVITY FOR MULTIPLICATION

Pooling A collections of size B and pooling that number of collections of size C gives a collection of the same resulting size as pooling the number of A collections of the size of the collection formed by pooling B collections of size C.	$\rightarrow$ Multiplying A times B and multiplying the result times C gives the same number as multiplying A times the result of multiplying B times C.

DISTRIBUTIVITY OF MULTIPLICATION OVER ADDITION

First, pool A collections of the size of the collection formed by adding a collection of size B to a collection of size C. This gives a collection of the same size as adding a collection formed by pooling A collections of size B to A collections of size C.	$\rightarrow$ First, multiply A times the sum of B plus C. This gives the same number as adding the product of A times B to the product of A times C.

MULTIPLICATIVE IDENTITY

Pooling one collection of size A results in a collection of size A.	$\rightarrow$ Multiplying one times A yields A
Pooling A collections of size one yields a collection of size A.	$\rightarrow$ Multiplying A times one yields A.

INVERSE OF MULTIPLICATION

Splitting a collection of size A into A subcollections yields subcollections of size one.	$\rightarrow$ Dividing A by A yields one.

In each case, a true statement about collections is projected by the metaphor in the pooling/splitting extension onto the domain of numbers, yielding a true statement about arithmetic. The same will work for iterative extension (i.e., repeated addition and repeated subtraction).

Thus, the Arithmetic Is Object Collection metaphor extends our understanding of number from the subitized numbers of innate arithmetic and from simple counting to the arithmetic of the natural numbers, grounding the extension of arithmetic in our everyday experience with groups of physical objects.

Zero

The Arithmetic Is Object Collection metaphor does, however, leave a problem. What happens when we subtract, say, seven from seven? The result cannot be understood in terms of a collection. In our everyday experience, the result of taking a collection of seven objects from a collection of seven objects is an absence of any objects at all—*not a collection of objects*. If we want the result to be a number, then in order to accommodate the Arithmetic Is Object Collection metaphor we must conceptualize the absence of a collection *as a collection*. A new conceptual metaphor is necessary. What is needed is a metaphor that creates something out of nothing: From the absence of a collection, the metaphorical mapping creates a unique collection of a particular kind—a collection with no objects in it.

THE ZERO COLLECTION METAPHOR

The lack of objects to form a collection	→	The empty collection

Given this additional metaphor as input, the Arithmetic Is Object Collection metaphor will then map the empty collection onto a number—which we call "zero."

This new metaphor is of a type common in mathematics, which we will call an *entity-creating metaphor*. In the previous case, the conceptual metaphor *creates* zero as an actual number. Although zero is an extension of the object-collection metaphor, it is not a natural extension. It does not arise from a correlation between the experience of collecting and the experience of subitizing and doing innate arithmetic. It is therefore an artificial metaphor, concocted ad hoc for the purpose of extension.

Once the metaphor Arithmetic Is Object Collection is extended in this way, more properties of numbers follow as entailments of the metaphor.

ADDITIVE IDENTITY

Adding the empty collection to a collection of size *A* yields a collection of size *A*.	→	Adding zero to *A* yields *A*.
Adding a collection of size *A* to the empty collection yields a collection of size *A*.	→	Adding *A* to zero yields *A*.

INVERSE OF ADDITION	
Taking a collection of size A away from a collection of size A → yields the empty collection.	Subtracting A from A yields zero.

These metaphors ground our most basic extension of arithmetic—from the innate cardinal numbers to the natural numbers plus zero. As is well known, this understanding of number still leaves gaps: It does not give a meaningful characterization of 2 minus 5 or 2 divided by 3. To fill those gaps we need further entity-creating metaphors, e.g., metaphors for the negative numbers. We will discuss such metaphors shortly.

At this point, we have explored only one of the basic grounding metaphors for arithmetic. There are three more to go. It would be unnecessarily repetitive to go into each in the full detail given above. Instead, we will sketch only the essential features of these metaphors.

Arithmetic As Object Construction

Consider such commonplaces of arithmetic as these: "Five is *made up of* two plus three." "You can *factor* 28 *into* 7 times 4." "If you *put* 2 and 2 *together*, you get 4."

How is it possible to understand a number, which is an abstraction, as being "made up," or "composed of," other numbers, which are "put together" using arithmetic operations? What we are doing here is conceptualizing numbers as wholes made up of parts. The parts are other numbers. And the operations of arithmetic provide the patterns by which the parts fit together to form wholes. Here is the metaphorical mapping used to conceptualize numbers in this way.

ARITHMETIC IS OBJECT CONSTRUCTION

Source Domain OBJECT CONSTRUCTION		*Target Domain* ARITHMETIC
Objects (consisting of ultimate parts of unit size)	→	Numbers
The smallest whole object	→	The unit (one)
The size of the object	→	The size of the number
Bigger	→	Greater
Smaller	→	Less
Acts of object construction	→	Arithmetic operations
A constructed object	→	The result of an arithmetic operation

A whole object	→	A whole number
Putting objects together with other objects to form larger objects	→	Addition
Taking smaller objects from larger objects to form other objects	→	Subtraction

As in the case of Arithmetic Is Object Collection, this metaphor can be extended in two ways via metaphorical blending: fitting together/splitting up and iterated addition and subtraction.

THE FITTING TOGETHER/SPLITTING UP EXTENSION

| The fitting together of A parts of size B to form a whole object of size C | → | Multiplication $(A \cdot B = C)$ |
| The splitting up of a whole object of size C into A parts of size B, a number that corresponds in the blend to an object of size A, which is the result | → | Division $(C \div B = A)$ |

THE ITERATION EXTENSION

| The repeated addition (A times) of A parts of size B to yield a whole object of size C | → | Multiplication $(A \cdot B = C)$ |
| The repeated subtraction of parts of size B from an initial object of size C until the initial object is exhausted. The result, A, is the number of times the subtraction occurs. | → | Division $(C \div B = A)$ |

Fractions are understood metaphorically in terms of the characterizations of division (as splitting) and multiplication (as fitting together).

Fractions

A part of a unit object (made by splitting a unit object into n parts)	→	A simple fraction ($1/n$)
An object made by fitting together m parts of size $1/n$	→	A complex fraction (m/n)

These additional metaphorical mappings yield an important entailment about number based on a truth about objects.

If you split a unit object into n parts and then you fit the n parts together again, you get the unit object back.	→	If you divide 1 by n and multiply the result by n, you get 1. That is, $1/n \cdot n = 1$.

In other words, $1/n$ is the multiplicative inverse of n.

As in the case of the object-collection metaphor, a special additional metaphor is needed to conceptualize zero. Since the lack of an object is not an object, it should not, strictly speaking, correspond to a number. The zero object metaphor is thus an artificial metaphor.

The Zero Object Metaphor

The Lack of a Whole Object	→	Zero

The object-construction metaphor is intimately related to the object-collection metaphor. The reason is that constructing an object necessarily requires collecting the parts of the object together. Every whole made up of parts is a collection of the parts, with the added condition that the parts are assembled according to a certain pattern. Since object construction is a more specific version of object collection, the metaphor of Arithmetic As Object Construction is a more specific version of the metaphor of Arithmetic As Object Collection. Accordingly, the object-construction metaphor has all the inferences of the object-collection metaphor—the inferences we stated in the previous section. It differs in that it is extended to characterize fractions and so has additional inferences—for example, $(1/n) \cdot n = 1$.

It also has metaphorical entailments that characterize the decomposition of numbers into parts.

| Whole objects are composites of their parts, put together by certain operations. | → | Whole numbers are composites of their parts, put together by certain operations. |

It is this metaphorical entailment that gives rise to the field of number theory, the study of which numbers can be decomposed into other numbers and operations on them.

The Measuring Stick Metaphor

The oldest (and still often used) method for designing buildings or physically laying out dimensions on the ground is to use a measuring stick or string—a stick or string taken as a unit. These are physical versions of what in geometry are called *line segments*. We will refer to them as "physical segments." A distance can be measured by placing physical segments of unit length end-to-end and counting them. In the simplest case, the physical segments are body parts: fingers, hands, forearms, arms, feet, and so on. When we put physical segments end-to-end, the result is another physical segment, which may be a real or envisioned tracing of a line in space.

In a wide range of languages throughout the world, this concept is represented by a classifier morpheme. In Japanese, for example, the word *hon* (literally, "a long, thin thing") is used for counting such long, thin objects as sticks, canes, pencils, candles, trees, ropes, baseball bats, and so on—including, of course, rulers and measuring tapes. Even though English does not have a single word for the idea, it is a natural human concept.

THE MEASURING STICK METAPHOR

Source Domain THE USE OF A MEASURING STICK		Target Domain ARITHMETIC
Physical segments (consisting of ultimate parts of unit length)	→	Numbers
The basic physical segment	→	One
The length of the physical segment	→	The size of the number
Longer	→	Greater
Shorter	→	Less
Acts of physical segment placement	→	Arithmetic operations
A physical segment	→	The result of an arithmetic operation

Putting physical segments together end-to-end with other physical segments to form longer physical segments	→	Addition
Taking shorter physical segments from larger physical segments to form other physical segments	→	Subtraction

As in the previous two metaphors, there are two ways of characterizing multiplication and division: fitting together/dividing up and iterated addition and subtraction.

THE FITTING TOGETHER/DIVIDING UP EXTENSION

The fitting together of A physical segments of length B to form a line segment of length C	→	Multiplication $(A \cdot B = C)$
The splitting up of a physical segment C into A parts of length B. A is a number that corresponds in the blend to a physical segment of length A, which is the result.	→	Division $(C \div B = A)$

THE ITERATION EXTENSION

The repeated addition (A times) of A physical segments of length B to form a physical segment of length C.	→	Multiplication $(A \cdot B = C)$
The repeated subtraction of physical segments of length B from an initial physical segment of length C until nothing is left of the initial physical segment. The result, A, is the number of times the subtraction occurs.	→	Division $(C \div B = A)$

As in the case of the object-construction metaphor, the physical segment metaphor can be extended to define fractions.

A part of a physical segment (made by splitting a single physical segment into n equal parts)	→	A simple fraction $(1/n)$
A physical segment made by fitting together (end-to-end) m parts of size $1/n$	→	A complex fraction (m/n)

Just as in the object-construction metaphor, this metaphor needs to be extended in order to get a conceptualization of zero.

The lack of any physical segment	→	Zero

Up to this point, the measuring stick metaphor looks very much like the object-construction metaphor: A physical segment can be seen as a physical object, even if it is an imagined line in space. But physical segments are very special "constructed objects." They are unidimensional and they are continuous. In their abstract version they correspond to the line segments of Euclidean geometry. As a result, the blend of the source and target domains of this metaphor has a very special status. It is a blend of line (physical) segments with numbers specifying their length, which we will call the Number/Physical Segment blend.

Moreover, once you form the blend, a fateful entailment arises. Recall that the metaphor states that Numbers Are Physical Segments, and that given this metaphor you can characterize natural numbers, zero, and positive complex fractions (the rational numbers) in terms of physical segments. That is, for every positive rational number, this metaphor (given a unit length) provides a unique physical segment. The metaphorical mapping is unidirectional. It does not say that for any line segment at all, there is a corresponding number.

But the blend of source and target domains goes beyond the metaphor itself and has new entailments. When you form a blend of physical segments and numbers, constrained by the measuring stick metaphor, then within the blend there is a one-to-one correspondence between physical segments and numbers. The fateful entailment is this: Given a fixed unit length, it follows that for every physical segment there is a number.

Now consider the Pythagorean theorem: In $A^2 + B^2 = C^2$, let C be the hypotenuse of a right triangle and A and B be the lengths of the other sides. Let $A = 1$ and $B = 1$. Then $C^2 = 2$. The Pythagoreans had already proved that C could not be expressed as a fraction—that is, that it could not be a rational number—

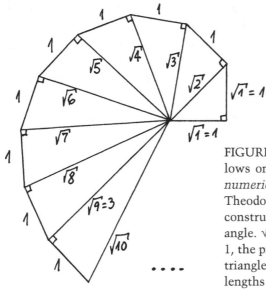

FIGURE 3.1 The measuring stick metaphor allows one to form physical segments of particular *numerical* lengths. In the diagram, taken from Theodorus of Cyrene (fourth century B.C.), √2 is constructed from the unit length 1 and a right triangle. √3 is then constructed from the unit length 1, the previously constructed length √2, and a right triangle. And so on. Without the metaphor, the lengths are just lengths, not numbers.

a ratio of physical lengths corresponding to integers. They assumed that only natural numbers and their ratios (the rational numbers) existed and that the length *C* was not a number at all; they called it an *incommensurable*—without ratio, that is, without a common measure.

But Eudoxus (c. 370 B.C.) observed, implicitly using the Number/Physical Segment blend, that corresponding to the hypotenuse in this triangle there must be a number: *C* = √2! This conclusion could not have been reached using numbers by themselves, taken literally. If you assume that only rational numbers exist and you prove that √2 cannot be a rational number, then it could just as well follow (as it did initially for the Pythagoreans) that √2 does not exist—that is, that 2 does not have any square root. But if, according to the Number/Physical Segment blend, there must exist a number corresponding to the length of every physical segment, then and only then must √2 exist as a number!

It was the measuring stick metaphor and the Number/Physical Segment blend that gave birth to the irrational numbers.

Arithmetic As Motion Along a Path

When we move in a straight line from one place to another, the path of our motion forms a physical segment—an imagined line tracing our trajectory. There is a simple relationship between a path of motion and a physical segment. The origin of the motion corresponds to one end of a physical segment; the endpoint of the motion corresponds to the other end of the physical segment; and the path of motion corresponds to the rest of the physical segment.

Given this correspondence between motions and physical segments, there is a natural metaphorical correlate to the measuring stick metaphor for arithmetic, namely, the metaphor that Arithmetic Is Motion Along a Path. Here is how that metaphor is stated.

ARITHMETIC IS MOTION ALONG A PATH

Source Domain MOTION ALONG A PATH		*Target Domain* ARITHMETIC
Acts of moving along the path	→	Arithmetic operations
A point-location on the path	→	The result of an arithmetic operation
The origin, the beginning of the path	→	Zero
Point-locations on a path	→	Numbers
The unit location, a point-location distinct from the origin	→	One
Further from the origin than	→	Greater than
Closer to the origin than	→	Less than
Moving from a point-location A away from the origin, a distance that is the same as the distance from the origin to a point-location B	→	Addition of B to A
Moving toward the origin from A, a distance that is the same as the distance from the origin to B	→	Subtraction of B from A

This metaphor can be extended to multiplication and division by means of iteration over addition and subtraction.

THE ITERATION EXTENSION

Starting at the origin, move A times in the direction away from the origin a distance that is the same as the distance from the origin to B.	→	Multiplication $(A \cdot B = C)$

| Starting at *C*, move toward the origin distances of length *B* repeatedly *A* times. | → | Division ($C \div B = A$) |

FRACTIONS

| Starting at 1, find a distance *d* such that by moving distance *d* toward the origin repeatedly *n* times, you will reach the origin. $1/n$ is the point-location at distance *d* from the origin. | → | A simple fraction ($1/n$) |
| Point-location reached moving from the origin a distance $1/n$ repeatedly *m* times. | → | A complex fraction (m/n) |

As we mentioned, the Arithmetic Is Motion metaphor corresponds in many ways to the measuring stick metaphor. But there is one major difference. In all the other metaphors that we have looked at so far, including the measuring stick metaphor, there had to be some entity-creating metaphor added to get zero. However, when numbers are point-locations on a line, the origin is by its very nature a point-location. When we designate zero as the origin, it is already a point-location.

Moreover, this metaphor provides a natural extension to negative numbers— let the origin be somewhere on a pathway extending indefinitely in both direc- tions. The negative numbers will be the point-locations on the other side of zero from the positive numbers along the same path. This extension was explicitly made by Rafael Bombelli in the second half of the sixteenth century. In Bombelli's extension of the point-location metaphor for numbers, positive num- bers, zero, and negative numbers are all point-locations on a line. This made it commonplace for European mathematicians to think and speak of the concept of a number *lying between* two other numbers—as in *zero lies between minus one and one*. Conceptualizing all (real) numbers metaphorically as point-loca- tions on the same line was crucial to providing a uniform understanding of number. These days, it is hard to imagine that there was ever a time when such a metaphor was not commonly accepted by mathematicians!

The understanding of numbers as point-locations has come into our language in the following expressions:

How *close* are these two numbers?
37 is *far away from* 189,712.

4.9 is *near* 5.

The result is *around* 40.

Count up *to* 20, without *skipping* any numbers.

Count *backward* from 20.

Count *to* 100, *starting at* 20.

Name all the numbers *from* 2 *to* 10.

The linguistic examples are important here in a number of respects. First, they illustrate how the language of motion can be recruited in a systematic way to talk about arithmetic. The conceptual mappings characterize what is systematic about this use of language. Second, these usages of language provide evidence for the existence of the conceptual mapping—evidence that comes not only from the words but also from what the words mean. The metaphors can be seen as stating generalizations not only over the use of the words but also over the inference patterns that these words supply from the source domain of motion, which are then used in reasoning about arithmetic.

We have now completed our initial description of the four basic grounding metaphors for arithmetic. Let us now turn to the relation between arithmetic and elementary algebra.

The Fundamental Metonymy of Algebra

Consider how we understand the sentence "When the pizza delivery boy comes, give him a good tip." The conceptual frame is Ordering a Pizza for Delivery. Within this frame, there is a role for the Pizza Delivery Boy, who delivers the pizza to the customer. In the situation, we do not know which *individual* will be delivering the pizza. But we need to conceptualize, make inferences about, and talk about that individual, whoever he is. Via the Role-for-Individual metonymy, the role "pizza delivery boy" comes to stand metonymically for the particular individual who fills the role—that is, who happens to deliver the pizza today. "Give him a good tip" is an instruction that applies to the individual, whoever he is.

This everyday conceptual metonymy, which exists outside mathematics, plays a major role in mathematical thinking: It allows us to go from concrete (case by case) arithmetic to general algebraic thinking. When we write "$x + 2 = 7$," x is our notation for a role, Number, standing for an individual number. "$x + 2 = 7$" says that whatever number x happens to be, adding 2 to it will yield 7.

This everyday cognitive mechanism allows us to state general laws like "$x + y = y + x$," which says that adding a number y to another number x yields the same result as adding x to y. It is this metonymic mechanism that makes the discipline of algebra possible, by allowing us to reason about numbers or other entities without knowing which particular entities we are talking about.

Clear examples of how we unconsciously use and master the Fundamental Metonymy of Algebra are provided by many passages in this very chapter. In fact, every time we have written (and every time you have read and understood) an expression such as "If collection A is the same size as collection B," or "adding zero to A yields A," we have been implicitly making use of the Fundamental Metonymy of Algebra. It is this cognitive mechanism that permits general proofs in mathematics—for example, proofs about any number, whatever it is.

The Metaphorical Meanings of One and Zero

The four grounding metaphors mentioned so far—Object Collection, Object Construction, the Measuring Stick, and Motion Along a Line—contain metaphorical characterizations of zero and one. Jointly, these metaphors characterize the symbolic meanings of zero and one. In the collection metaphor, zero is the empty collection. Thus, zero can connote *emptiness*. In the object-construction metaphor, zero is either thc lack of an object, the absence of an object or, as a result of an operation, the destruction of an object. Thus, zero can mean *lack*, *absence*, or *destruction*. In the measuring stick metaphor, zero stands for the *ultimate in smallness*, the lack of any physical segment at all. In the motion metaphor, zero is the origin of motion; hence, zero can designate an *origin*. Hence, zero, in everyday language, can symbolically denote emptiness, nothingness, lack, absence, destruction, ultimate smallness, and origin.

In the collection metaphor, one is the collection with a lone member and, hence, symbolizes *individuality* and *separateness* from others. In the object-construction metaphor, one is a whole number and, by virtue of this, signifies *wholeness*, *unity*, and *integrity*. In the measuring stick metaphor, one is the length specifying the unit of measure. In this case, one signifies a *standard*. And in the motion metaphor, one indicates the first step in a movement. Hence, it symbolizes a *beginning*. Taken together, these metaphors give one the symbolic values of individuality, separateness, wholeness, unity, integrity, a standard, and a beginning. Here are some examples.

- *Beginning:* One small step for a man; one great step for mankind.
- *Unity:* E pluribus unum ("From many, one").
- *Integrity:* Fred and Ginger danced as one.
- *Origin:* Let's start again from zero.

- *Emptiness:* There's zero in the refrigerator.
- *Nothingness:* I started with zero and made it to the top.
- *Destruction:* This nullifies all that we have done.
- *Lack (of ability):* That new quarterback is a big zero.

These grounding metaphors therefore explain why zero and one, which are literally numbers, have the symbolic values that they have. But that is not the real importance of these metaphors for mathematics. The real importance is that they explain how innate arithmetic gets extended systematically to give arithmetic distinctive properties that innate arithmetic does not have. Because of their importance we will give the four grounding metaphors for arithmetic a name: *the 4Gs*. We now turn to their implications.

4

Where Do the Laws
of Arithmetic Come From?

The Significance of the 4Gs

Innate arithmetic, as we saw, is extremely limited: It includes only subitizing, addition, and subtraction up to the number 4 at most. The 4Gs each arise via a conflation in everyday experience. Take object collection. Young children form small collections, subitize them, and add and take away objects from them, automatically forming additions and subtractions within the subitizable range. The same is true when they make objects, take steps, and later use sticks, fingers, and arms to estimate size. These correlations in everyday experience between innate arithmetic and the source domains of the 4Gs give rise to the 4Gs. The metaphors—at least in an automatic, unconscious form—arise naturally from such conflations in experience.

   The significance of the 4Gs is that they allow human beings, who have an innate capacity to form metaphors, to extend arithmetic beyond the small amount that we are born with, while preserving the basic properties of innate arithmetic. The mechanism is as follows: In each conflation of innate arithmetic with a source domain, the inferences of innate arithmetic fit those of the source domain (say, object collection). Just as $3 - 1 = 2$ abstractly, if you take one object from a collection of three objects, you get a collection of two objects. In other words, the inferences of abstract innate arithmetic hold when it is conceptually blended with object collection.

This may seem so obvious as to hardly be worth mentioning, but it is the basis for the extension of innate arithmetic way beyond its inherent limits. Because innate arithmetic "fits" object collection and construction, motion, and manipulation of physical segments, those four domains of concrete experience are suitable for metaphorical extensions of innate arithmetic that preserve its properties. Taking one step after taking two steps gets you to the same place as taking three steps, just as adding one object to a collection of two objects yields a collection of three objects.

Thus, the properties of innate arithmetic can be seen as "picking out" these four domains for the metaphorical extension of basic arithmetic capacities beyond the number 4. Indeed, the reason that these four domains all fit innate arithmetic is that there are structural relationships across the domains. Thus, object construction always involves object collection; you can't build an object without gathering the parts together. The two experiences are conflated and thereby neurally linked. Putting physical segments end-to-end is similar to object construction (think of legos here). When you use a measuring stick to mark off a distance, you are mentally constructing a line segment out of parts—a "path" from the beginning of the measurement to the end. A path of motion from point to point corresponds to such an imagined line segment. In short, there are structural correspondences between

- object collection and object construction
- the construction of a linear object and the use of a measuring stick to mark off a line segment of certain length
- using a measuring stick to mark off a line segment, or "path," and moving from location to location along a path.

As a result of these structural correspondences, there are *isomorphisms* across the 4G metaphors—namely, the correlations just described between the source domains of those metaphors.

That isomorphism defines a one-to-one correlation between metaphoric definitions of arithmetic operations—addition and multiplication—in the four metaphors. For there to be such an isomorphism, the following three conditions must hold:

- There is a one-to-one mapping, M, between elements in one source domain and elements in the other source domain—that is, the "images" under the "mapping."

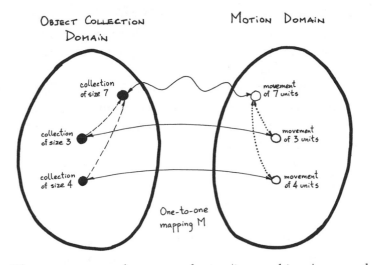

FIGURE 4.1 There are structural correspondences (isomorphisms) across the source domains of the four grounding metaphors for arithmetic (4Gs). The diagram depicts the isomorphism between the source domains of Object Collection and Motion Along a Path. There is a one-to-one mapping between the elements of the two source domains. Moreover, the images of sums correspond to the sums of images, and the images of products correspond to the products of images. For example, for a collection of sizes 3 and 4 there are unique movements of 3 and 4 units, respectively (solid lines). Besides, the result of "adding collections of size 3 and 4" (segmented lines) has a unique image, that is, the image of sums: "a movement of 7 units" (wavy line). This is equivalent to the sum of images: Adding the image of "a collection of size 3" (i.e., "a movement of 3 units") and the image of "a collection of size 4" (i.e., "a movement of 4 units") yields "a movement of 7 units" (dotted lines).

- M preserves sums: $M(x + y) = M(x) + M(y)$; that is, the images of sums correspond to the sums of images.
- M preserves products: $M(x \cdot y) = M(x) \cdot M(y)$; that is, the images of products correspond to the products of images.

Consider, for example, the source domains of object collection and motion, which appear quite dissimilar. There is such an isomorphism between those two source domains. First, there is a one-to-one correspondence, M, between sizes of collections and distances moved. For example, a collection of size three is uniquely mapped to a movement of three units of length, and conversely. It should be obvious from Figure 4.1 that the last two conditions are also met.

This is something very special about the conceptual system of mathematics: The source domains of all four basic grounding metaphors for the arithmetic of natural numbers are isomorphic in this way! Note that there are no numbers in these source domains; there are only object collections, motions, and so on. But given how they are mapped onto the natural numbers, the relevant inferential structures of all these domains are isomorphic.

Aside from the way they are mapped onto the natural numbers, these four source domains are not isomorphic: Object construction characterizes fractions but not zero or negative numbers, whereas motion along a path characterizes zero and negative numbers. In other words, if you look at the complete domains in isolation, you will not see an isomorphism across the source domains. What creates the isomorphism is the collection of mappings from these source domains of the 4Gs onto natural numbers. And what grounds the mappings onto natural numbers are the experiences we have, across the four domains, with innate arithmetic—with subitizing and counting in such early experiences as forming collections, putting things together, moving from place to place, and so on.

Numbers Are Things

In each of the 4Gs, numbers are things that exist in the world: collections of objects, complex objects with parts, physical segments, and locations. These four metaphors thus induce a more general metaphor, that *Numbers Are Things in the World*. Though they can function as quantifiers in everyday language (e.g., "five apples"), numbers in arithmetic statements function like things; the names of numbers (e.g., "five") are proper nouns and go in noun positions, as in "Two is less than four," "Divide 125 by 5," and so on.

The metaphor Numbers Are Things in the World has deep consequences. The first is the widespread view of mathematical Platonism. If objects are real entities out there in the universe, then understanding Numbers metaphorically as Things in the World leads to the metaphorical conclusion that numbers have an objective existence as real entities out there as part of the universe. This is a metaphorical inference from one of our most basic unconscious metaphors. As such, it seems natural. We barely notice it. Given this metaphorical inference, other equally metaphorical inferences follow, shaping the intuitive core of the philosophy of mathematical Platonism:

- Since real objects in the world have unique properties that distinguish them from all other entities, so there should be a uniquely true mathematics. Every mathematical statement about numbers should be absolutely true or false. There should be no equally valid alternative forms of mathematics.

- Numbers should not be products of minds, any more than trees or rocks or stars are products of minds.
- Mathematical truths are discovered, not created.

What is particularly ironic about this is that *it follows from the empirical study of numbers as a product of mind that it is natural for people to believe that numbers are* not *a product of mind!*

Closure

The metaphor that Numbers Are Things in the World has a second important consequence for both the structure and practice of mathematics. In most of our everyday experience, when we operate on actual physical entities, the result is another physical entity. If we put two objects together, we get another object. If we combine two collections, we get another collection. If we start moving at one location, we wind up at another location. Over much of our experience, a general principle holds:

- An operation on physical things yields a physical thing of the same kind.

The metaphor that Numbers Are Things yields a corresponding principle:

- An operation on numbers yields a number of the same kind.

The name for this metaphorical principle in mathematics is *closure*. Closure is not a property of innate arithmetic. "Subitizable 3" plus "subitizable 4" does not produce a subitizable number; we don't normally subitize 7. But closure does arise naturally from the grounding metaphors.

Closure is a central idea in mathematics. It has led mathematicians to extend number systems further and further until closure is achieved—and to stop with closure. Thus, the natural numbers had to be extended in various ways to achieve closure relative to the basic arithmetic operations (addition, subtraction, multiplication, division, raising to powers, and taking roots):

- Because, say, $5 - 5$ is not a natural number, zero had to be added.
- Because, say, $3 - 5$ is not a natural number, negative numbers had to be added.

- Because, say, $3 \div 5$ is not a natural number, rational numbers (fractions) had to be added.
- Because, say, $\sqrt{2}$ is not a rational number, irrational numbers had to be added to form the "real numbers."
- Because, say, $\sqrt{-1}$ is not a real number, the "imaginary numbers" had to be added to form the "complex numbers."

Extending the natural numbers to the complex numbers finally achieved closure relative to the basic operations of arithmetic. As the fundamental theorem of algebra implies, any arithmetic operation on any complex numbers yields a complex number (except for division by zero).

The notion of closure is central to all branches of mathematics and is an engine for creating new mathematics. Given any set of mathematical elements and a set of operations on them, we can ask whether that set is "closed" under those operations—that is, whether those operations always yield members of that set. If not, we can ask what other elements need to be added to that set in order to achieve closure, and whether closure is even achievable. A very large and significant part of mathematics arises from issues surrounding closure.

Numbers and Numerals

Numbers

Via the Arithmetic Is Object Construction metaphor, we conceptualize numbers as wholes put together out of parts. The operations of arithmetic provide the patterns by which the parts are arranged within the wholes. For example, every natural number can be conceptualized uniquely as a product of prime numbers. Thus, 70 equals 2 times 5 times 7, and so can be conceptualized as the unique sequence of primes (2, 5, 7), which uniquely picks out the number 70.

Similarly, every natural number can be conceptualized as a polynomial—that is, a sum of integers represented by simple numerals times powers of some integer B. B is called the base of the given number system. In the binary system, B is two. In the octal system, B is eight. In the system most of the world now uses, B is ten. Thus, 8,307 is eight times ten to the third power plus three times ten to the second power, plus zero times ten to the first power, plus seven times ten to the zeroth power.

$$8,307 = (8 \cdot 10^3) + (3 \cdot 10^2) + (0 \cdot 10^1) + (7 \cdot 10^0)$$

When the natural numbers are extended to the reals, this metaphorical representation of numbers is extended to include infinite decimals—sums of the same sort, where the powers of ten can be negative numbers and products of all negative powers of ten are included. Thus, π is understood as an infinite sum:

$$\pi = (3 \cdot 10^0) + (1 \cdot 10^{-1}) + (4 \cdot 10^{-2}) + (1 \cdot 10^{-3}) + \ldots = 3.141\ldots$$

Numerals

There is a big difference between numbers, which are concepts, and numerals, which are written symbols for numbers. In innate arithmetic, there are numbers but no numerals, since newborn children have not learned to symbolize numbers using numerals. The difference can be seen in Roman numerals versus Arabic numerals, where the same *number* is represented by different *numerals*; for example, the number fourteen is represented by XIV in Roman numerals and by 14 in Arabic numerals with base ten. The Arabic numeral system, now used throughout the world, is based on the metaphorical conceptualization of numbers as sums of products of small numbers times powers of ten.

Suppose we replaced Arabic numerals with Roman numerals, where, for example, 3 = III, 5 = V, 4 = IV, 6 = VI, 50 = L, 78 = LXXVIII, and 1998 = MCMXCVIII. With Roman numerals instead of Arabic numerals, nothing would be changed about *numbers*. Only the *symbols* for numbers would be changed. Every property of numbers would remain the same.

The Roman notation is also based on the Arithmetic Is Object Construction metaphor, and the Roman notation uses not only addition (e.g., VI + I = VII) but subtraction (e.g., X – I = IX) for certain cases. Is arithmetic the same with both notations, even when the properties of numbers are all the same?

Yes and no. Conceptually it would remain the same. But *doing* arithmetic by calculating with numerals would be very different. The notation for numbers is part of the mathematics of calculation. To see this, consider systems of notations with different bases. The binary number system uses a different version of the metaphor that numbers are sums of products of other numbers. In the binary version, the base is two and every number is a sum of products of powers of two. Thus, fourteen in the binary system is one times two to the third power plus one times two to the second power, plus one times two to the first power plus zero times two to the zeroth power. In binary numerals, that sum would be represented as 1110.

The decimal, binary, octal, and other base-defined notations are all built on various versions of the metaphor that numbers are sums of products of small numbers times powers of some base. Thus, cognitively, it is important to make a three-way distinction:

- *The number* (e.g., thirteen)
- *The conceptual representation of the number:* the sum of products of powers adding up to that number (e.g., one times ten to the first power plus three times ten to the zero[th] power)
- *The numeral that symbolizes the number* by, in turn, symbolizing the sum of products of powers (e.g., 13).

From a cognitive perspective, bidirectional conceptual mappings are used to link conceptual representations to numerals. Here are the mappings for decimal and binary systems (where n is an integer):

THE DECIMAL NUMERAL-NUMBER MAPPING

Numeral Sequences $x_i (i = \text{'0'}, \dots, \text{'9'})$		Sums of Products of Powers of Numbers $\dots (x_n \cdot 10^n) + (x_{n-1} \cdot 10^{n-1}) \dots$
	$\leftrightarrow$	

THE BINARY NUMERAL-NUMBER MAPPING

Numeral Sequences $x_i (i = \text{'0'}, \text{'1'})$		Sums of Products of Powers of Numbers $\dots (x_n \cdot 2^n) + (x_{n-1} \cdot 2^{n-1}) \dots$
	$\leftrightarrow$	

Any facts about arithmetic can be expressed in any base, because all conceptual representations of numbers in terms of sums of products of powers and all symbolic representations of numbers in terms of sequences of digits refer to the same system of numbers. Thirteen is a prime number whether you conceptualize and write it 1101 in binary, 111 in ternary, 15 in octal, 13 in decimal, 10 in base thirteen, or even XIII in Roman numerals.

Though the numbers are the same, the numerical representations of them are different. The tables for numbers are the same, but the numerical representations of the tables are different. For example, here are the addition tables for numbers between zero and five in binary and decimal notations:

ADDITION TABLE IN BINARY NOTATION

+	0	1	10	11	100	101
0	0	1	10	11	100	101
1	1	10	11	100	101	110
10	10	11	100	101	110	111
11	11	100	101	110	111	1000
100	100	101	110	111	1000	1001
101	101	110	111	1000	1001	1010

ADDITION TABLE IN DECIMAL NOTATION

+	0	1	2	3	4	5
0	0	1	2	3	4	5
1	1	2	3	4	5	6
2	2	3	4	5	6	7
3	3	4	5	6	7	8
4	4	5	6	7	8	9
5	5	6	7	8	9	10

The *numbers* represented in these tables are the same. 1010 in the lower right-hand cell of the binary table represents the number ten, just as 10 does in the decimal table. The numerals are different. And the way of conceptualizing the numbers using the numerals are different. Thus, in the binary table, numbers are conceptualized as well as numerically represented in terms of sums of products of powers of two.

Calculation

A system of calculation based on the Roman numerals would be very different from any system we presently have. From a cognitive perspective, it would be prohibitively difficult. Parsing a long Roman numeral is simply harder than parsing an equivalently long decimal notation, since the notation is not strictly *positional*; that is, it is not the case that one symbol stands for one multiple of ten to some power. For example, in XI (eleven), you are adding I to X (one to ten). But in XIX (nineteen), the I is not being added to the first X; rather, it is being subtracted from the second (one taken from ten), with the result added to the first (ten plus nine). This demands more cognitive activity for human beings than using decimal notation. For beings like us, positional notation requires less cognitive effort, not just in recognition but especially in calculation. Procedures for adding, subtracting, multiplying, and dividing require less cognitive effort in positional notations than in nonpositional notations like Roman numerals. Imagine doing long division with Roman numerals!

Not that it isn't doable. One could program a computer to do arithmetic using Roman numerals, and given the speed of contemporary computers, we probably wouldn't notice the difference in computation time, for any normal computation. But we don't use Roman numerals, and we will never go back to them because of the cognitive load they place on us. Moreover, we don't use binary notation, even though computers do, because our ten fingers make it easier for us to use base 10.

Our mathematics of calculation and the notation we do it in is chosen for bodily reasons—for ease of cognitive processing and because we have ten fingers and learn to count on them. But our bodies enter into the very idea of a linearly ordered symbolic notation for mathematics. Our writing systems are linear partly because of the linear sweep of our arms and partly because of the linear sweep of our gaze. The very idea of a linear symbol system arises from the peculiar properties of our bodies. And linear symbol systems are at the heart of mathematics. Our linear, positional, polynomial-based notational system is an optimal solution to the constraints placed on us by our bodies (our arms and our gaze), our cognitive limitations (visual perception and attention, memory, parsing ability), and possibilities given by conceptual metaphor.

Calculation Without Understanding

As we have seen, the mathematics of calculation, including the tables and algorithms for arithmetic operations, is all defined in terms of numerals, not numbers. Using the algorithms, we can manipulate the numerals correctly without having contact with numbers and without necessarily knowing much about numbers as opposed to numerals. The algorithms have been explicitly created for such efficient calculation, not for understanding. And we can know how to use the algorithms without much understanding of what they mean.

When we learn procedures for adding, subtracting, multiplying, and dividing, we are learning algorithms for manipulating symbols—numerals, not numbers. What is taught in grade school as arithmetic is, for the most part, not ideas about numbers but automatic procedures for performing operations on numerals—procedures that give consistent and stable results. Being able to carry out such operations does not mean that you have learned meaningful content about the nature of numbers, even if you always get the right answers!

There is a lot that is important about this. Such algorithms minimize cognitive activity while allowing you to get the right answers. Moreover, these algorithms work generally—for all numbers, regardless of their size. And one of the best things about mathematical calculation—extended from simple arithmetic to higher forms of mathematics—is that the algorithm, being freed from meaning and understanding, can be implemented in a physical machine called a computer, a machine that can calculate everything perfectly without understanding anything at all.

Equivalent Result Frames and the Laws of Arithmetic

Part of our knowledge about voting is that there are two equivalent ways to vote. You can write away for an absentee ballot, fill it in at home, and send it in before the election. Or you can go to your polling place, show your identification, get a ballot, fill it in, and leave it at the polling place on the day of the election. That is, given an election, an election day, a ballot, and the procedure of filling in a ballot, there are two ways to achieve the result of voting. Similarly, we have knowledge about lots of other equivalent ways to achieve desired results. You can buy a product by shopping at a store, using a mail-order catalogue and telephoning your order, or placing your order over the Internet. Familiarity with the various processes that achieve an identical result is an important part of our overall knowledge. From a cognitive perspective, such knowledge is represented in a conceptual frame within Charles Fillmore's theory of frame semantics (Fillmore, 1982, 1985). An *Equivalent Result Frame* (hereafter, ERF) includes

- a desired result,
- essential actions and entities, and
- a list of alternative ways of performing those actions with those entities to achieve the result.

For example, an important property of collections of objects can be stated in terms of the following ERF:

THE *Associative ERF* FOR COLLECTIONS

Desired result:	A collection N
Entities:	Collections A, B, and C
Operation:	"add to"

Equivalent alternatives:
- A added to [the collection resulting from adding B to C] yields N
- [the collection resulting from adding A to B] added to C yields N

The metaphor Arithmetic Is Object Collection maps this ERF onto a corresponding ERF for arithmetic:

THE *Associative ERF* FOR ARITHMETIC

Desired result:	A number N
Entities:	Numbers A, B, and C
Operation:	"+"

Equivalent alternatives:
- $A + (B + C) = N$
- $(A + B) + C = N$

This ERF expresses what we understand the associative law for arithmetic to mean. From a cognitive perspective, this ERF *is* the cognitive content of the associative law for arithmetic: $A + (B + C) = (A + B) + C$. Here we can see a clear example of how the grounding metaphors for arithmetic yield the basic laws of arithmetic, when applied to the ERFs for the source domains of collections, construction, motion, and so on.

Note, incidentally, that the associative law does not hold for innate arithmetic, where all numbers must be subitizable—that is, less than 4. The reason, of course, is that closure does not hold for innate arithmetic. Thus, if we let $A = 1$, $B = 2$, and $C = 3$, then $A + B = 3$, $B + C = 5$, and $A + B + C = 6$. Since 6 is beyond the usual range of what is subitizable, this assignment of these subitizable results to A, B, and C yields a result outside of innate arithmetic. This means that the associative law cannot arise in innate arithmetic and must arise elsewhere. As we have just seen, it arises in the source domains of the four grounding metaphors.

Why Calculation with Numerals Works

Why does calculation using symbolic numerals work? There is nothing magical about it. Suppose you have to add eighty-three to the sum of seventeen and thirty-nine. The numeral-number mapping, which underlies our understanding of the numeral-number relationship, maps this problem into the symbolization:

$$83 + (17 + 39) = ?$$

By the associative law, this is equivalent to

$$(83 + 17) + 39 = ?$$

Since we know that $83 + 17 = 100$, it is clear that the right answer is 139. But why does this work? Is the associative law god-given? Not at all. It works for the following reason:

- The 4Gs ground our understanding of arithmetic and extend it from innate arithmetic.
- The source domains of the 4Gs are object collection, object construction, physical segmentation, and motion. Each of these is part of our understanding of the real world.
- The associative equivalent result frame is true of each physical source domain.

- The 4Gs map those equivalent result frames onto the conceptual content of the associative law.
- The numeral-number mapping maps associative ERF for arithmetic onto the symbolized form of the associative law: $A + (B + C) = (A + B) + C$. This is used in calculation to replace an occurrence of "$A + (B + C)$" by "$(A + B) + C$" in the calculation.
- Because the symbolized equation corresponds to the cognitive content of the equivalent result frame, the symbolic substitution yields an equivalent conceptual result.

Note what a cognitive account such as this does *not* say: It does not say that the reason the calculation works is that the associative law is an *axiom*. Rather, it says that there is a reason following from our embodied understanding of what arithmetic is. It is our embodied understanding that allows blind calculation with numerals to work in arithmetic.

Up to this point we have looked at the 4Gs and how they extend innate arithmetic and enrich it with properties like closure and the basic laws of arithmetic. We have also considered the cognitive mechanism for the symbolization of numbers, which makes arithmetic calculable. We now need to flesh out the grounding metaphors discussed so far.

Stretching the 4Gs

The 4Gs, as we have stated them so far, are grounding metaphors that arise naturally from experience that conflates innate arithmetic with one of the domains. These natural metaphors extend innate arithmetic considerably. They also allow for extensions from natural numbers to other numbers. For example, the Arithmetic Is Motion Along a Path metaphor allows the path to be extended indefinitely from both sides of the origin, permitting zero and negative numbers to be conceptualized as point-locations on the path and therefore to be seen as numbers just like any other numbers. Fractions can then be conceptualized as point-locations between the integers, at distances determined by division.

Given that such numbers have natural interpretations in this metaphor, one might think that this metaphor as it stands would be sufficient to ground arithmetic for zero, negative numbers, and fractions. It is not. No single natural metaphor permits closure in arithmetic. To achieve closure, the metaphors must be extended, or "stretched."

Let us begin with addition and multiplication for negative numbers. Let negative numbers be point-locations on the path on the side opposite the origin from positive numbers. The result in the source domain of the path is symme-

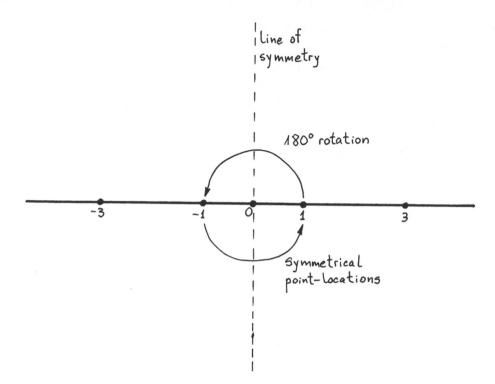

FIGURE 4.2   Mental rotation is a natural cognitive operation. Given that numbers are metaphorically conceptualized as point-locations on a line, rotation gives us a natural way of relating positive and negative numbers: A 180° rotation around zero maps the positive numbers to the corresponding negative numbers, and vice versa. This cognitive operation provides grounding for a metaphor for multiplication by negative numbers: Rotation by 180° Is Multiplication by –1.

try: For every point-location at a given distance on one side of the origin, there is a unique point-location at the same distance on the other side; let us call it the "symmetrical point." Thus, –5 is the symmetrical point of +5, and +5 is the symmetrical point of –5.

Mapping point-locations to numbers in the target domain, we get a symmetry in the number system: For every positive number, there is a unique negative number, and conversely. Since positive and negative numbers are symmetric, we need to distinguish them by picking an orientation: The usual choice is that positive numbers are on the right of the origin and negative numbers are on the left (see Figure 4.2).

The metaphorical mapping for addition must now be changed slightly. Addition of positive numbers will now be conceptualized as moving toward the right, whereas addition of negative numbers will be moving toward the left. Thus, 3 + (−5) will straightforwardly be −2. The metaphor for subtraction must be changed accordingly. Subtraction of positive numbers will be moving toward the left, while subtraction of negative numbers will be moving toward the right. Thus, (−4) − (−6) will be +2, straightforwardly. It follows from these metaphors that the result of adding a negative number is the same as the result of subtracting a positive number:

$$A + (-B) = A - B.$$

The converse is also entailed:

$$A - (-B) = A + B.$$

Multiplication, however, is not quite so straightforward. Multiplication by positive numbers requires performing an action—moving—a certain number of times. Multiplication of a negative number $-B$ by a positive number A is no problem: You perform repeated addition of $-B$ A times; that is, starting from the Origin, you move B units to the left A times.

But multiplying *by* a negative number is not a simple conceptual extension of multiplying by a positive number. In the source domain of motion, doing something a negative number of times makes no sense. A different metaphor is needed for multiplication by negative numbers. That metaphor must fit the laws of arithmetic—the most basic entailments of the four most basic grounding metaphors. Otherwise, the extension will not be consistent. For example, 5 · (−2) does have a metaphorical meaning, as we just saw, and it gives the result that 5 · (−2) = −10. The commutative law therefore constrains any metaphor for multiplication by negative numbers to the following result: (−2) · 5 = −10.

The symmetry between positive and negative numbers motivates a straightforward metaphor for multiplication by $-n$: First, do multiplication by the positive number n and then move (or "rotate" via a mental rotation) to the symmetrical point—the point on the other side of the line at the same distance from the origin. This meets all the requirements imposed by all the laws. Thus, (−2) · 5 = −10, because 2 · 5 = 10 and the symmetrical point of 10 is −10. Similarly, (−2) · (−5) = 10, because 2 · (−5) = −10 and the symmetrical point of −10 is 10. Moreover, (−1) · (−1) = 1, because 1 · (−1) = −1 and the symmetrical point of −1 is 1.

The process we have just described is, from a cognitive perspective, another metaphorical blend. Given the metaphor for multiplication by positive numbers, and given the metaphors for negative numbers and for addition, we form a blend in which we have both positive and negative numbers, addition for both, and multiplication for only positive numbers. To this conceptual blend we add the new metaphor for multiplication by negative numbers, *which is formulated in terms of the blend!* That is, to state the new metaphor, we must use

- negative numbers as point-locations to the left of the origin,
- addition for positive and negative numbers in terms of movement, and
- multiplication by positive numbers in terms of repeated addition a positive number of times, which results in a point-location.

Only then can we formulate the new metaphor for multiplication by negative numbers using the concept of moving (or rotating) to the symmetrical point-location.

MULTIPLICATION BY −1 IS ROTATION

Source Domain SPACE		Target Domain ARITHMETIC
Rotation to the symmetry point of n	$\rightarrow$	$-1 \cdot n$

Given this metaphor, $-n \cdot a = (-1 \cdot n) \cdot a = -1 \cdot (n \cdot a)$, which is conceptualized as a mental rotation to the symmetrical point of $(n \cdot a)$.

What Is Stretched

The metaphor Arithmetic Is Motion Along a Path has a source domain of natural motions: movement in one direction or the other, and iterated movements. Mental rotations around a center are also cognitively natural. Thus the metaphor for multiplication of negative numbers uses a natural cognitive mechanism. However, it does not arise from subitizing and counting as part of some natural activity. Instead, it is added to fit the needs of closure, which are given higher priority than consistency with subitizing or counting.

What is important to understand is the difference between the four basic grounding metaphors and extensions like this one. The 4Gs do arise naturally for natural activities correlated with subitizing and counting. *They are what make the arithmetic of natural numbers natural!* The laws of arithmetic for natural numbers are entailments of those metaphors. None of this is true of the extensions of those metaphors.

Further Stretching

We can characterize division by negative numbers by stretching the 4Gs even further. Here's how.

Consider three cases. Case 1 involves dividing a negative by a negative. This is just like division of a positive by a positive, but done on the negative side of the line. It is simply repeated subtraction, moving in the opposite direction. And the result, which is a number of times, is a positive number. Thus, division of a negative by a negative gives the same result as the corresponding division of a positive by a positive.

Now consider case 2: $(-A) \div B$, where A and B are positive. What answer do we need to get? From the entailments of the four basic metaphors, we know that division is the inverse of multiplication: Whereas multiplication is repeated addition, division is repeated subtraction. Therefore, what we need to get is a C such that $C \cdot B = -A$. From the metaphors for multiplication by negatives, we know that C must be negative. We can now stretch the motion metaphor further and metaphorically define $(-A) \div B$: Perform the simple division of positive by positive for $A \div B$. Then rotate to the symmetrical point. This will give us a consistent result, which is a negative number.

Finally consider case 3: $A \div (-B)$. For the same reason, the answer has to be a C which is a negative number. The stretched metaphor for division will therefore be the same in this case as in the last: Perform the simple division of positive by positive for $A \div B$. Then rotate to the symmetrical point. This will give us a consistent result, which is a negative number.

It is interesting that such metaphorical stretching yields consistent results. Compare this with the situation of division by zero. There is no possible consistent stretching of either the motion metaphor or any of the other grounding metaphors to allow division by zero. The requirements of closure say that the operations of arithmetic should give determinate results for any operation on numbers. But there is no way to metaphorically conceptualize division by zero to get a determinate answer.

Let's look at this in, say, the object-collection metaphor. Here is the metaphor for division (see Chapter 3):

- Division $(C \div B = A)$ Is the Splitting Up of a Collection of Size C into A Subcollections of Size B.

Take a collection of a particular size C. To divide C by zero, you would have to split up the collection into subcollections of size zero—that is, into a unique, determinate number of empty subcollections exhausting the collection of size *C*. But there is no such unique, determinate number. Moreover, there is no consistent way to extend this metaphor to get such a unique, determinate number. The same situation holds in all the other grounding metaphors.

This *is* an interesting result. The lack of divisibility by zero is a consequence of the lack of any extensions of the four basic grounding metaphors consistent with the laws of arithmetic and that would meet the requirement of closure that the result be a unique, determinate number.

In short, the four basic grounding metaphors are natural and are constitutive of our fundamental understanding of arithmetic. Stretched versions of the 4Gs get more and more contrived as one stretches more and more. But sometimes it is impossible to stretch in a sensible way to fit the requirements of closure and consistency reflected in the notational system.

Metaphoric Blends and the Effectiveness of Arithmetic in the World

We have now identified the four grounding metaphors that allow us to extend innate arithmetic. We have seen how they lead to the requirements of closure, and we have seen how much of that closure (not including the reals and the imaginaries) is realized via the stretching of those grounding metaphors. Given this conceptual structure, metaphoric blends in which source and target domain are both activated arise naturally. These are blends of arithmetic with object collections, object construction, measuring, and motion. We use such blends when we employ arithmetic in each of those activities. It should not be surprising, therefore, that arithmetic "works" for each of those activities.

What makes the arithmetic of natural numbers effective in the world are the four basic grounding metaphors used to extend innate arithmetic and the metaphoric blends that arise naturally from those metaphors.

Summary

Let us review how arithmetic is "put together" cognitively. We can get an overall picture in terms of answers to a series of questions of the form "Where does *X* come from cognitively?" where *X* includes all the following:

- The natural numbers
- The concept of closure—the idea that the operations of arithmetic always work on numbers to produce other numbers

- The laws of arithmetic
- Fractions, zero, and negative numbers
- The notation of arithmetic
- Generalizations: Laws of arithmetic work for numbers in general, not just in specific cases. For example, $a + b = b + a$, no matter what a and b are.
- Symbolization and calculation
- The special properties of arithmetic: precision, consistency, and stability across time and communities
- The fact that arithmetic "works" in so much of our experience.

In showing where each of these properties comes from, we answer the big question—namely, Where does arithmetic as a whole come from? Here is a brief summary of the answers we have given.

Where Do the Natural Numbers and the Laws of Arithmetic Come From?

The natural numbers and the laws governing them arise from the following sources:

1. *Subitizing Plus.* We have innate capacities, instantiated neurally in our brains, for subitizing and innate arithmetic—limited versions of addition and subtraction (up to three). We have basic capacities—grouping, ordering, focusing attention, pairing, and so on—that can be combined to do primitive counting.

2. *Primary Experiences.* We have primary experiences with object collection, object construction, physical segmentation, and moving along a path. In functioning in the world, each of these primary experiences is correlated with subitizing, innate arithmetic, and simple counting, as when we automatically and largely unconsciously subitize the number of objects in a collection or automatically count a small number of steps taken.

3. *The Conflation of Subitizing Plus with Primary Experiences.* Such correlations form the experiential basis of the four basic grounding metaphors for arithmetic, in which the structure of each domain (e.g., object collection) is mapped onto the domain of number as structured by subitizing, innate arithmetic, and basic counting experiences. This is what makes the arithmetic of natural numbers natural.

4. *Conflation Among the Primary Experiences.* Object construction (putting objects together) always involves object collection. Placing physical

segments end to end is a form of object construction and, hence, of object collection. Trajectories from one point to another are imagined physical segments. Each of these is a conflation of experiences. From a neural perspective, they involve coactivations of those areas of the brain that characterize each of the experiences. And constant coactivation presumably results in neural links. As a consequence, an isomorphic structure emerges across the source domains of the 4Gs. That isomorphic structure *is independent of numbers themselves* and lends stability to arithmetic.

5. *Subitizing and Innate Arithmetic.* Within their range, subitizing and innate arithmetic are precise, consistent, stable, and common to all normal human beings. In those cultures where subitizing, innate arithmetic, and basic counting are correlated with object collection, object construction, and so on, to form the experiential basis of the basic grounding metaphors, those metaphors bring precision, stability, consistency, and universality to arithmetic.

6. *Laws from Entailments of Grounding Metaphors.* The laws of arithmetic (commutativity, associativity, and distributivity) emerge, first, as properties of the source domains of the 4Gs, then as properties of numbers via those metaphors, since the metaphors are inference-preserving conceptual mappings. For example, associativity is a property of the physical addition of the contents of one collection to another. The same collection results from either order. The Numbers Are Object Collections metaphor maps that knowledge onto the associative law for arithmetic.

7. *The Equivalent Result Frames.* They give the commutative and associative laws.

Why Does Arithmetic Fit the World?

The metaphoric blends of the source and target domains of the 4Gs associate the arithmetic of natural numbers with a huge range of experiences in the world: experiences of collections, structurings of objects, the manipulation of physical segments, and motion. This is the basis of the link between arithmetic and the world as we experience it and function in it. It forms the basis of an explanation for why mathematics "works" in the world.

Where Do Fractions, Zero, and Negative Numbers Come From?

The nonisomorphic portions of the 4Gs provide groundings for other basic concepts in arithmetic: Fractions make sense as parts of wholes in the object-construction metaphor. Zero and negative numbers make sense as point-locations in the motion metaphor.

Why Do We Think of Numbers As Things in the World?

The Numbers Are Things in the World metaphor arises as a generalization over the 4Gs, since in each one, numbers are things (collections; wholes made from parts; physical segments, like sticks; and locations). This metaphor has important entailments, many of which fit the so-called Platonic philosophy of mathematics:

- Since things in the world have an existence independent of human minds, numbers have an existence independent of human minds. Hence, they are not creations of human beings and they and their properties can be "discovered."
- There is only one true arithmetic, since things in the world have determinate properties.
- Arithmetic is consistent. It has to be, if its objects and their properties have an objective existence.
- Regarding closure, the Numbers Are Things in the World metaphor maps (a) onto (b): (a) Operations on things in the world yield unique, determinate other things in the world of the same kind. (b) Operations on numbers yield unique, determinate other numbers.

In other words, the system of numbers should be closed under arithmetic operations. If it appears not to be, then there must be numbers still to be discovered that would produce this result. Thus, given the natural numbers and their arithmetic, there should be fractions (e.g., 1 divided by 2), zero (e.g., 3 − 3), negative numbers (e.g., 3 − 5), irrational numbers (e.g., $\sqrt{2}$), and imaginary numbers (e.g., $\sqrt{-1}$).

Symbolization and Calculation

Symbolization and calculation are central features of arithmetic. Embodied cognition and metaphor play important roles in both.

- Linear notation is motivated by the sweep of our gaze and our arms.
- Positional notation is motivated by memory constraints and cognitive ease of symbolic calculation, which reduces memory load.

- The numeral-number mappings build on linear and positional notation to provide a notation for the natural numbers by means of a one-to-one conceptual mapping that links Arabic notation with the conceptualization of numbers as polynomial sums.
- The ERFs characterizing the laws of arithmetic, together with the numeral-number mapping, characterize equations that can be used in defining purely symbolic algorithms to mirror rational processes.

Where Does Generalizability in Arithmetic Come From?

Cognitive frames work at a general level within human conceptual systems. They specify general roles that can be filled by a member of a category. For example, in the travel frame, the role of traveler can be filled by any person physically able to travel. In mathematics this is accomplished through the Fundamental Metonymy of Algebra (see Chapter 3).

Conceptual metaphor, as it functions across the entire human conceptual system, also operates at a general level. For example, in the Love Is a Journey metaphor, where lovers are conceptualized as travelers, *any* lovers (not just particular ones) can be conceptualized as travelers. Similarly, inferences about travelers in general apply to lovers in general.

These facts about conceptual frames and metaphors have a profound consequence in arithmetic. The basic grounding metaphors for natural numbers apply to *all* natural numbers. The laws of arithmetic, which are entailments of those basic grounding metaphors, therefore apply to *all* natural numbers. It is this property of metaphor, together with the Fundamental Metonymy of Algebra, that makes arithmetic generalizable, since metaphor is used to ground arithmetic in everyday experience.

The Centrality of Calculation

Calculation is the backbone of mathematics. It has been carefully designed for overall consistency. The algorithmic structure of arithmetic has been carefully put together to mirror rational processes and to be usable when those processes are disengaged. The mechanisms required are

- Equivalent Result Frames, true of the physical source domains
- the 4Gs, which map ERFs from physical source domains onto arithmetic

- the numeral-number mapping, which maps ERFs in arithmetic onto equations in symbolic form, which can be used in calculations, where one side of the equation is replaced by the other, yielding an equivalent result.

The Superstructure of Basic Arithmetic

The elements required for the conceptual system of basic arithmetic are as follows:

- The innate neural structures that allow us to subitize and do basic concrete addition and subtraction up to about four items
- Basic experiences that form the basis of metaphors, like subitizing while (a) forming collections, (b) putting objects together, (c) manipulating physical segments (fingers, arms, feet, sticks, strings), (d) moving in space, (e) stretching things
- Grounding metaphors, like the 4Gs, and their inferences, such as the basic laws of arithmetic and closure
- Extensions of metaphors to satisfy the requirements of closure
- Basic arithmetic concepts, like number; the operations of addition, subtraction, multiplication, and division; identities and inverses under operations; decomposition of numbers into wholes and parts (factoring); Equivalent Result Frames (equations) characterizing which combinations of operations on the same elements yield the same results
- Symbolization: The mapping of symbols to concepts. For example, 0 maps to the additive identity, 1 maps to the multiplicative identity, $-n$ maps to the additive inverse of n, $1/n$ maps to the multiplicative inverse of n, + maps to addition, and so on.
- Numerals: Particular symbols for particular numbers
- Calculations: Algorithms using symbols and numerical ERF equations for performing calculations.

So far we have discussed just basic arithmetic. When we start discussing more advanced forms of mathematics, additional cognitive mechanisms will enter the picture. The most prominent are:

- Linking metaphors, which conceptualize one domain of mathematics in terms of another

- Arithmetization metaphors—a special case of linking metaphors, those that conceptualize ideas in other domains of mathematics in terms of arithmetic
- Foundational metaphors, which choose one domain of mathematics (say, set theory or category theory) as fundamental and, via metaphor, "reduce" other branches of mathematics to that branch.

The ideas discussed here will recur throughout the book. But before we move on, we need to bring up an important issue.

How Do We Know?

To those unfamiliar with the methodology of cognitive linguistics, it will not be obvious how we arrived at the metaphorical mappings given throughout the last chapter and this one, and how we know that they work.

The various branches of cognitive science use a wide range of methodologies. In cognitive neuroscience and neuropsychology, there are PET scans and fMRIs, the study of the deficiencies of brain-damaged patients, animal experimentation, and so on. In developmental psychology, there is, first of all, careful observation, but also experiments where such things as sucking rates and staring are measured. In cognitive psychology, there is model building plus a relatively small number of experimental paradigms (e.g., priming studies) to gather data. In cognitive linguistics, the main technique is building models that generalize over the data. In the study of one's native language, the data may be clear and overwhelmingly extensive, although common techniques of data gathering (e.g., texts, recordings of discourse) may supplement the data. In the study of other languages, there are field techniques and techniques of text gathering and analysis. As in other sciences, one feels safe in one's conclusions when the methodologies give convergent results.

In this study, we are building on results about innate mathematics from neuroscience, cognitive psychology, and developmental psychology. The analyses given in this book must also mesh with what is generally known about the human brain and mind. In particular, these results must mesh with findings about human conceptual systems in general. As we discussed in Chapter 2, the human conceptual system is known to be embodied in certain ways. The general mechanisms found in the study of human conceptual systems are radial categories, image schemas, frames, conceptual metaphors, conceptual blends and so on.

Much of this book is concerned with conceptual metaphors. In Chapter 2, we cited ten sources of convergent evidence for the existence of conceptual metaphor in everyday thought and language. These metaphor studies mesh with studies showing that the conceptual system is embodied—that is, shaped by the structure of our brains, our bodies, and everyday interactions in the world. In particular they show that abstract concepts are grounded, via metaphorical mappings, in the sensory-motor system and that abstract inferences, for the most part, are metaphorical projections of sensory-motor inferences.

Our job in this chapter and throughout the book is to make the case that human mathematical reason works in roughly the same way as other forms of abstract reason—that is, via sensory-motor grounding and metaphorical projection. This is not an easy job. We must propose plausible ultimate embodied groundings for mathematics together with plausible metaphorical mappings. The hypothesized groundings must have just the right inferential structure and the hypothesized metaphors must have just the right mapping structure to account for *all* the relevant mathematical inferences and all the properties of the branch of mathematics studied. Those are the data to be accounted for. For example, in this chapter, the data to be accounted for include the basic properties of arithmetic (given by the laws) and all the computational inferences of arithmetic. That is a huge, complex, and extremely precise body of data.

We made use of the model-building methodology to account for this data. The constraints on the models used were as follows:

1. The grounding metaphors must be plausible; that is, they arise via conflation in everyday experience (see Chapter 2).
2. Given the sensory-motor source domain and the mappings, all the properties and computational inferences about the mathematical target domain must follow.
3. In the case of arithmetic, the analysis must fit and extend what is known about innate arithmetic.
4. The models must be maximally general.
5. The models must accord with what is generally known about embodied cognition; that is, they must be able to fit with what is known about human brains and minds.

These constraints are so demanding that, at first, we found it difficult to come up with any models at all that fit within them.

Indeed, in studying arithmetic, for example, we depended on the prior research of Ming Ming Chiu (1996). Chiu's dissertation set out some first approximations that met a number of these constraints. Starting there, we made many successive revisions until the constraints were met.

To the novice in metaphor theory it may not be obvious that not just any metaphorical mapping will meet these constraints. Indeed, hardly any do. For example, arithmetic is not an apple. The structure of an apple—growing on trees; having a skin, flesh, and a core with seeds; being about the size to fit in your hand, having colors ranging from green to red; and so on—does not map via any plausible metaphor (or any at all) onto arithmetic. The inferences about apples just do not map onto the inferences about numbers. Apples won't do as a source domain, nor will couches, churches, clouds, or most things we experience in the everyday physical world.

As we have seen, there are four plausible grounding domains—forming collections, putting objects together, using measuring sticks, and moving through space. Each of them forms just the right kind of conflation with innate arithmetic to give rise to just the right metaphorical mappings so that the inferences of the source domains will map correctly onto arithmetic—almost. Only two of them have an equivalent to zero.

But that isn't good enough. The constraints on an analysis are so stringent that an additional metaphor is necessary to extend the system to negative numbers. That metaphor makes use of a known cognitive mechanism—mental rotation. Moreover, these metaphors must have just the right properties to account for the properties of arithmetic—its ability to fit experience, its stability over time, its universal accessibility, its combinatorial possibilities, and so on.

In short, mathematics provides a formidable challenge to the model-building methodology of cognitive science, because the constraints on the possible models are so severe. The biggest surprise in our research to date is that we have been able to get as far as we have.

The metaphors given so far are called grounding metaphors because they directly link a domain of sensory-motor experience to a mathematical domain. But as we shall see in the chapters to come, abstract mathematics goes beyond *direct* grounding. The most basic forms of mathematics are directly grounded. Mathematics then uses other conceptual metaphors and conceptual blends to link one branch of mathematics to another. By means of linking metaphors, branches of mathematics that have direct grounding are extended to branches that have only indirect grounding. The more indirect the grounding in experience, the more "abstract" the mathematics is. Yet ultimately the entire edifice of mathematics does appear to have a bodily grounding, and the mechanisms linking abstract mathematics to that experiential grounding are conceptual metaphor and conceptual blending.

In addition, a certain aspect of our linguistic capacities is used in mathematics—namely, the capacity for symbolizing. By this we mean the capacity for associating written symbols (and their phonological representations) with mathematical ideas. This is just one aspect of our linguistic capacities, but it is anything but trivial. As in natural languages, mathematical symbols can be polysemous; that is, they can have multiple, systematically associated meanings. For example, + is used not only for addition but for set union and other algebraic operations with the properties of addition. 1 and 0 are sometimes used to mean true and false—and as we shall see, it is no accident that 1 is used for true and 0 for false and not the reverse. Similarly, it is no accident that + rather than, say, √ is used for set union. As we shall also see, these uses of symbols accord with the metaphorical structure of our system of mathematical ideas.

We are now in a position to move from basic arithmetic to more sophisticated mathematics.

Part II

Algebra, Logic, and Sets

5

Essence and Algebra

Have you ever wondered why axioms are so important in mathematics? Or why it is so important for mathematicians to find the smallest number of independent axioms for a subject matter? Or why it was assumed for over two millennia (from Euclid until Gödel) that a whole mathematical subject matter should follow from a small number of postulates or axioms?

Furthermore, why did these ideas begin with Greek mathematics? They were not present in Babylonian, Egyptian, or Mayan mathematics. What was it about ancient Greece that made such ideas seem natural? And what is it about European conceptual history that made these ideas flourish into the twentieth century?

By putting together what we know both from cognitive science and from history, we can approach an answer to these questions. The contribution from cognitive science is the notion of a *folk theory*—a cognitive structure characterizing a typically unconscious, informal "theory" about some subject matter.

Essence and Axioms

We can find the answer in pre-Socratic Greek philosophy, where the idea of "essence" was central. Indeed, the concept of essence is still with us, in the form of an unconscious folk theory about the world that most people in Western culture still take for granted. Here is that folk theory.

The Folk Theory of Essences

- *Every specific thing is a kind of thing.* For example, there are kinds of animals: lions, elephants, dogs, and so on. Each specific animal—say, your pet Fido—is a kind of animal, in this case a dog.

- *Kinds are categories, which exist as entities in the world.* In other words, the category *dog* is as much an entity as Fido is.
- *Everything has an essence—a collection of essential properties—that makes it the kind of thing it is.* Each elephant, for example, has certain features that are *essential* to elephanthood: a trunk, large floppy ears, stumplike legs, and so on. Those features constitute the "essence" of each elephant. They are what make every animal with those features the same kind of thing—namely, an elephant.
- *Essences are causal; essences—and only essences—determine the natural behavior of things.* For example, elephants eat the way they do because they have a trunk and because their legs cannot be used to pick up food. All natural behavior follows from the essences. Moreover, the essences cannot follow from one another—or they wouldn't be essences.
- *The essence of a thing is an inherent part of that thing.*

There are three basic metaphors for characterizing what an essence is. They are:

- Essences Are Substances.
- Essences Are Forms.
- Essences Are Patterns of Change.

The folk theory of essences is part of what constitutes our everyday "common sense" about physical objects; that is, it is part of the unconscious conceptual system that governs our everyday reasoning. Take a particular tree, for example. We understand that tree to be an instance of the general kind, Tree. The general kind is seen as having an existence of its own. When we say that there are trees in the world, we don't just mean the particular trees that happen to exist now. What is it that makes a tree a tree?

- *Substance.* To start with, it is made of wood. If it were made of plastic, it wouldn't be a real tree. Thus, substance counts as part of its essence.
- *Form.* Next, it has a form: trunk, bark on the trunk, roots, branches, leaves (or needles) on the branches, the roots underground, the trunk oriented relatively perpendicular to the ground, the branches extending out from the trunk. Without such a form, it wouldn't be a tree.
- *Pattern of Change.* Finally, it has a pattern of change: It grows out of a seed, matures, dies.

Consider, for example, the smell of the eucalyptus tree. The tree contains a substance, eucalyptus oil. That substance is part of what makes a eucalyptus a eucalyptus; that is, it makes the tree the kind of thing it is. Eucalyptus oil is thus part of the essence of the eucalyptus tree. It is the causal source of the smell of the eucalyptus. It is part of the natural behavior of a eucalyptus tree to have that smell. Thus, the essence of the eucalyptus is the causal source of that natural behavior.

We apply this everyday folk theory of essences, which works very well for things like trees, to abstract cases as well. We conceptualize people as having essences: Their personality is their emotional essence. Their character is their moral essence. Our personality and character are seen as part of what makes us the kinds of people we are. A person may be friendly or mean, have a heart of gold, or be rotten to the core. Depending on such judgments about their essence, we generate expectations about how people will behave.

In its expert version, the theory of essences fits together with the classical theory of categories, which goes back to Aristotle. In the classical theory, a category is defined by a set of necessary and sufficient conditions: a list of inherent properties that each member has. That category-defining set of properties is an essence. All and only the members of the category have exactly that essence. It was Aristotle who defined "definition" in terms of essences: A definition is a list of properties that are both *necessary* and *sufficient* for something to be the kind of thing it is, and from which all its natural behavior flows.

Aristotle's expert version of the folk theory of essences is at the heart of a great deal of scientific practice, though not all. Physics, for example, seeks the essential properties of matter and other physical phenomena—the properties that make a thing the kind of thing it is and allow us to predict its behavior. Though the folk theory of essences has informed expert scientific theories, it is not true that the theory of essences, in either its folk or expert version, fits the physical world. The notion of a species in biology is a case where it fails: A species cannot be defined by necessary and sufficient conditions (see Mayr, 1984; Lakoff, 1987). Indeed, in biology, the folk theory of essences has interfered with the practice of science.

Euclid brought the folk theory of essences into mathematics in a big way. He claimed that only five postulates characterized the essence of plane geometry as a subject matter. He believed that from this essence all other geometric truths could be derived by deduction—by reason alone! From this came the idea that every subject matter in mathematics could be characterized in terms of an essence—a short list of axioms, taken as truths, from which all other truths about the subject matter could be deduced.

Thus, the axiomatic method is the manifestation in Western mathematics of the folk theory of essences inherited from the Greeks.

Essences Within Mathematics

Mathematics as a discipline has followed Aristotle's definition of "definition" as an essence: a collection of necessary and sufficient conditions that make an entity the kind of thing it is, and from which all its natural properties follow.

Numbers and other mathematical entities (e.g., triangles, groups, and topological spaces) are conceptualized as objects and therefore are assumed to have essences. It is the essence of a triangle to be a polygon with three angles. It is the essence of real numbers to have the following properties (entailed by the four basic metaphors for arithmetic)—namely, commutativity (for addition and multiplication), associativity (for addition and multiplication), and distributivity (of multiplication over addition). Axiom systems are taken as defining the essence of each mathematical subject matter.

This is something special about the conceptual system of mathematics. In virtually all other domains of human conceptual systems (e.g., love, cooking, politics, social life, religion), categories are not usually characterized by necessary and sufficient conditions (see Lakoff, 1987). The folk theory of essences may be used—but not in its detailed, expert version, where there is a conscious, explicit, and passionate effort to distinguish what is part of an essence as opposed to what follows from that essence.

Essence and Algebra

Algebra is about essence. It makes use of the same metaphor for essence that Plato did—namely, Essence Is Form.

Algebra is the study of mathematical form or "structure." Since form (as the Greek philosophers assumed) is taken to be abstract, algebra is about abstract structure. Since it is about essence, it has a special place within mathematics. It is that branch of mathematics that is conceptualized as characterizing essences in other branches of mathematics. In other words, mathematics implicitly assumes a particular metaphor for the essences of mathematical systems: The Essence of a Mathematical System Is an Abstract Algebraic Structure.

Historically, algebra began with basic arithmetic structures (e.g., the operations of addition and multiplication over the integers, the rational numbers, the real numbers, the complex numbers). Via the Fundamental Metonymy of Algebra (see Chapter 3), generalizations over particular numbers could be stated as precisely as sums of particular numbers.

Algebra has also come to include substructures of arithmetic (e.g., the numbers [0, 1, 2] with the operation of addition modulo 3). Algebra asks what the essence of each such structure is, where the essence of an arithmetic structure is taken to include

- the elements in the structure,
- the number and type of operations used on those elements (e.g., two binary operations, one unary operation, and so on), and
- the essential properties of the operations (i.e., the axioms governing the operations).

Notice that this list does not include numbers specifically; it mentions only "elements." Nor does it include arithmetic operations like addition and multiplication; it mentions only "operations." In other words, algebra is "abstract"; it is about mathematical essences in general, *across* mathematical domains, not just about essences *within* particular mathematical domains.

Since the domain of mathematics from which algebra originally came is arithmetic, algebraic operations are often written "+" and "·" and are called "addition" and "multiplication" (again, by virtue of the Fundamental Metonymy of Algebra). This sometimes creates confusion, since algebra is its own separate mathematical domain.

Let us consider a simple example: the numbers 1, 2, and 3 under the operation of addition modulo 3—that is, addition with multiples of 3 subtracted out when the result is greater than or equal to 3. Thus, 2 + 1 would normally be 3, but with 3 subtracted out, 2 + 1 = 0. Similarly, 2 + 2 would normally equal 4, but with 3 subtracted out, 2 + 2 = 1. Here is the addition table for addition modulo 3:

+ TABLE

+	0	1	2
0	0	1	2
1	1	2	0
2	2	0	1

In this table, 0 is the identity element, since $X + 0 = X$, for each element X. For each element X, there is an inverse element Y, such that $X + Y = 0$. Thus, 0 is its own inverse, since $0 + 0 = 0$, and 1 is the inverse of 2, since $2 + 1 = 0$. The following laws hold with respect to this table:

- *Closure:* The sum of every two elements is an element.
- *Associativity:* $X + (Y + Z) = (X + Y) + Z$.
- *Commutativity:* $X + Y = Y + X$.
- *Identity:* $X + 0 = X$.
- *Inverse:* For each X, there is a Y such that $X + Y = 0$.

This set of elements {1, 2, 3} with this addition table is in the subject matter of arithmetic with addition modulo 3. The essence of this arithmetic structure is the set of elements, the binary operation of addition modulo 3, and the stated arithmetic laws governing addition over these elements.

Algebra is not about particular essences (like arithmetic essences or geometric essences) but, rather, about *general essences*, considered as things in themselves. These general essences are called "abstract"—abstract elements, abstract operations, abstract laws. Here is an example of a set of abstract elements, an abstract operation, and a set of abstract laws.

The elements: {I, A, B}
The operation: *, defined over the elements by the following table:

* TABLE

*	I	A	B
I	I	A	B
A	A	B	I
B	B	I	A

The laws holding with respect to this "abstract" table:

- *Closure:* The operation * on any two elements of the set is an element of the set.
- *Associativity:* $X * (Y * Z) = (X * Y) * Z$.
- *Commutativity:* $X * Y = Y * X$.
- *Identity:* There is an element I, such that $X * I = X$.
- *Inverse:* For each X, there is a Y such that $X * Y = I$.

The abstract elements, the abstract operation *, and the abstract laws are independent of arithmetic. They are part of the subject matter of abstract algebra, and they form an abstract structure called the *commutative group with three elements*.

The study of such structures is interesting in itself, but its appeal does not end there. What makes algebra a central discipline in mathematics is its relationship to other branches of mathematics, in which algebraic structures are conceptualized as the essences of other mathematical structures in other mathematical domains. This requires a collection of what we will call Algebraic Essence metaphors (AE metaphors for short)—conceptual mappings from algebraic structures to structures in other mathematical domains.

The most simpleminded example of an Algebraic Essence metaphor is the mapping from the commutative group of three elements to addition modulo 3. Here is the AE metaphor:

ADDITION MODULO 3 AS A COMMUTATIVE GROUP WITH THREE ELEMENTS

Source Domain ALGEBRA: GROUPS		*Target Domain* MODULAR ARITHMETIC
The commutative group with 3 elements	$\rightarrow$	The essence of addition modulo 3
The set {I, A, B} of abstract elements	$\rightarrow$	The set {0, 1, 2} of numbers
The abstract binary operation $\star$	$\rightarrow$	The addition operation +
The algebraic-identity element I	$\rightarrow$	The arithmetic-identity element 0
The abstract element A	$\rightarrow$	The number 1
The abstract element B	$\rightarrow$	The number 2
Abstract closure law	$\rightarrow$	Arithmetic closure law
Abstract associative law	$\rightarrow$	Arithmetic associative law
Abstract commutative law	$\rightarrow$	Arithmetic commutative law
Abstract identity: There is an element I, such that $X \star I = X$.	$\rightarrow$	Arithmetic identity: There is an element 0, such that $X + 0 = X$.
Abstract inverse: For each X, there is a Y such that $X \star Y = I$.	$\rightarrow$	Arithmetic inverse: For each X, there is a Y such that $X + Y = 0$.
$\star$ Table	$\rightarrow$	+ Table

This is an AE metaphor. That is, it is a conceptual mapping by means of which we can conceptualize addition modulo 3 in terms of an abstract algebraic structure, the commutative group with three elements.

This is a thorough explication of a largely unconscious metaphorical mapping used by mathematicians to assign an algebraic essence to an arithmetic structure. Mathematicians tend to think of the algebraic structure as being "in" the arithmetic structure. Mathematicians speak of addition modulo 3 as *being* literally a commutative group with three elements, or as "having the structure of a commutative group with three elements." But from a cognitive perspective, this is a metaphorical idea. Algebra and arithmetic are different domains, and numbers are different from abstract elements.

We can see this even more clearly with an additional example. Consider an equilateral triangle, with vertices marked 1, 2, 3 as in Figure 5.1. Consider the following three things you can do to this triangle:

- Rotate it around its center by 120 degrees (small rotation).
- Rotate it around its center by 240 degrees (middle rotation).
- Rotate it around its center by 360 degrees (full rotation).

All we have here so far is a triangle and a bunch of rotations. A full rotation returns the triangle to its original orientation. The rotations can be combined in a sequence of two rotations. A small rotation followed by a middle rotation, or a middle rotation followed by a small rotation, also returns the triangle to its original orientation. A small rotation followed by a small rotation leaves it in the same state as a single middle rotation. And a middle rotation followed by another middle rotation leaves it in the same state as a single small rotation.

Here is a table summarizing the effects of the various combinations of two successive rotations, with @ symbolizing two successive rotations:

ROTATION TABLE

@	Full	Small	Middle
Full	Full	Small	Middle
Small	Small	Middle	Full
Middle	Middle	Full	Small

Thus, for example, the cell at the lower right is marked "Small" because a middle rotation (240°) followed by another middle rotation (240°) is equivalent to a small rotation (120°).

According to this table, the following properties of the operation @ hold:

- *Closure:* The operation @, which combines any two successive rotations, yields the equivalent of another rotation.
- *Associativity:* $X @ (Y @ Z) = (X @ Y) @ Z$.
- *Commutativity:* $X @ Y = Y @ X$.

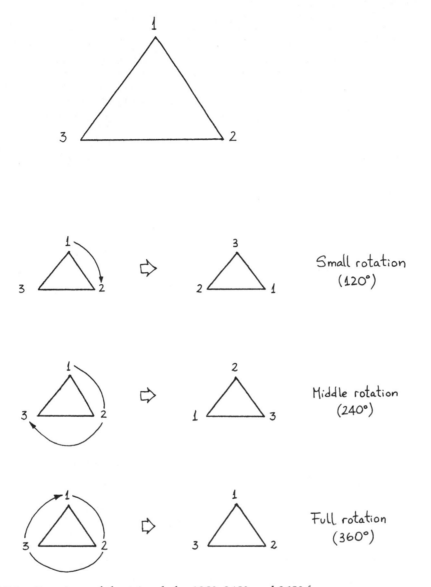

FIGURE 5.1   Rotations of the triangle by 120°, 240°, and 360° form a group.

- *Identity:* There is a rotation—namely, full rotation—such that X @ full rotation = X.
- *Inverse:* For each X, there is a Y such that X @ Y = full rotation.

The rotations of this triangle can be conceptualized as a commutative group via the following metaphor.

THE ROTATION GROUP METAPHOR

Source Domain ALGEBRA: GROUPS		*Target Domain* GEOMETRY: ROTATIONS
The commutative group with 3 elements	→	The essence of the system of rotations of an equilateral triangle through multiples of 120°
The set {I, A, B} of abstract elements	→	The set {Full, Small, Middle} of rotations of the triangle
The abstract binary operation *	→	The combination of two successive rotations @
The algebraic-identity element I	→	The rotational-identity element full rotation
The abstract element A	→	The small rotation
The abstract element B	→	The middle rotation
Abstract closure law	→	Rotational closure Law
Abstract associative law	→	Rotational associative law
Abstract commutative law	→	Rotational commutative law
Abstract identity: There is an element I, such that $X * I = X$.	→	Rotational identity: There is a rotation, Full, such that for any rotation X, X @ Full = X.
Abstract inverse: For each X, there is a Y such that $X * Y = I$.	→	Rotational inverse: For each rotation X, there is a rotation Y such that X @ Y = Full.
* Table	→	Rotation table

This metaphor maps the "* Table" onto the domain of triangle rotations, with the abstract elements I, A, and B mapped onto the rotations full, small, and middle; and it maps the abstract * operator onto the rotation operator @.

What makes this a conceptual metaphor? Rotations of a triangle in themselves are just rotations of a triangle. In order to conceptualize what you are doing to that triangle as having a group structure, you need a very special and not particularly obvious way of conceptualizing rotations, sequences of rotations, and their results in terms of an abstract group structure. From the cognitive perspective—the perspective from which you are conceptualizing one kind of thing in terms of a very different kind of thing—what you need is exactly the peculiar metaphorical mapping just given. From a cognitive perspective, the abstract group structure is *not* inherently *in* the rotations you are performing on the triangle. Conceptually, the group structure is imposed by the metaphorical mapping. It takes extraordinary ingenuity and imagination by mathematicians to construct such metaphors imposing algebraic structures of all kinds in domains outside arithmetic. This use of Algebraic Essence metaphors is one of the great accomplishments of the mathematical imagination.

The two metaphors we have just discussed are used by mathematicians to conceptualize addition modulo 3 and rotations of equilateral triangles as having the same mathematical structure, *the same abstract essence:* that of a commutative group with three elements. Mathematicians usually think of the algebraic structure as *inhering* in both equilateral-triangle rotations and addition modulo 3. They form metaphorical blends for each metaphor—speaking, for example, of a "rotation group," in which the group structure is superimposed on the rotation structure to yield a very specific algebraic-geometric conceptual blend; the algebraic structure is conceptualized together with what it gives structure to.

Such AE metaphors and their corresponding conceptual blends are an essential tool for seeing identical structures in very different mathematical domains and for formulating them precisely. Since the AE metaphors, like all conceptual metaphors, preserve inferential structure, anything proved about the algebraic structure will also hold for the mathematical structure it is mapped onto. Thus, for example, both addition modulo 3 and rotations of equilateral triangles have all the entailed properties of commutative groups with three elements.

Algebraic Essence Metaphors in General

When a mathematician claims that a particular mathematical system "has" or "forms" a particular algebraic structure, that claim must be substantiated via a complex mapping from the algebraic structure to the mathematical system. From a cognitive perspective, such a mapping is an AE metaphor that characterizes the essence of the mathematical system in terms of a structure in the domain of algebra. All AE metaphors have the same form:

THE GENERAL FORM OF ALGEBRAIC ESSENCE METAPHORS	
Source Domain ALGEBRA	*Target Domain* A MATHEMATICAL DOMAIN *D* OUTSIDE OF ALGEBRA
An algebraic structure $\rightarrow$	The essence of a mathematical system in Domain *D*

A set S_A of abstract elements	$\rightarrow$	A set S_D of elements of Domain D
Abstract operations	$\rightarrow$	Operations in Domain D
Algebraic identity element(s)	$\rightarrow$	Identity element(s) in Domain D
Other abstract element(s)	$\rightarrow$	Other element(s) in Domain D
Abstract laws	$\rightarrow$	Laws governing Domain D
Tables for abstract operation(s) over the Set S_A	$\rightarrow$	Tables for operation(s) in Domain D over the set S_D

Summary

We can now answer the questions we began with.

- Why are axioms so important in mathematics?
- Why do mathematicians seek to find the smallest number of logically independent axioms for a subject matter?
- Why was it assumed for over two millennia (from Euclid until Gödel) that a whole mathematical subject matter should follow from a small number of postulates or axioms?
- Why did these ideas begin with Greek mathematics?
- Why are algebraic structures seen as essences of other mathematical entities?

Once we see the link between the axiomatic method and the folk theory of essences, the answers to these questions become clear.

Historically, the axiomatic approach to mathematics was just one of many approaches (compare the Mayan, Babylonian, and Indian approaches). It became dominant in Europe beginning with Euclid and shaped the history of mathematics in the West. The axiomatic method arose in Greek mathematics and extended to European mathematics because the folk theory of essences was central to Greek philosophy, which formed the basis of European philosophy.

Axiomatic mathematics is built on the folk theory of essences as applied to mathematical systems. It assumes that each mathematical system can be completely characterized by an essence—that is, a relatively small number of essential properties that are independent of one another. Axioms minimally characterize those properties, and a system of axioms represents the essence of a mathematical system. Written symbolically, a collection of axioms symbolically represents, in compact form, the essence of an entire mathematical system.

In the folk theory of essences, essences are causal; that is, they are the causal source of all natural behavior. In axiomatic mathematics, the causal relationship becomes a deductive one: The axioms become the deductive source of all the natural properties of the system (the theorems).

The field of algebra begins with the Fundamental Metonymy of Algebra and builds on it, adding the metaphor Essences Are Forms—a metaphor inherited directly from Greek philosophy, which saw form as abstract and independent of substance. This is the source of the idea that abstract algebraic structures characterize the essential forms of particular mathematical systems.

From the perspective of cognitive science, the field of abstract algebra is a magnificent case study of the use of fundamental cognitive mechanisms in mathematical cognition. It begins with a metonymy and is based on a folk theory—the folk theory of essences—and a metaphor—The Essence of a Mathematical System Is an Abstract Algebraic Structure. It is this metaphor that links algebra to other branches of mathematics.

What we learn from this is not any new algebra but, rather, what algebra *is*, from a cognitive perspective. The analytic activity we have engaged in within this chapter is what we have been calling *mathematical idea analysis*. It tells us things that traditional descriptions of algebra do not:

- The folk theory of essences lies behind the very idea of abstract algebra. It tells us how we normally think about algebra relative to other mathematical systems. But the folk theory of essences is not an account of mathematical cognition.
- The cognitive structure of an algebraic entity (e.g., a group) is not an essence that inheres in other cognitive structures of mathematical entities (e.g., a collection of rotations). Rotations are conceptualized independently of groups, and groups are conceptualized independently of rotations.

Here is an example of how the folk theory of essences applies to algebra from a cognitive perspective. A group structure is ascribed to a collection of rotations when the collection of rotations can be metaphorically conceptualized as a group. This requires a metaphorical mapping from groups to rotations of the sort just described.

The metaphor given above—namely, The Essence of a Mathematical System Is an Abstract Algebraic Structure—attributes essence to algebraic structures (e.g., the group) that are mapped onto other mathematical structures (e.g., collections of rotations).

The moral here is one that will be repeated throughout this book:

- Mathematical symbolism is *not* an analysis of mathematical ideas.
- Mathematical notation must be understood in terms of mathematical ideas.
- A *mathematical idea analysis* is required to make precise the ideas implicit in mathematical notations.

6

Boole's Metaphor:
Classes and Symbolic Logic

S YMBOLIC LOGIC AND THE LOGIC OF CLASSES are branches of mathematics. There is a widespread belief that these forms of mathematics characterize reason itself. Even cognitive psychology as recently as the early 1970s largely took this to be true. But recent cognitive science, concerned with the embodiment of mind, has found that these branches of mathematics are far from adequate for the characterization of human reason, which must include prototypes and image schemas, as well as conceptual frames, metaphors, and blends (see Chapter 2). From our perspective, symbolic logic and the logic of classes are just two mathematical subject matters whose conceptual structure can be studied precisely through mathematical idea analysis. The questions we ask are like the questions we asked in the previous chapters. How are these subject matters conceptualized by human beings, how were they created, and how is our understanding of them grounded in our experience?

A number of questions immediately arise for many people who come to the study of these subject matters. What really *is* the empty set, why is there only one such set, and why is it a subset of every set? And where did this idea come from, anyway? Why should symbolic logic have such strange implications as "*P*, therefore either *P* or *Q*"? Where did the *Q* come from? And what about "*P* and not *P*, therefore *Q*"? Why should anything at all follow from a contradiction, much less everything? Generations of students have found these to be disturbing questions. Formal definitions, axioms, and proofs do not answer these questions. They just raise further questions, like "Why these axioms and not others?"

We believe that mathematical idea analysis can provide insight into these matters.

Boole's Classes

When we draw illustrations of Container schemas, as we did in Chapter 2, we find that they look rather like Venn diagrams for Boolean classes. This is by no means an accident. The reason is that classes are normally conceptualized in terms of Container schemas. We think of elements as being *in* or *out* of a class. Venn diagrams are visual instantiations of Container schemas. The reason that Venn diagrams work as symbolizations of classes is that classes are usually metaphorically conceptualized as containers—that is, as bounded regions in space.

As we saw in Chapter 2, Container schemas have a logic that is very much like Boolean logic—a logic that appears to arise from the structure of part of our visual and imaging system (Regier, 1996), adapted for more general use. It can be used both for structuring space and for more abstract reason, and is projected onto our everyday conceptual system by the Categories (or Classes) Are Containers metaphor. This accounts for part (by no means all!) of our reasoning about conceptual categories. Boolean logic, as we shall see shortly, also arises from our capacity to perceive the world in terms of Container schemas and to form mental images using them.

Since the time of Aristotle, the logic of propositions and predications has been directly linked with what have variously been called categories, classes, or sets. Aristotle conceptualized a predication like "Socrates is mortal" as a category statement of the form "Socrates is in the category (or class or set) of mortals." Boole's logic of thought was in this tradition, and his initial formulation of what developed into symbolic logic has been interpreted as being a logic of classes. The technical distinction in modern mathematics between "classes" and "sets" was introduced much later and will be discussed in the next chapter. For the time being, we will use the term "class" as a convenience, generally meaning a collection of any sort.

The intuitive premathematical notion of classes is conceptualized in terms of Container schemas. In other words, a class of entities is conceptualized in terms of a bounded region of space, with all members of the class *inside* the bounded region and nonmembers *outside* the bounded region. From a cognitive perspective, intuitive classes are thus metaphorical conceptual containers, characterized cognitively by a metaphorical mapping: a grounding metaphor, the Classes Are Containers metaphor.

CLASSES ARE CONTAINERS

Source Domain CONTAINER SCHEMAS		Target Domain CLASSES
Interiors of Container schemas	→	Classes
Objects in interiors	→	Class members
Being an object in an interior	→	The membership relation
An interior of one Container schema within a larger one	→	A subclass in a larger class
The overlap of the interiors of two Container schemas	→	The intersection of two classes
The totality of the interiors of two Container schemas	→	The union of two classes
The exterior of a Container schema	→	The complement of a class

This is our natural, everyday unconscious conceptual metaphor for what a class is. It is a *grounding metaphor*. It grounds our concept of a class in our concept of a bounded region in space, via the conceptual apparatus of the image schema for containment. This is the way we conceptualize classes in everyday life.

Boole's Metaphor

The English mathematician George Boole (1815–1864) noticed certain structural similarities between arithmetic and classes, much like the relationships between numbers and collections of objects we noted in Chapters 3 and 4. At the time, reasoning was understood in terms of Aristotelian logic, which depended on a central metaphor that Aristotle had introduced more than two millennia earlier (see Lakoff & Johnson, 1999, ch. 18). For Aristotle, a predication like "Socrates is mortal" is a class-membership statement: Socrates is a member of the class of mortals.

ARISTOTLE'S PREDICATION METAPHOR

Source Domain CLASSES		Target Domain PREDICATION
A class	→	A predicate
A member of a class	→	An entity the predicate applies to

| The membership relation (between class and member) | $\rightarrow$ | The predication relation (between predicate and subject) |

For Boole, writing in the 1850s, Aristotle's predication metaphor meant that providing algebraic principles governing classes would be tantamount to formulating the laws of thought. Moreover, providing a symbolic calculus of classes would be providing a symbolic calculus of thought.

Boole saw a way to use the similarities he perceived between arithmetic and classes to formulate an algebra of classes so that algebraic laws and arithmetic-like notation could be used to construct a calculus of classes, which would provide a notation for "laws of thought." Boole achieved this (with some modifications by his countryman Augustus De Morgan) by introducing a new conceptual metaphor, which we refer to as "Boole's metaphor."

What Boole proposed, from a cognitive perspective, was a historic metaphor that allowed one to conceptualize classes as having an algebraic structure. He developed this metaphor in stages. The first stage was a partial metaphorical understanding of classes in terms of arithmetic. Boole observed that if you conceptualized classes as numbers, and operations on classes (union and intersection) as operations on numbers (addition and multiplication), then the associative, commutative, and distributive laws of arithmetic would hold for classes. In cognitive terms, he constructed a *linking metaphor* between arithmetic and classes, mapping numbers to classes, arithmetic operations to class operations, and arithmetic laws to "laws of thought"—that is, the laws governing operations on classes.

To complete the first-stage metaphor, Boole needed to map the identity elements for arithmetic—zero and one—onto classes. But in the intuitive notion of classes described here via the Classes Are Containers metaphor, there are no natural correlates of zero and one. Boole invented them to flesh out his arithmetic metaphor for classes. Corresponding to zero, Boole had to invent a class (he called it "nothing") that when "added" to another class did not change it, just as adding zero to a number does not change the number. Zero was to map onto this "empty" class. However, zero is a unique number. Therefore, the empty class that zero had to map onto had to be a unique entity. Moreover, an important arithmetic law for zero had to be maintained—namely, "X times 0 yields 0, for all numbers X." This law would have to map onto a new corresponding "law" for classes: "A intersect the empty class yields the empty class, for all classes A," which implies that the empty class is a subclass of every class.

In addition, there is no natural correlate to 1 in our ordinary understanding of classes. Boole invented such a correlate, which he called "universe"—the universal class containing everything in our "universe of discourse." Thus, the law of arithmetic "*X* times 1 yields *X*, for all numbers *X*" would be mapped onto "*A* intersect the universal class yields *A*."

Here is the resulting metaphor.

BOOLE'S FIRST-STAGE METAPHOR

Source Domain ARITHMETIC		Target Domain CLASSES
Numbers	$\rightarrow$	Classes
Addition	$\rightarrow$	Union, symbolized by '$\cup$'
Multiplication	$\rightarrow$	Intersection, symbolized by '$\cap$'
Commutative law for addition	$\rightarrow$	Commutative law for union
Commutative law for multiplication	$\rightarrow$	Commutative law for intersection
Associative law for addition	$\rightarrow$	Associative law for union
Associative law for multiplication	$\rightarrow$	Associative law for intersection
Distributive law for multiplication over addition	$\rightarrow$	Distributive law for intersection over union
0	$\rightarrow$	The empty class, symbolized by $\varnothing$
1	$\rightarrow$	The universal class, symbolized by I
Identity for addition: 0	$\rightarrow$	Identity for union: $\varnothing$
Identity for multiplication: 1	$\rightarrow$	Identity for intersection: I
$A + 0 = A$	$\rightarrow$	$A \cup \varnothing = A$
$A \cdot 1 = A$	$\rightarrow$	$A \cap I = A$
$A \cdot 0 = 0$	$\rightarrow$	$A \cap \varnothing = \varnothing$

This is a *linking metaphor*, not a *grounding metaphor*. It links one branch of mathematics (the logic of classes) to another branch of mathematics (arithmetic).

Boole was entirely conscious of this mapping. It enabled him to pick out certain "resemblances" between arithmetic and classes. Some resemblances are obvious: The laws have the same "shape." Boole saw this clearly through the use of symbols. If you let + stand for "union," · for "intersection," and *A* and *B* for classes, the commutative, associative, and distributive laws for arithmetic can be seen as holding for classes.

Commutative law for addition	$A + B = B + A$
Commutative law for multiplication	$A \cdot B = B \cdot A$
Associative law for addition	$A + (B + C) = (A + B) + C$
Associative law for multiplication	$A \cdot (B \cdot C) = (A \cdot B) \cdot C$
Distributive law for multiplication over addition	$A \cdot (B + C) = (A \cdot B) + (A \cdot C)$

There were further resemblances. For disjoint classes (those sharing no members), union is like addition, in that forming a union is intuitively like adding members of one class to the members of the other to form a larger class.

Intersection can be seen, for certain cases, as corresponding to multiplication by fractions between zero and one. Consider the class of cows and two subclasses, the brown cows and the free-range cows (those that range freely to feed). Assume that 1/2 of all cows are brown, and that 1/4 of all cows are free-range and equally distributed among the brown and nonbrown cows. Take the intersection of the two classes: the class of brown, free-range cows. What percentage of all cows are both brown and free-range? 1/2 times 1/4 equals 1/8. 1/8 of all cows are both brown and free-range. In such cases, intersection of classes corresponds to multiplication by fractions. This intuition lies behind Boole's idea of conceptualizing intersection as multiplication in his first-stage metaphor.

However, Boole realized that his first-stage metaphor would not work as it stood. Consider the two arithmetic equations $X + X = 2X$ and $X \cdot X = X^2$. According to the first-stage metaphor, these should correspond to the union of a class with itself and the intersection of a class with itself. But the union of a class A with itself is A. No new members are added. And the intersection of class A with itself is A. Every member of A is in A. In symbolic terms, we should have what are called the *idempotent laws: $A + A = A$ and $A \cdot A = A$.* Boole observed that "$A \cdot A = A$" would work for only two numbers, 1 and 0, and that "$A + A = A$" would work for only one number, 0. The only way that Boole could preserve his first-stage metaphor was by limiting his numbers to 1 and 0, and by inventing a new addition table for his "arithmetic"—a table in which $1 + 1$ is not 2, but 1!

Boole's New Addition Table

+	0	1
0	0	1
1	1	1

With this new table for "addition," and with numbers limited to 1 and 0, Boole could make his first-stage metaphor work—kind of. It could be made to work for a universe with only two classes: the universal class and the empty class—the classes invented by Boole to make the metaphor itself work! It could not work at all for a universe with ordinary classes. Yet, Boole bore in mind, the similarities between arithmetic and classes were unmistakable. Since Boole was an algebraist, he turned to algebra as a source domain for his metaphor. In algebra, he was not limited to exactly the arithmetic properties of numbers. In algebra he could find abstract symbols that could obey the laws he observed—those laws that were true of numbers and other laws that were not.

For his new metaphor, Boole devised a new algebraic source domain—what has come to be called Boolean algebra. It is an abstract structure with the following properties.

The Properties of a Boolean Algebra

- It has abstract elements, not numbers.
- It is closed under two abstract binary operators, $\oplus$ and $\otimes$. These are not addition and multiplication of numbers but abstract operators operating on abstract elements.
- It has algebraic analogs of the associative and commutative laws of arithmetic.
- It has idempotent laws: $A \oplus A = A$ and $A \otimes A = A$.
- It has identity elements for both binary operators, 0 and 1, which are not numbers but designated abstract elements.
- It obeys two inverse laws: For each element A, there is an element A' such that $A \oplus A' = 1$ and $A \otimes A' = 0$.
- It obeys a distributive law like the distributive law of arithmetic: $A \otimes (B \oplus C) = (A \otimes B) \oplus (A \otimes C)$.
- It has a second distributive law, unlike anything in arithmetic: $A \oplus (B \otimes C) = (A \oplus B) \otimes (A \oplus C)$.

An important special case of a Boolean algebra consists of only the two elements 0 and 1, with these tables:

$\oplus$	0	1
0	0	1
1	1	1

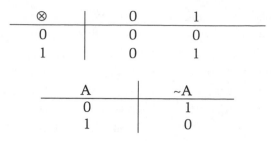

With these adjustments, taking algebra as his source domain, Boole was able to state his metaphor for conceptualizing classes in terms of an algebraic structure of the sort required. Here is the final metaphor:

BOOLE'S METAPHOR

Source Domain ALGEBRA		Target Domain CLASSES
Abstract elements	→	Classes
An abstract operation $\oplus$	→	Union, symbolized by $\cup$
An abstract operation $\otimes$	→	Intersection, symbolized by $\cap$
Commutative law for $\oplus$	→	Commutative law for union
Commutative law for $\otimes$	→	Commutative law for intersection
Associative law for $\oplus$	→	Associative law for union
Associative law for $\otimes$	→	Associative law for intersection
Distributive laws for $\otimes$ and $\oplus$	→	Distributive laws for intersection and union
0	→	The empty class, symbolized by $\emptyset$
1	→	The universal class, symbolized by I
Identity for $\oplus$: 0	→	Identity for union: $\emptyset$
Identity for $\otimes$: 1	→	Identity for intersection: I
$A \oplus 0 = A$	→	$A \cup \emptyset = A$
$A \otimes 1 = A$	→	$A \cap I = A$
$A \otimes 0 = 0$	→	$A \cap \emptyset = \emptyset$
A unary abstract operator $\sim$, such that $A \oplus \sim A = 1$ and $A \otimes \sim A = 0$	→	Complement of A: A' such that $A \cup A' = I$ and $A \cap A' = \emptyset$
Closure for $\oplus$, $\otimes$, and $\sim$	→	Closure for $\cup$, $\cap$, and $'$

This metaphor does everything Boole needs. It shows the inherent similarities between classes and arithmetic. It creates additional similarities by the metaphorical creation of an empty class and a universal class. And it introduces the algebraic laws (e.g., idempotent laws and an additional distributive law) needed to fit classes.

The Symbolization of Operations on Classes

Boole not only extended our ordinary, premathematical concept of a class to a mathematical one but also introduced a symbolization and a calculus for classes. From a cognitive perspective, any notation for classes requires a cognitive mapping between the conceptual domain of classes and the domain of symbols. The symbols that have become traditional and that are used here are those of De Morgan, not Boole.

THE SYMBOLIZATION MAPPING FOR CLASSES

Conceptual Domain CLASSES		*Symbolic Domain* SYMBOLS
Classes A, B, . . .	↔	Symbols A, B, . . .
The empty class	↔	$\emptyset$
The universal class	↔	I
Union	↔	$\cup$
Intersection	↔	$\cap$
Complement	↔	$'$
The classes A and B have the same members.	↔	$A = B$

When we learn such a symbolization, what we are learning is how to match purely symbolic expressions with the content specified in the conceptual domain. Mathematicians often introduce such symbolization mappings with expressions like "Let '$\cup$' be the binary operation of union."

Boole saw himself as having accomplished two things: an understanding for classes in terms of algebra, and an algebraic notation and calculus for classes. Since he accepted Aristotle's predication metaphor, he assumed that he had achieved a "scientific formulation" of the "laws of thought," as well as a notation for keeping track of reasoning.

Boole from a Cognitive Perspective

Why is it important to know the cognitive structure of Boole's "laws of thought"? Why should one care about the metaphorical nature of his algebraic "analog" of classes, and the cognitive mapping required for his symbolic calculus? In short, why does the cognitive science of Boolean classes matter besides simply characterizing the cognitive mechanisms involved?

Perhaps the most important thing is to understand that Boole did not achieve what he thought he had achieved and what is still commonly taught as his achievement—a rigorous calculus for the laws of everyday human thought. There are two major differences between what he achieved and what he thought he had achieved.

First, our everyday notion of classes does not include Boole's metaphorical inventions: the empty class and the universal class. Without these, his calculus does not work. Second, Boole had thought that classes of the kind he was describing fit everyday language. They do not. The meanings of ordinary category terms are far more complex than that; they include prototype structures of many kinds (typical cases, ideal cases, social stereotypes, salient exemplars, and more), radial categories, frame structure, metaphoric structure, and so on.

This does not lessen Boole's creative achievement, but it does cast it in a different light. Boole created a concept of class that had enormous mathematical utility. It became the basis for the mathematical notion of a "set," which we will discuss in the next chapter. Moreover, Boole's notation and calculus were the precursors of the development of symbolic logic as a mathematical discipline, taken in the twentieth century by those in the Foundations movement as being foundational for all mathematics.

In addition, Boole's ideas are central to twentieth-century logical philosophy, which includes metaphysical assumptions about the world that Boole himself did not endorse. In logical philosophy, the logic of classes is sometimes represented as providing the rational structure of the universe: Things in the world are assumed to come in natural classes that have a Boolean structure. In this philosophical interpretation of the theory of classes, the empty class is a feature of the universe, and it is a basic truth about the universe that the empty class is a subclass of every class. It is important to understand that this is a peculiar philosophical interpretation of the theory of classes. The empty class is an extremely useful metaphorical invention for a branch of mathematics, but it is not a feature of the universe any more than it is a feature of ordinary everyday human thought.

The universal class is another mathematically indispensable metaphorical creation of Boole's. Without the universal class, Boole's theory of classes does not work. It is introduced only to be the metaphorical equivalent of 1. There is no universal class in our everyday conceptual systems. And there is no reason to believe that the universal class has any objective existence at all. That is, there is no scientifically valid reason to believe that the physical entities in the universe form a subclass of an objectively existing universal class. When we conceptualize physical objects as members of the universal class, we are implicitly using Boole's metaphor, through which the universal class is created.

Symbolic Logic

Propositional logic is a mathematical system built on Boole's algebra of classes and its notation. The relationship between propositional logic and Boole's algebra of classes is given by a linking metaphor, which we will call the Propositional Logic metaphor.

The trick here is to conceptualize propositions in terms of classes, in just such a way that Boole's algebra of classes will map onto propositions, preserving inferences. This is done in the semantics of propositional logic by conceptualizing each proposition P as *being* (metaphorically of course) the class of world states in which that proposition P is true. For example, suppose the proposition P is *It is raining in Paris*. Then the class of world states A, where P is true, is the class of all the states of the world in which it is raining in Paris. Similarly, if Q is the proposition *Harry's dog is barking*, then B is the class of states of the world in which Harry's dog is barking. The class $A \cup B$ is the class of world states where either P or Q is true—that is, where it is raining in Paris or Harry's dog is barking, or both.

The way this trick works from a cognitive perspective is by having the source domain of the metaphor be a conceptual blend of classes and propositions as defined above. Each proposition is conceptualized metaphorically in terms of that blend—that is, in terms of the class of world states in which the proposition itself is true.

THE PROPOSITIONAL LOGIC METAPHOR

Source Domain CLASSES OF WORLD STATES		Target Domain PROPOSITIONS: P, Q, ...
A class of world states in which proposition P is true	$\rightarrow$	Proposition P
Union, symbolized by $\cup$	$\rightarrow$	Disjunction *"or"*
Intersection, symbolized by $\cap$	$\rightarrow$	Conjunction *"and"*

Complement of A: $A' =$ the class of world states where P is not true	$\rightarrow$	The negation of P: *not P*
B is a subclass of A (where B is the class of world states where P is true and A is the class of world states where Q is true)	$\rightarrow$	The material implication of Q by P: *if P, then Q*
Class identity: "is the same class as"	$\rightarrow$	Logical equivalence: *if and only if*
Commutative law for union	$\rightarrow$	Commutative law for disjunction
Commutative law for intersection	$\rightarrow$	Commutative law for conjunction
Associative law for union	$\rightarrow$	Associative law for disjunction
Associative law for intersection	$\rightarrow$	Associative law for conjunction
Distributive laws for intersection and union	$\rightarrow$	Distributive laws for conjunction and disjunction
The empty class of world states (where contradictions are true), symbolized by $\emptyset$	$\rightarrow$	Contradictions: propositions whose truth value is always false
The universal class of world states (where tautologies are true), symbolized by I	$\rightarrow$	Tautologies: propositions whose truth value is always true
Identity for union: $\emptyset$	$\rightarrow$	*Q and not Q*
Identity for intersection: I	$\rightarrow$	*Q or not Q*
$A \cup \emptyset = A$	$\rightarrow$	*P or (Q and not Q) if and only if P*
$A \cap I = A$	$\rightarrow$	*P and (Q or not Q) if and only if P*
$A \cap \emptyset = \emptyset$	$\rightarrow$	*P and (Q and not Q) if and only if (Q and not Q)*

The target domain of this metaphor is a conceptual structure—namely, propositional logic. To get a symbolic logic—in this case, propositional calculus—we need to map the conceptual structure onto symbols. Here is the cognitive mapping:

THE SYMBOLIC LOGIC MAPPING

Conceptual Domain PROPOSITIONAL LOGIC		*Symbolic Domain* PROPOSITIONAL CALCULUS
Propositions P, Q, . . .	$\leftrightarrow$	Symbols 'P', 'Q', . . .
Disjunction *"or"*	$\leftrightarrow$	$\vee$
Conjunction *"and"*	$\leftrightarrow$	$\wedge$
The negation of P: *not P*	$\leftrightarrow$	$\neg P$
The material implication of Q by P: *if P, then Q*	$\leftrightarrow$	$P \supset Q$
if and only if	$\leftrightarrow$	$\Leftrightarrow$
Commutative law for disjunction	$\leftrightarrow$	$P \vee Q \Leftrightarrow Q \vee P$
Commutative law for conjunction	$\leftrightarrow$	$P \wedge Q \Leftrightarrow Q \wedge P$
Associative law for disjunction	$\leftrightarrow$	$P \vee (Q \vee R) \Leftrightarrow (P \vee Q) \vee R$
Associative law for conjunction	$\leftrightarrow$	$P \wedge (Q \wedge R) \Leftrightarrow (P \wedge Q) \wedge R$
Distributive laws for conjunction and disjunction	$\leftrightarrow$	$P \wedge (Q \vee R) \Leftrightarrow (P \wedge Q) \vee (P \wedge R)$ $P \vee (Q \wedge R) \Leftrightarrow (P \vee Q) \wedge (P \vee R)$
Q and not Q	$\leftrightarrow$	$Q \wedge \neg Q$
Q or not Q	$\leftrightarrow$	$Q \vee \neg Q$
P or (Q and not Q) if and only if P	$\leftrightarrow$	$P \vee (Q \wedge \neg Q) \Leftrightarrow P$
P and (Q or not Q) if and only if P	$\leftrightarrow$	$P \wedge (Q \vee \neg Q) \Leftrightarrow P$
P and (Q and not Q) if and only if (Q and not Q)	$\leftrightarrow$	$P \wedge (Q \wedge \neg Q) \Leftrightarrow (Q \wedge \neg Q)$

At this point we can see the relationships among the following:

- Container schemas (everyday cognitive mechanisms; grounding for classes)
- Everyday classes (conceptualized via the Classes Are Containers metaphor)
- Boolean algebra (adapted from the algebraic structure of arithmetic)
- Boolean classes (constructed from everyday classes and Boole's metaphor)

- Propositional logic (constructed via the Propositional Logic metaphor, which builds on Aristotle's Predication metaphor)
- Propositional calculus (a cognitive mapping between propositional logic and symbols).

Here we can see that propositional calculus (the simplest form of symbolic logic) is ultimately grounded in Container schemas in the visual system. That is why, when we see Venn diagrams employed to illustrate propositional calculus, we can immediately understand the symbols in terms of those diagrams, which are visual representations of Container schemas. The mappings used are:

- The Classes Are Containers metaphor (maps Container schemas to everyday classes)
- Boole's metaphor (maps Boolean algebra to Boolean classes)
- The Propositional Logic metaphor (maps a blend of Boolean classes and propositions to propositions)
- The symbolic-logic mapping (maps propositions to symbolic expressions).

Why is this important? First, it allows us to see that symbolic logic, far from being purely abstract or part of the logic of the universe, is actually grounded in the body, partly via Container schemas in the visual system and partly via the grounding of arithmetic. Second, it allows us to see how the inferential laws of Boolean classes, propositional calculus, and symbolic logic are also grounded in the body. This is important for philosophy in general and the philosophy of mathematics in particular. Before we discuss what that importance is, let us first look in detail at how these inferential laws are grounded.

The Mapping of Inferential Laws

We are going to consider four fundamental inferential laws of logic and their source in conceptual image schemas, in particular, container schemas. As we saw in Chapter 2, Container schemas appear to be realized neurally using such brain mechanisms as topographic maps of the visual field, center-surround receptive fields, gating circuitry, and so on (Regier, 1996). The structural constraints on Container schemas give them an inferential structure, which we will call *Laws of Container Schemas*. These so-called laws are conceptual in nature and are reflections at the cognitive level of brain structures at the neural level.

The four inferential laws we will discuss are Container schema versions of classical logic laws: excluded middle, modus ponens, hypothetical syllogism, and modus tollens. These laws, as we shall see, are cognitive entities and, as such, are embodied in the neural structures that characterize Container schemas. They are part of the body. Since they do not transcend the body, they are not laws of any transcendent reason.

Inferential Laws of Embodied Container Schemas

- *Excluded Middle:* Every object X is either *in* Container schema A or *outside* Container schema A.
- *Modus Ponens:* Given two Container schemas A and B and an object X, if A is *in* B and X is *in* A, then X is *in* B.
- *Hypothetical Syllogism:* Given three Container schemas A, B, and C, if A is *in* B and B is *in* C, then A is *in* C.
- *Modus Tollens:* Given two Container schemas A and B and an object Y, if A is *in* B and Y is *outside* B, then Y is *outside* A.

Metaphors allow the inferential structure of the source domain to be used to structure the target domain. The Classes Are Containers metaphor maps the inferential laws given above for embodied Container schemas onto conceptual classes. These include both everyday classes and Boolean classes, which are metaphorical extensions (via Boole's metaphor) of everyday classes.

Inferential Laws for Classes Mapped from Embodied Container Schemas

- *Excluded Middle:* Every element X is either *a member of* class A or *not a member of* class A.
- *Modus Ponens:* Given two classes A and B and an element X, if A is *a subclass of* B and X is *a member of* A, then X is *a member of* B.
- *Hypothetical Syllogism:* Given three classes A, B, and C, if A is *a subclass of* B and B is *a subclass of* C, then A is *a subclass of* C.
- *Modus Tollens:* Given two classes A and B and an element Y, if A is *a subclass of* B and Y is *not a member of* B, then Y is *not a member of* A.

Via the Propositional Logic metaphor, the inferential laws for classes are mapped onto the following inferential laws for propositions:

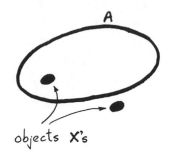

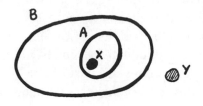

(a)   (b)

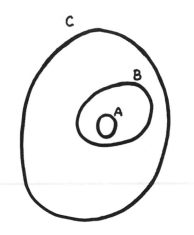

(c)

FIGURE 6.1   The implicit logic of Container schemas. The first drawing (a) depicts the implicit excluded middle, with an entity X either in the interior or the exterior of the Container
schema A. Drawing (b) depicts both modus ponens and modus tollens. Implicit modus ponens can be seen in the fact that A is in B, X is in A, and therefore X is in B. Implicit modus
tollens can be seen in the fact that A is in B, Y is outside B, and therefore Y is outside A. Finally, drawing (c) depicts the implicit hypothetical syllogism, where A is in B, B is in C, and
therefore A is in C.

Inferential Laws for Propositions Mapped from Classes

- *Excluded Middle:* Every proposition P is either true or not true.
- *Modus Ponens:* If P implies Q is true, and P is true, then Q is true.
- *Hypothetical Syllogism:* If P implies Q is true, and Q implies R is true, then P implies R is true.
- *Modus Tollens:* If P implies Q is true, and Q is not true, then P is not true.

Finally, the inferential laws for propositions are mapped onto inferential laws in symbolic form, "rules" usable in mindless algorithms that simply manipulate symbols. These symbolic rules are understood via the cognitive mechanisms we have just given. The reason that such rules for manipulating symbols "preserve truth" is that they have been constructed using these mappings so that this will be the case. It is the inference-preserving capacity of conceptual metaphors that permits these rules of inference to "preserve truth."

Inferential Laws in Symbolic Form Mapped from Propositions

- *Excluded Middle:* $P \lor \neg P$
- *Modus Ponens:* $((P \supset Q) \land P) \supset Q$
- *Hypothetical Syllogism:* $((P \supset Q) \land (Q \supset R)) \supset (P \supset R)$
- *Modus Tollens:* $((P \supset Q) \land \neg Q) \supset \neg P$

Each of these is a tautology in symbolic form. As such, each can always be introduced into a line of a proof without affecting the validity of the proof. Moreover, modus ponens, hypothetical syllogism, and modus tollens can be put in the forms of rules of inference that allow one to move from line to line in a proof:

- *Modus Ponens:* $P \supset Q$, P, $\therefore Q$
- *Hypothetical Syllogism:* $P \supset Q$, $Q \supset R$, $\therefore P \supset R$
- *Modus Tollens:* $P \supset Q$, $\neg Q$, $\therefore \neg P$

By virtue of such rules for getting one line of a proof from previous lines, propositional calculus can be mechanized, so that by the purely mechanical manipulation of symbols we can be sure (given the mappings just discussed) that truth will be preserved.

Overview

Our job as cognitive scientists is to ask how mathematics, including mathematical logic, is conceptualized and understood in terms of normal cognitive mechanisms—for example, image schemas, conceptual metaphors, conceptual blends, and cognitive mappings between concepts and symbols. It is also our job to ask how our understanding is grounded in the body—in the structure of the brain and in bodily and interpersonal experience.

The argument in this chapter takes the following form:

- There is evidence from cognitive linguistics, cognitive psychology, neuroscience, and computational neural modeling that Container schemas are grounded in the sensory-motor system of the brain, and that they have inferential structures like those just discussed. These include Container schema versions of the four inferential laws of classical logic.
- We know from extensive studies of conceptual metaphor in cognitive linguistics, cognitive psychology, and computational neural modeling that conceptual metaphors are cognitive cross-domain mappings that preserve inferential structure.
- From evidence in cognitive linguistics, we know that there is a Classes Are Containers metaphor. This grounds our understanding of classes, by mapping the inferential structure of embodied Container schemas to classes as we understand them.
- Boole's metaphor and the Propositional Logic metaphor have been carefully crafted by mathematicians to mathematicize classes and map them onto propositional structures.
- The symbolic-logic mapping was also crafted by mathematicians, so that propositional logic could be made into a symbolic calculus governed by "blind" rules of symbol manipulation.
- Thus, our understanding of symbolic logic traces back via metaphorical and symbolic mappings to the inferential structure of embodied Container schemas.

We have discussed only propositional logic and Boole's logic of classes here. There are extremely restricted cases and artificially constructed cases in natural language that do fit these logics. And over the years far more sophisticated logics have been developed. But real meanings of natural-language expressions are far more complex than any logicians have yet approached or are likely to in the foreseeable future. In fact, writers of logic textbooks usually have to search far and wide for examples that fit their formulas and proofs.

Real reasoning includes phenomena like graded categories, various kinds of prototypes (typical, ideal, nightmare, salient examples, social stereotypes), radial categories, frames, metaphors, and so on. Symbolic logic, no matter how sophisticated its present form, is nowhere near characterizing real everyday human reason. The developments in mathematical logic include a great many interesting—even spectacular—results, often with important implications for other fields like computer science and economics. But they are *mathematical* results, not (for the most part) results about the nature of everyday human reason.

Moreover, no matter how sophisticated formal logics get, they remain mathematical tools conceptualized, created, and understood by human beings with human brains and bodies. It is the responsibility of the cognitive science of mathematics to understand the cognitive mechanisms by which new mathematical logics are constructed, conceptualized, and understood.

What we have done in this chapter has been to construct a mathematical idea analysis of Boole's logic of classes and of propositional logic. In addition, we have explicitly indicated how the mathematical ideas are symbolized. Such idea analyses (and their symbolizations) make explicit what is implicit in normal mathematical notation. We have been able to add precision at the level of ideas and symbolization to mathematics, an enterprise that has maximal precision as one of its major goals.

We are now ready to discuss sets.

7

Sets and Hypersets

Some Conceptual Issues in Set Theory

We have been looking at mathematics from a cognitive perspective, seeking to understand the cognitive and ultimately the bodily basis of mathematics. Set theory is at the heart of modern mathematics. It is therefore incumbent upon us to understand the ideas implicit in set theory as well as we can.

In our discussion of Boolean classes in Chapter 6, we saw that the concept of a class is experientially grounded via the metaphor that Classes Are Containers—mental containers that we use in perceiving, conceptualizing, and reasoning about our experience. As we saw there, notions like the empty class and the technical notion of subclass (in which a class is a subclass of itself) are introduced via Boole's metaphor.

Modern set theory begins with these basic elements of our grounding metaphor for classes and with Boole's metaphor. That is, it starts with the notion of a class as a containerlike entity; the ideas of intersection, union, and complement; the associative, commutative, and distributive laws; the empty set; and the idea of closure. Sets are more sophisticated than Boolean classes.

Take a simple case. In Boolean algebra, classes can be subclasses of other classes but not *members* of those classes. Contemporary set theory allows sets to be members, not just subclasses, of other sets. What is the grounding for the intuitive conceptual distinction between sets as subsets and sets as members? As subsets, intuitive sets are Container schemas, mental containers organizing objects into groups. For sets to be *members* in this conceptualization, they, too, must be conceptualized as *objects* on a par with other objects and not just mental Container schemas providing organizations for objects. To conceptualize sets as objects, we need a conceptual metaphor providing such a conceptualization. We will call this metaphor Sets Are Objects.

Via this metaphor, the empty set can be viewed not just as a subset but as a unique entity in its own right—an entity that may or may not be a *member* of a given set, even though it is a *subset* of every set. The Sets Are Objects metaphor allows us to conceptualize *power sets*, in which all the *subsets* of a given set A are made into the *members* of another set, $P(A)$. For example, consider the set $A = \{a, b, c\}$. The following are the subsets of A: $\varnothing$, $\{a\}$, $\{b\}$, $\{c\}$, $\{a, b\}$, $\{b, c\}$, $\{a, c\}$, $\{a, b, c\}$. The power set, $P(A)$ has these subsets of A as its members. $P(A) = \{\varnothing, \{a\}, \{b\}, \{c\}, \{a, b\}, \{b, c\}, \{a, c\}, \{a, b, c\}\}$. Note that the set $\{a, b, c\}$ is a member of its power set.

Incidentally, as long as we conceptualize sets as Container schemas, it will be impossible to conceptualize sets as members of themselves. The reason is that a member is properly included in the interior of a Container schema, and no Container schema can be in its own interior.

The Ordered Pair Metaphor

Once we conceptualize sets as objects that can be members of another set, we can construct a metaphorical definition for ordered pairs. Intuitively, an ordered pair is conceptualized nonmetaphorically as a subitized pair of elements (by what we will call a Pair schema) structured by a Path schema, where the source of the path is seen as the first member of the pair and the goal of the path is seen as the second member. This is simply our intuitive notion of what an ordered pair is.

With the addition of the Sets Are Objects metaphor, we can conceptualize ordered pairs metaphorically, not in terms of Path and Pair schemas but in terms of sets:

THE ORDERED PAIR METAPHOR

Source Domain SETS		Target Domain ORDERED PAIRS
The Set $\{\{a\}, \{a, b\}\}$	$\rightarrow$	The Ordered Pair (a, b)

Using this metaphorical concept of an ordered pair, one can go on to metaphorically define relations, functions, and so on in terms of sets. One of the most interesting things that one can do with this metaphorical ordered-pair definition is to metaphorically conceptualize the natural numbers in terms of sets that have other sets as members, as the great mathematician John von Neumann (1903–1957) did.

THE NATURAL NUMBERS ARE SETS METAPHOR

Source Domain SETS		Target Domain NATURAL NUMBERS
The empty set ∅	→	Zero
The set containing the empty set {∅} (i.e., {0}).	→	One
The set {∅, {∅}} (that is, {0, 1})	→	Two
The set {∅, {∅}, {∅, {∅}}} (i.e., {0, 1, 2})	→	Three
The set of its predecessors (built following the above rule)	→	A natural number

Here, the set with no members is mapped onto the number zero, the set containing the empty set as a member (i.e., it has one member) is mapped onto the number one, the set with two members is mapped onto the number two, and so on. By virtue of this metaphor, every set containing three members is in a one-to-one correspondence with the number three. Using this metaphor, you can metaphorically construct the natural numbers out of nothing but sets.

From a cognitive perspective, this is a metaphor that allows us to conceptualize numbers, which are one kind of conceptual entity, in terms of sets, which are a very different kind of conceptual entity. This is a *linking metaphor*—a metaphor that allows one to conceptualize one branch of mathematics (arithmetic) in terms of another branch (set theory). Linking metaphors are different from grounding metaphors in that both the source and target domains of the mapping are within mathematics itself. Such linking metaphors, where the source domain is set theory, are what is needed conceptually to "reduce" other branches of mathematics to set theory in order to satisfy the Formal Foundations metaphor. (For a discussion of the Formal Foundations program and the metaphors needed to achieve it, see Chapter 16.)

Cantor's Metaphor

Our ordinary everyday conceptual system includes the concepts Same Number As and More Than. They are based, of course, on our experience with finite, not infinite, collections. Among the criteria characterizing our ordinary everyday versions of these concepts are:

- *Same Number As:* Group *A* has the same number of elements as group *B* if, for every member of *A*, you can take away a corresponding member of *B* and not have any members of *B* left over.

- *More Than:* Group *B* has more objects than group *A* if, for every member of *A*, you can take away a member of *B* and still have members left in *B*.

It is important to contrast our everyday concept of Same Number As with the German mathematician Georg Cantor's concept of pairability—that is, capable of being put into a one-to-one correspondence. Finite collections with the same number of objects are pairable, since two finite sets with the same number of objects can be put in a one-to-one correspondence. This does not mean that Same Number As and pairability are the same idea; the ideas are different in a significant way, but they happen to correlate precisely for finite sets. The same is not true for infinite sets.

Compare the set of natural numbers and the set of even integers. As Cantor (1845–1918) observed, they are pairable; that is, they can be put into a one-to-one correspondence. Just multiply the natural numbers by two, and you will set up the one-to-one correspondence (see Figure 7.1). Of course, these two sets do not have the same number of elements according to our everyday criterion. If you take the even numbers away from the natural numbers, there are still all the odd numbers left over. According to our usual concept of "more than," *there are more natural numbers than even numbers*, so the concepts Same Number As and pairability are different concepts.

Our everyday concepts of Same Number As and More Than are, of course, linked to other everyday quantitative concepts, like How Many or As Many As, and Size Of, as well as to the concept Number itself—the basic concept in arithmetic. In his investigations into the properties of infinite sets, Cantor used the concept of pairability in place of our everyday concept of Same Number As. In doing so, he established a conceptual metaphor, in which one concept (Same Number As) is conceptualized in terms of the other (pairability).

CANTOR'S METAPHOR

Source Domain MAPPINGS		Target Domain NUMERATION
Set *A* and set *B* can be put into one-to-one correspondence	→	Set *A* and set *B* have the same number of elements

That is, our ordinary concept of having the same number of elements is metaphorically conceptualized, especially for infinite sets, in terms of the very different concept of being able to be put in a one-to-one correspondence.

This distinction has never before been stated explicitly using the idea of conceptual metaphor. Indeed, because the distinction has been blurred, generations of students have been confused. Consider the following statement made by

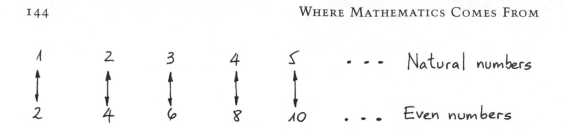

FIGURE 7.1   The one-to-one correspondence between the natural numbers and the positive even numbers.

many mathematics teachers: *"Cantor proved that there are just as many positive even integers as natural numbers."*

Given our ordinary concept of "As Many As," Cantor proved no such thing. He proved only that the sets were pairable. In our ordinary conceptual system, there are *more* natural numbers than there are positive even integers. It is only by use of Cantor's metaphor that it is correct to say that he proved that there are, metaphorically, "just as many" even numbers as natural numbers.

The same comment holds for other proofs of Cantor's. According to our ordinary concept of "More Than," there are more rational numbers than natural numbers, since if you take the natural numbers away from the rational numbers, there will be lots left over. But Cantor did prove that the two sets are pairable, and hence they can be said (via Cantor's metaphor) to metaphorically have the "same number" of elements. (For details, see Chapter 10.)

Cantor's invention of the concept of pairability and his application of it to infinite sets was a great conceptual achievement in mathematics. What he did in the process was create a new technical mathematical concept—pairability—and with it, new mathematics. But Cantor also intended pairability to be a *literal* extension of our ordinary notion of Same Number As from finite to infinite sets. There Cantor was mistaken. From a cognitive perspective, it is a metaphorical rather than literal extension of our everyday concept. The failure to teach the difference between Cantor's technical metaphorical concept and our ordinary concept confuses generation after generation of introductory students.

Axiomatic Set Theory and Hypersets

Axiomatic Set Theory

In Chapter 16, we will discuss the approach to mathematics via formal systems of axioms. We will see there that such an approach requires a linking metaphor for conceptualizing complex and sophisticated mathematical ideas in terms of much simpler mathematical ideas, those of set theory and the theory of formal systems. For example, there must be a metaphor for characterizing what it means for a mathematical structure to "fit" a collection of axioms that have no inherent meaning. Whereas Euclid understood his postulates for geometry to be meaningful to human beings, formal systems of axioms are taken as mind-free sequences of symbols. They are defined to be free of human conceptual systems and human understanding.

On the formalist view of the axiomatic method, a "set" is any mathematical structure that "satisfies" the axioms of set theory as written in symbols. The traditional axioms for set theory (the Zermelo-Fraenkel axioms) are often taught as being about sets conceptualized as containers. Many writers speak of sets as "containing" their members, and most students think of sets that way. Even the choice of the word "member" suggests such a reading, as do the Venn diagrams used to introduce the subject. But if you look carefully through those axioms, you will find nothing in them that characterizes a container. The terms "set" and "member of" are both taken as undefined primitives. In formal mathematics, that means that they can be anything that fits the axioms. Here are the classic Zermelo-Fraenkel axioms, including the axiom of choice; together they are commonly called the ZFC axioms.

- *The axiom of Extension:* Two sets are equal if and only if they have the same members. In other words, a set is uniquely determined by its members.
- *The axiom of Specification:* Given a set A and a one-place predicate $P(x)$ that is either true or false of each member of A, there exists a subset of A whose members are exactly those members of A for which $P(x)$ is true.
- *The axiom of Pairing:* For any two sets, there exists a set that they are both members of.
- *The axiom of Union:* For every collection of sets, there is a set whose members are exactly the members of the sets of that collection.
- *The axiom of Powers:* For each set A, there is a set $P(A)$ whose members are exactly the subsets of set A.
- *The axiom of Infinity:* There exists a set A such that (1) the empty set is a member of A, and (2) if X is a member of A, then the successor of X is a member of A.
- *The axiom of Choice:* Given a disjoint set S whose members are non-empty sets, there exists a set C which has as its members one and only one element from each member of S.

There is nothing in these axioms that explicitly requires sets to be containers. What these axioms do, collectively, is to *create* entities called "sets," first from elements and then from previously created classes and sets. The axioms do not say explicitly how sets are to be conceptualized.

The point here is that, within formal mathematics, where all mathematical concepts are mapped onto set-theoretical structures, the "sets" used in these structures are not technically conceptualized as Container schemas. They do not have Container schema structure with an interior, boundary, and exterior. Indeed, within formal mathematics, there are no concepts at all, and hence sets are not conceptualized as anything in particular. They are undefined entities whose only constraints are that they must "fit" the axioms. For formal logicians and model theorists, sets are those entities that fit the axioms and are used in the modeling of other branches of mathematics.

Of course, most of us do conceptualize sets in terms of Container schemas, and that is perfectly consistent with the axioms given above. However, when we conceptualize sets as Container schemas, a constraint follows automatically: *Sets cannot be members of themselves*, since containers cannot be inside themselves. Strictly speaking, this constraint does not follow from the axioms but from our metaphorical understanding of sets in terms of containers. The axioms do not rule out sets that contain themselves. However, an extra axiom was proposed by von Neumann that does rule out this possibility.

- *The axiom of Foundation:* There are no infinite descending sequences of sets under the membership relation; that is, $\ldots S_{i+1} \in S_i \in \ldots \in S$ is ruled out.

Since allowing sets to be members of themselves would result in such a sequence, this axiom has the indirect effect of ruling out self-membership.

Hypersets

Technically, within formal mathematics, model theory has nothing to do with everyday understanding. Model theorists do not depend upon our ordinary container-based concept of a set. Indeed, certain model theorists have found that our ordinary grounding metaphor that Classes Are Container schemas gets in the way of modeling kinds of phenomena they want to model, especially recursive phenomena. For example, take expressions like

$$x = 1 + \cfrac{1}{1 + \cfrac{1}{1 + \ldots}}$$

If we observe carefully, we can see that the denominator of the main fraction has in fact the value defined for x itself. In other words, the above expression is equivalent to

$$x = 1 + \frac{1}{x}$$

Such recursive expressions are common in mathematics and computer science. The possibilities for modeling these expressions using "sets" are ruled out if the only kind of "sets" used in the modeling must be ones that cannot have themselves as members. Set theorists have realized that a new noncontainer metaphor is needed for thinking about sets, and they have explicitly constructed one: hyperset theory (Barwise & Moss, 1991).

The idea is to use graphs, not containers, for characterizing sets. The kinds of graphs used are Accessible Pointed Graphs, or APGs. "Pointed" denotes an asymmetric relation between nodes in the graph, indicated visually by an arrow pointing from one node to another—or from one node back to that node itself. "Accessible" means that one node is linked to all other nodes in the graph and can therefore be "accessed" from any other node.

From the axiomatic perspective, hyperset theorists have replaced the axiom of Foundation with an Anti-Foundation axiom. From a cognitive point of view, the implicit conceptual metaphor they have used is this:

THE SETS ARE GRAPHS METAPHOR

Source Domain ACCESSIBLE POINTED GRAPHS		*Target Domain* SETS
An APG	→	The membership structure of a set
An arrow	→	The membership relation
Nodes that are tails of arrows	→	Sets
Decorations on nodes that are heads of arrows	→	Members
APGs with no loops	→	Classical sets with the Foundation axiom
APGs with or without loops	→	Hypersets with the Anti-Foundation axiom

The effect of this metaphor is to eliminate the notion of containment from the concept of a "set." The graphs have no notion of containment built into them at all. And containment is not modeled by the graphs.

Graphs that have no loops satisfy the ZFC axioms and the axiom of Foundation. They thus work just like sets conceptualized as containers. But graphs that do have loops model sets that can have themselves as members; these graphs do not work like sets that are conceptualized as containers, and they do not satisfy the axiom of Foundation.

A *hyperset* is an APG that may or may not contain loops. Hypersets thus fit not the axiom of Foundation but, rather, another axiom with the opposite intent:

- *The Anti-Foundation axiom:* Every APG pictures a unique set.

The fact that hypersets satisfy the Zermelo-Fraenkel axioms confirms what we said above: *The Zermelo-Fraenkel axioms for set theory—the ones generally accepted in mathematics—do not define our ordinary concept of a set as a container at all!* That is, the axioms of "set theory" are not, and were never meant to be, about what we ordinarily call "sets," which we conceptualize in terms of containers.

Numbers Are Sets, Which Are Graphs

If we adjust the von Neumann metaphor for numbers in terms of sets to the new graph-theoretical models, zero (the empty set) becomes a node with no arrows leading from it. One becomes the graph with two nodes and one arrow leading from one node to the other. Two and three are represented by the graphs in Figures 7.2(a) and 7.2(b), respectively. And where an arrow bends back on itself, as in Figure 7.2(c), we get recursion—the equivalent of an infinitely long chain of nodes with arrows not bending back on themselves, as in Figure 7.2(d).

Here we see the power of conceptual metaphor in mathematics. Sets, conceptualized in everyday terms as containers, do not have the right properties to model everything needed. So we can now metaphorically redefine "sets" to exclude containment by using certain kinds of graphs. The only confusing thing is that this special case of graph theory is still called "set theory" for historical reasons.

It is sometimes said that the theory of hypersets is "a set theory in which sets can contain themselves." But this is misleading, because it is not a theory of "sets" as we ordinarily understand them in terms of containment. The reason that these graph-theoretical objects are called "sets" is a functional one: They play the role in modeling axioms that classical sets with the axiom of Foundation used to play.

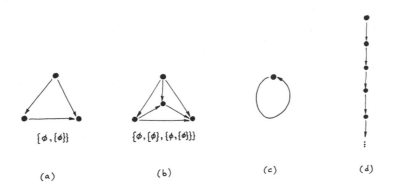

FIGURE 7.2 Hypersets: sets conceptualized as graphs, with the empty set as the graph having no arrows leading from it. The set containing the empty set is a graph whose root has one arrow leading to the empty set. The metaphor Numbers Are Sets is taken for granted, with zero being the empty set and each natural number being the set of all its predecessors. Drawing (a) depicts the number 2, since it is a graph whose root (the top node) has two arrows (membership relations): one leading to the empty set, or zero (the lower right node), and the other leading to the set containing the empty set, or one (the subgraph whose root is the node at the lower left). Drawing (b) depicts the number 3, since its root has three arrows leading to root nodes for subgraphs for the numbers 2, 1, 0, as depicted in (a). Drawing (c) is a graph representing a set that is a "member" of itself, under the Sets Are Graphs Metaphor. And drawing (d) is an infinitely long chain of nodes in an infinite graph, which is equivalent to (c). (Adapted from Barwise & Moss, 1991.)

What Are Sets, Really?

As we have just seen, mathematics has two quite inconsistent metaphorical conceptions of sets—one in terms of Container schemas and one in terms of graphs. Is one of these conceptions right and the other wrong? There is a perspective from which one might think so—a perspective that says there is only one correct notion of a "set." There is another perspective, the most common one, which recognizes that these two distinct notions of "set" define different and mutually inconsistent subject matters, conceptualized via radically different metaphors. This situation is much more common in mathematics than most people realize.

The Roles Played by Conceptual Metaphors

In this chapter, as in previous chapters, conceptual metaphors have played several roles.

- First, there are *grounding metaphors*—metaphors that ground our understanding of mathematical ideas in terms of everyday experience. Examples include the Classes Are Container schemas and the four grounding metaphors for arithmetic.
- Second, there are *redefinitional metaphors*—metaphors that impose a technical understanding replacing ordinary concepts. Cantor's metaphor is a case in point.
- Third, there are *linking metaphors*—metaphors within mathematics itself that allow us to conceptualize one mathematical domain in terms of another mathematical domain. Examples include Boole's metaphor, von Neumann's Natural Numbers Are Sets metaphor, and the Sets Are Graphs metaphor.

The linking metaphors are in many ways the most interesting of these, since they are part of the fabric of mathematics itself. They occur whenever one branch of mathematics is used to model another, as happens frequently. Moreover, linking metaphors are central to the creation not only of new mathematical concepts but often of new branches of mathematics. As we shall see, such classical branches of mathematics as analytic geometry, trigonometry, and complex analysis owe their existence to linking metaphors.

But before we go on, we should point out an important phenomenon arising from the fact that different linking metaphors can have the same source domain.

Metaphorically Ambiguous Sets

Recall for a moment two central metaphors that are vital to the set-theoretical foundations of mathematics: The Ordered Pair metaphor and the Natural Numbers Are Sets metaphor.

THE ORDERED PAIR METAPHOR

Source Domain SETS		Target Domain ORDERED PAIRS
The set $\{\{a\}, \{a, b\}\}$	$\rightarrow$	The ordered pair (a, b)

THE NATURAL NUMBERS ARE SETS METAPHOR

Source Domain SETS		Target Domain NATURAL NUMBERS
The empty set $\varnothing$	$\rightarrow$	Zero

The set containing the empty set {∅} (i.e., {0}).	→	One
The set {∅, {∅}} (i.e., {0, 1})	→	Two
The set {∅, {∅}, {∅, {∅}}} (i.e., {0, 1, 2})	→	Three
The set of its predecessors (built following the above rule)	→	A natural number

Within formal mathematics, these metaphors are called "definitions." The purpose of these definitions is to reduce arithmetic to set theory. An "eliminativist" is someone who takes this reduction literally—that is, who believes that these metaphorical definitions eliminate numbers and ordered pairs from the ontology of mathematics, replacing them by sets with these structures. The eliminativist claims that numbers and ordered pairs do not really exist as separate kinds of things from sets; rather, they are sets—sets of the form given in the two metaphors. In the eliminativist interpretation, there is no number zero separate from the empty set. Rather, the number zero in reality *is* the empty set. And there is no number one separate from the set containing the empty set. Rather, the set containing the empty set *is* the number one. Similarly, no ordered pairs exist as ordered pairs distinct from sets. Rather, on the eliminativist view, the ordered pair (a, b) really *is* a set, the set {{a}, {a, b}}.

Let us consider the following two questions:

1) Exactly what set *is* the ordered pair of numbers (0, 1)?
2) Exactly what set *is* the set containing the numbers 1 and 2 as members?

According to the two metaphorical definitions, the answers should be as follows:

1) The ordered pair (0, 1) does not exist as such. It is the set: {{∅}, {∅, {∅}}}.
2) The set containing the numbers 1 and 2 as members does not exist as such. It is the set: {{∅}, {∅, {∅}}}.

The answer is the same.

The ordered pair (0, 1) is a conceptual entity different from the set {1, 2}. One might expect that the reduction of ordered pairs and natural numbers to sets would preserve such conceptual distinctions. It is clear from this example that it does not. Is this a problem for eliminativists? It depends on whether they have such an expectation or not.

From a cognitive point of view, this is no problem. It is common for different concepts to have the same metaphorical source. For example, "You're warm" can have at least two metaphorical meanings—You are affectionate or You are getting close to the answer. In mathematics, the Ordered Pair metaphor and the Natural Numbers Are Sets metaphor serve the purposes of set theory. Since they preserve inference, inferences about sets will be mapped appropriately onto both numbers and ordered pairs.

The moral is simple: Mathematics is not literally reducible to set theory in a way that preserves conceptual differences. However, ingenious metaphors linking ordered pairs to sets and numbers to sets have been explicitly constructed and give rise to interesting mathematics. It is important to distinguish a literal definition from a metaphorical one. Indeed, there are so many conceptual metaphors used in mathematics that it is extremely important to know just what they are and to keep them distinct.

Mathematical idea analysis of the sort we have just done clarifies these issues. Where there are metaphorical ambiguities in a mere formalist symbolization like {{∅}, {∅, {∅}}}, it reveals those ambiguities because it makes explicit what is implicit. It also clarifies the concept "same size" for a set, distinguishing the everyday notion from that defined by Cantor's metaphor. And it makes simple and comprehensible the idea of hypersets as an instance of the Sets Are Graphs metaphor.

Part III

The Embodiment of Infinity

8

The Basic Metaphor of Infinity

Infinity Embodied

One might think that if any concept cannot be embodied, it is the concept of infinity. After all, our bodies are finite, our experiences are finite, and everything we encounter with our bodies is finite. Where, then, does the concept of infinity come from?

A first guess might be that it comes from the notion of what is finite—what has an end or a bound—and the notion of negation: something that is *not* finite. But this does not give us any of the richness of our conceptions of infinity. Most important, it does not characterize infinite *things*: infinite sets, infinite unions, points at infinity, transfinite numbers. To do this, we need not just a negative notion ("not finite") but a positive notion—a notion of infinity as an entity-in-itself.

What is required is a *general* mathematical idea analysis of all the various concepts of infinity in mathematics. Such an analysis must answer certain questions: What do the various forms of infinity have to do with each other? That is, What is the relationship among ideas like infinite sets, points at infinity, and mathematical induction, where a statement is true for an infinity of cases. How are these ideas related to the idea of a limit of an infinite sequence? Or an infinite sum? Or an infinite intersection? And finally, how are infinitely large things conceptually related to infinitely small things?

To begin to see the embodied source of the idea of infinity, we must look to one of the most common of human conceptual systems, what linguists call the *aspectual system*. The aspectual system characterizes the structure of event-concepts—events as we conceptualize them. Some actions, for example, are inherently iterative, like tapping or breathing. Others are inherently continuous, like moving. Some have inherent beginning and ending points, like jumping. Some just have ending points, like arriving. Others just have starting points, like leaving or embarking on a trip. Those that have ending points also have resulting states. In addition, some actions have their completions conceptualized as part of the action (e.g., landing is part of jumping), while others have their completions conceptualized as external to the actions (e.g., we do not normally conceptualize landing as part of flying, as when a comet flies by the earth).

Of course, in life, hardly anything one does goes on forever. Yet we conceptualize breathing, tapping, and moving as *not having completions*. This conceptualization is called *imperfective aspect*. As we saw in Chapter 2, the concept of aspect appears to be embodied in the motor-control system of the brain. Narayanan (1997), in a study of computational neural modeling, showed that the neural computational structure of the aspectual system is the same as that found in the motor-control system.

Given that the aspectual system is embodied in this way, we can see it as the fundamental source of the concept of infinity. Outside mathematics, a process is seen as infinite if it continues (or iterates) indefinitely without stopping. That is, it has imperfective aspect (it continues indefinitely) without an endpoint. This is *the literal concept of infinity* outside mathematics. It is used whenever one thinks of perpetual motion—motion that goes on and on forever.

Continuative Processes Are Iterative Processes

A process conceptualized as not having an end is called an imperfective process—one that is not "perfected," that is, completed. Two of the subtypes of imperfective processes are *continuative* (those that are continuous) and *iterative* (those that repeat and have an intermediate endpoint and an intermediate result). In languages throughout the world, continuous processes are conceptualized as if they were iterative processes. The syntax used is commonly that of conjunction. Consider a sentence like *John jumped and jumped again, and jumped again.* Here we have an iteration of three jumps. But *John jumped and jumped and jumped* is usually interpreted not as three jumps but as an open-ended, indefinite number.

Now, "jump" is an inherently perfective verb: Each jump has an endpoint and a result. But verbs like *swim, fly,* and *roll* are imperfective, with no indicated endpoint. Consider sentences indicating iteration via the syntactic device of conjunction: *John swam and swam and swam. The eagle flew and flew and*

flew. This sentence structure, which would normally indicate indefinite iteration with perfective verbs, here indicates a continuous process of swimming or
flying. The same is true in the case of aspectual particles like *on* and *over.* For
example, *John said the sentence over* indicates a single iteration of the sentence. But *John said the sentence over and over and over* indicates ongoing repetition. Similarly, *The barrel rolled over and over* indicates indefinitely
continuous rolling, and *The eagle flew on and on* indicates indefinitely continuous flying. In these sentences, the language of iteration for perfectives (e.g.,
verb *and* verb *and* verb; *over and over and over*) is used with imperfectives to
express something quite different—namely, an indefinitely continuous process.
In short, the idea of iterated action is being used in various syntactic forms to
express the idea of continuous action. This can be characterized in cognitive
terms by the metaphor Indefinite Continuous Processes Are Iterative Processes.

There is a cognitive reason why such a metaphor should exist. Processes in general are conceptualized metaphorically in terms of motion via the event structure
metaphor, in which processes are extended motions (see Lakoff & Johnson, 1999).
Indefinitely continuous motion is hard to visualize, and for extremely long periods it is impossible to visualize. What we do instead is visualize short motions
and then repeat them, thus conceptualizing indefinitely continuous motion as repeated motion. Moreover, everyday continuous actions typically require iterated
actions. For example, continuous walking requires repeatedly taking steps; continuous swimming requires repeatedly moving the arms and legs; continuous flying by a bird requires repeatedly flapping the wings. This conflation of continuous
action and repeated actions gives rise to the metaphor by which continuous actions are conceptualized in terms of repeated actions.

Why is this metaphor important for infinity? The reason is that we commonly
apply it to infinitely continuous processes. Continuous processes without end—
infinite continuous processes—are conceptualized via this metaphor as if they
were infinite iterative processes, processes that iterate without end but in which
each iteration has an endpoint and a result. For example, consider infinitely continuous motion, which has no intermediate endpoints and no intermediate locations where the motion stops. Such infinitely continuous motion can be
conceptualized metaphorically as iterated motion with intermediate endings to
motion and intermediate locations—but with infinitely many iterations.

This metaphor is used in the conceptualization of mathematics to break
down continuous processes into infinitely iterating step-by-step processes, in
which each step is discrete and minimal. For example, the indefinitely continuous process of reaching a limit is typically conceptualized via this metaphor as
an infinite sequence of well-defined steps.

Actual Infinity

The kind of infinity we have just seen—ongoing processes or motions without end—was called *potential infinity* by Aristotle, who distinguished it from *actual infinity,* which is infinity conceptualized as a realized "thing." Potential infinity shows up in mathematics all the time: when you imagine building a series of regular polygons with more and more sides, when you imagine writing down more and more decimals of √2, and so on (see Figure 8.1). But the interesting cases of infinity in modern mathematics are cases of actual infinity— cases that go beyond mere continuous or iterative processes with no end. These include, for example, infinite sets (like the set of natural numbers) and points at infinity—mathematical entities characterized by infiniteness.

We hypothesize that the idea of actual infinity in mathematics is metaphorical, that the various instances of actual infinity make use of the ultimate metaphorical *result* of a process without end. Literally, there is no such thing as the result of an endless process: If a process has no end, there can be no "ultimate result." But the mechanism of metaphor allows us to conceptualize the "result" of an infinite process—in the only way we have for conceptualizing the result of a process—that is, in terms of a process that does have an end.

We hypothesize that all cases of actual infinity—infinite sets, points at infinity, limits of infinite series, infinite intersections, least upper bounds—are special cases of a single general conceptual metaphor in which processes that go on indefinitely are conceptualized as having an end and an ultimate result. We call this metaphor the *Basic Metaphor of Infinity,* or the BMI for short. The target domain of the BMI is the domain of processes without end—that is, what linguists call imperfective processes. The effect of the BMI is to add a metaphorical completion to the ongoing process so that it is seen as having a result—an infinite *thing.*

The source domain of the BMI consists of an ordinary iterative process with an indefinite (though finite) number of iterations with a completion and resultant state. The source and target domains are alike in certain ways:

- Both have an initial state.
- Both have an iterative process with an unspecified number of iterations.
- Both have a resultant state after each iteration.

In the metaphor, the initial state, the iterative process, and the result after each iteration are mapped onto the corresponding elements of the target domain. But the crucial effect of the metaphor is *to add to the target domain the completion of the process and its resulting state.* This metaphorical addition is

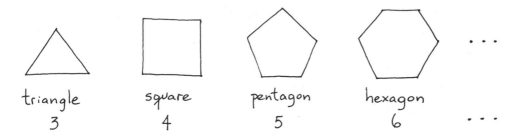

FIGURE 8.1 A case of potential infinity: the sequence of regular polygons with n sides, starting with $n = 3$. This is an unending sequence, with no polygon characterizing an ultimate result.

indicated in boldface in the statement of the metaphor that follows. It is this last part of the metaphor that allows us to conceptualize the ongoing process in terms of a completed process—and so to produce the concept of actual infinity.

THE BASIC METAPHOR OF INFINITY

Source Domain COMPLETED ITERATIVE PROCESSES		*Target Domain* ITERATIVE PROCESSES THAT GO ON AND ON
The beginning state	→	The beginning state
State resulting from the initial stage of the process	→	State resulting from the initial stage of the process
The process: From a given intermediate state, produce the next state.	→	The process: From a given intermediate state, produce the next state.
The intermediate result after that iteration of the process	→	The intermediate result after that iteration of the process
The final resultant state	→	**"The final resultant state"** **(actual infinity)**
Entailment *E:* The final resultant state is unique and follows every nonfinal state.	→	**Entailment *E*:** **The final resultant state** **is unique and follows every** **nonfinal state.**

Notice that the source domain of the metaphor has something that does not correspond to anything in the literal target domain—namely, a final resultant state.

The conceptual mapping imposes a "final resultant state" on an unending process. The literal unending process is given on the right-hand side of the top three arrows. The metaphorically imposed final resultant state (which characterizes an actual infinity) is indicated in boldface on the right side of the fifth line of the mapping.

In addition, there is a crucial entailment that arises in the source domain and is imposed on the target domain by the metaphor. In any completed process, the final resultant state is *unique*. The fact that it is the *final* state of the process means that:

- There is no earlier final state; that is, there is no distinct previous state within the process that both follows the completion stage of the process yet precedes the final state of the process.
- Similarly, there is no later final state of the process; that is, there is no other state of the process that both results from the completion of the process and follows the final state of the process. Any such putative state would have to be "outside the process" as we conceptualize it.

Thus, the uniqueness of the final state of a complete process is a product of human cognition, not a fact about the external world. That is, it follows from the way we conceptualize completed processes. This entailment is also shown in boldface.

The Basic Metaphor of Infinity maps this uniqueness property for final resultant states of processes onto actual infinity. Actual infinity, as characterized by *any given application of the BMI*, is unique. As we shall see later, the existence of degrees of infinity (as with transfinite numbers) requires multiple applications of the BMI.

What results from the BMI is a metaphorical creation that does not occur literally: a process that goes on and on indefinitely and yet has a unique final resultant state, a state "at infinity." This metaphor allows us to conceptualize "potential" infinity, which has neither end nor result, in terms of a familiar kind of process that has a unique result. Via the BMI, infinity is converted from an open-ended process to a specific, unique entity.

We have formulated this metaphor in terms of a simple iterative step-by-step process—namely: From a given state, produce the next state. From an initial state, the process produces intermediate resultant states. The metaphorical process has an infinity of such intermediate states and a metaphorical final, unique, resultant state.

Notice that the nature of the process is unspecified in the metaphor. The metaphor is general: It covers any kind of process. Therefore, we can form special cases of this general metaphor by specifying what process we have in mind. The process could be a mechanism for forming sets of natural numbers resulting in an infinite set. Or it could be a mechanism for moving further and further along a line until the "point at infinity" is reached. Our formulation of the metaphor is sufficiently precise and sufficiently general so that we can fill in the details of a wide variety of different kinds of infinity in different mathematical domains. Indeed, this will be our strategy, both in this chapter and in chapters to come. We will argue that a considerable number of infinite processes in mathematics are special cases of the BMI that can be arrived at by specifying what the iterative process is in detail. We believe that *all* notions of infinity in mathematics can be seen as special cases of the BMI, and we will discuss what we hope is a wide enough range of examples to make our case.

The Origin of the BMI Outside Mathematics

The Basic Metaphor of Infinity is a general cognitive mechanism. It can occur by itself, as when one speaks of "the infinite" as a thing. But most people don't go around speaking of "the infinite" as a thing; the notion occurs usually only in philosophical or spiritual contexts—or in special cases in which some particular process is under discussion.

There is a long history to the use of this metaphor in special cases, a history that goes back at least to pre-Socratic philosophy. Greek philosophy from its earliest beginnings held that any particular thing is an instance of a higher category. For example, a particular goat is an instance of the category Goat. Moreover, every such category was assumed to be itself a thing in the world. That is, not only were individual goats in the world but the category Goat—a natural kind—was also seen as an entity in the world. This meant that it, too, was an instance of a still higher category, and so on.

Thus there was taken to be an indefinitely long ascending hierarchy of categories. It was further assumed that this indefinitely ascending hierarchy had an end—the category of Being, which encompassed everything (see Lakoff & Johnson, 1999, chs. 16–18). The reasoning implicit in this move from an indefinitely ascending hierarchy to a highest, all-encompassing point in the hierarchy can be seen as an instance of the Basic Metaphor of Infinity, in which the result of an indefinitely iterative process of higher categorization results in a highest category. The following table shows, for each element of the target domain of the BMI, the corresponding element of the special case (the formation of ascending categories). The symbol "$\Rightarrow$" indicates the relation between target domain elements and their corresponding special cases.

THE HIGHEST CATEGORY—BEING

Target Domain ITERATIVE PROCESSES THAT GO ON AND ON		*Special Case of the Target Domain* THE FORMATION OF ASCENDING CATEGORIES
The beginning state	⇒	Specific entities with no categories
State resulting from the initial stage of the process	⇒	Categories of specific entities: The lowest level of categories
The process: From a prior intermediate state, produce the next state.	⇒	From categories at a given level, form categories at the next highest level that encompass lower-level categories.
The intermediate result from that iteration of the process	⇒	Intermediate-level categories formed by each iteration of the process
"The final resultant state" **(actual infinity)**	⇒	**"The final resultant state":** **The category of Being**
Entailment *E*: The final resultant state is unique and follows every nonfinal state.	⇒	**Entailment *E*: The category of Being is unique and encompasses all other categories.**

In this case, the process specified in the BMI is the process of forming higher categories. The first step is to form a category from specific cases and make that category a thing. The iterative process is to form higher categories from immediately lower categories. And the final resultant state is the highest category, the category of Being.

What we have just done is consider a special case of the BMI outside mathematics, and we have shown that by considering the category of Being as produced by the BMI, we get all of its requisite properties. Hereafter, we will use the same strategy *within* mathematics to produce special cases of actual infinity, and we will use similar tables to illustrate each special case.

This metaphorical concept of infinity as a unique entity—the highest entity—was extended naturally to religion. In Judaism, the Kabbalistic concept of God is *Ein Sof*—that is, "without end," a single, unique God. The uniqueness entailment applied to deities yields monotheism. In Christianity, God is *all*-powerful: Given an ascending hierarchy of power, that hierarchy is assumed to have a unique highest point, something all-powerful—namely, the power of God. It is no coincidence that Georg Cantor believed that the study of infinity in mathematics had a theological importance. He sought, in his mathematical concept of the infinite, a way to reconcile mathematics and religion (Dauben, 1983).

Processes As Things

Processes are commonly conceptualized as if they were static things—often containers, or paths of motion, or physical objects. Thus we speak of being *in the middle of* a process or at the *tail end* of it. We see a process as having a *length*—*short* or *long*—able to be *extended* or *attenuated, cut short*. We speak of the *parts* of a process, as if it were an object with parts and with a size. This extends to infinite processes as well: Some infinities are *bigger* than other infinities.

As we saw in Chapter 2, one of the most important cognitive mechanisms for linking processes and things is what Talmy (1996, 2000) has called "fictive motion," cases in which an elongated path, object, or shape can be conceptualized metaphorically as a process tracing the length of that path, object, or shape. A classic example is *The road runs through the forest,* where *runs* is used metaphorically as a mental trace of the path of the road.

Processes, as we ordinarily think of them, extend over time. But in mathematics, processes can be conceptualized as atemporal. For example, consider Fibonnacci sequences, in which the $n + 2^{nd}$ term is the sum of the nth term and the $n + 1^{st}$ term. The sequence can be conceptualized either as an ongoing infinite process of producing ever more terms or as a thing, an infinite sequence that is atemporal. This dual conceptualization, as we have seen, is not special to mathematics but part of everyday cognition.

Throughout this chapter, we will speak of infinite processes in mathematics. Sometimes we conceptualize them as extending over time, and sometimes we conceptualize them as static. For our purpose here, the difference will not matter, since we all have conceptual mechanisms (Talmy's fictive-motion schemas) for going between static and dynamic conceptualizations of processes. For convenience, we will be using the dynamic characterization throughout.

What Is Infinity?

Our job in this chapter is to answer a general question, *What is infinity?* Or, put another way, *How do we, mere human beings, conceptualize infinity?* This is not an easy question to answer, because of all the constraints on it. First, the answer must be biologically and cognitively plausible. That is, it must make use of normal cognitive and neural mechanisms. Second, the answer must cover all the cases in mathematics, which on the surface are very different from one another: mathematical induction, points meeting at infinity, the infinite set of natural numbers, transfinite numbers, limits (functions get infinitely close to them), things that are infinitely small (infinitesimal numbers and points), and so on. Third, the answer must be sufficiently precise so that it can be demonstrated that these and other concepts of the infinite in mathematics can indeed be characterized as special cases of a general cognitive mechanism.

Our answer is that that general cognitive mechanism is the Basic Metaphor of Infinity, the BMI. It has both neural and cognitive plausibility. We will devote the remainder of this chapter to showing how it applies in a wide variety of cases. Chapters 9 through 14 extend this analysis to still more cases.

The "Number" ∞

One of the first things that linguists, psychologists, and cognitive scientists learn is that when there are explicit culturally sanctioned warnings not to do something, you can be sure that people are doing it. Otherwise there would be no point to the warnings. A marvelous example of such a warning comes from G. H. Hardy's magnificent classic text, *A Course of Pure Mathematics* (1955; pp. 116–117):

> Suppose that n assumes successively the values 1, 2, 3, ... The word "successively" naturally suggests succession in time, and we may suppose n, if we like, to assume these values at successive moments of time. . . . However large a number we may think of, a time will come when n has become larger than this number. . . . This however is a mere matter of convenience, the variation of n having usually nothing to do with time.
>
> The reader cannot impress upon himself too strongly that when we say that n "tends to ∞" we mean simply that n is supposed to assume a series of values which increases beyond all limit. *There is no number "infinity"*: Such an equation as
>
> $$n = \infty$$
>
> is as it stands *meaningless*: A number n cannot be equal to ∞, because "equal to ∞" means nothing. . . . [T]he symbol ∞ means nothing at all except in the phrase "tends to ∞," the meaning of which we have explained above.

Hardy is going through all this trouble to keep the reader from thinking of infinity as a number just because people do tend to think of infinity as a number, which is what the language "tends to infinity" and "approaches infinity" chosen by mathematicians indicates. Hardy suggests that it is a mistake to think of infinity as a number—a mistake that many people make. If people are, mistakenly or not, conceptualizing infinity as a number, then it is our job as cognitive scientists to characterize the cognitive mechanism by which they are making that "mistake." And if we are correct in suggesting that a single cognitive mechanism—namely, the BMI—is used for all conceptions of infinity, then we have to show how the BMI can be used to conceptualize infinity as a number, whether it is a "mistake" to do so or not.

Cognitive science is, after all, *de*scriptive, not *pre*scriptive. And it must explain, as well, why people think as they do. The "mistake" of thinking of infinity as a number is not random. "∞" is usually used with a precise meaning—as a number in an enumeration, not as a number in a calculation. In "1, 2, 3, . . . , ∞" ∞ is taken as an endpoint in an enumeration, larger than any finite number and beyond all of them. But people do not use ∞ as a number in calculations: We see no cases of "17 times ∞, minus 473," which is of course a meaningless expression, as Hardy correctly points out. The moral here is that there are, cognitively, different uses for numbers—enumeration, comparison, and calculation. As a number, ∞ is used in enumeration and comparison but not in calculation. Even mathematicians use infinity as a number in enumeration, as in the sum of a sequence a_n from $n = 1$ to $n = \infty$:

$$\sum_{n=1}^{\infty} a_n$$

When Hardy warns us not to assume that ∞ is a number, it is because mathematicians have devised notions and ways of thinking, talking, and writing, in which ∞ is a number with respect to enumeration, though not calculation.

Indeed, the idea of ∞ as a number can also be seen as a special case of the BMI. Note that the BMI does not have any numbers in it. Suppose we apply the BMI to the integers used to indicate order of enumeration. The inherent structure of the target domain, independent of the metaphor, has a potential infinity, an unending sequence of ordered integers. The effect of the BMI is to turn this into an actual infinity with a largest "number" ∞.

THE BMI FOR ENUMERATION

Target Domain ITERATIVE PROCESSES THAT GO ON AND ON		*Special Case* THE UNENDING SEQUENCE OF INTEGERS USED FOR ENUMERATION
The beginning state	⇒	No integers
State resulting from the initial stage of the process	⇒	The integer 1

The process: From a prior intermediate state, produce the next state.	$\Rightarrow$	Given integer $n-1$, form the next largest integer n.
The intermediate result after that iteration of the process.	$\Rightarrow$	$n > n-1$
"The final resultant state" (actual infinity)	$\Rightarrow$	**The "integer" ∞**
Entailment *E:* The final resultant state is unique and follows every nonfinal state.	$\Rightarrow$	**Entailment *E:* The "integer" ∞ is unique and larger than every other integer.**

The BMI itself has no numbers. However, the unending sequence of integers used for enumeration (but not calculation) can be a special case of the target domain of the BMI. As such, the BMI produces ∞ as the largest integer used for enumeration. This is the way most people understand ∞ as a number. It cannot be used for calculation. It functions exclusively as an *extremity*. Its operations are largely undefined. Thus, $\infty/0$ is undefined, as is $\infty \cdot 0$, $\infty - \infty$, and ∞/∞. And thus ∞ is not a full-fledged number, which was Hardy's point. For Hardy, an entity either was a number or it wasn't, since he believed that numbers were objectively existing entities. The idea of a "number" that had one of the functions of a number (enumeration) but not other functions (e.g., calculation) was an impossibility for him. But it is not an impossibility from a cognitive perspective, and indeed people do use it. ∞ as *the extreme natural number* is commonly used with the implicit or explicit sequence "1, 2, 3, . . . , ∞" in the characterization of infinite processes. Each such use involves a hidden use of the BMI to conceptualize ∞ as the extreme natural number.

Mathematicians use "1, . . . , n, . . . , ∞" to indicate the terms of an infinite sequence. Although the BMI in itself is number-free, it will be notationally convenient from here on to index the elements of the target domain of the BMI with the integers "ending" with metaphorical "∞." For mathematical purists we should note that when an expression like $n-1$ is used, it is taken only as a notation indexing the stage previous to stage n, and not as the result of the operation of subtraction of 1 from n.

Target Domain
ITERATIVE PROCESSES THAT GO ON AND ON

The beginning state (0)
State (1) resulting from the initial stage of the process

The process: From a prior intermediate state $(n-1)$,
    produce the next state (n).
The intermediate result after that iteration of the
    process (the relation between n and $n-1$)
"The final resultant state" (actual infinity "∞")
Entailment *E:* The final resultant state ("∞") is unique
    and follows every nonfinal state.

The utility of this notation will become apparent in the next section.

Projective Geometry: Where Parallel Lines Meet at Infinity

In projective geometry, there is an axiom that *all parallel lines meet at infinity.*
From a cognitive perspective, this axiom presents the following problems. (1)
How can we conceptualize what it means for there to be a point "at infinity?"
(2) How can we conceptualize parallel lines as "meeting" at such a point? (3)
How can such a conceptualization use the same general mechanism for com-
prehending infinity that is used for other concepts involving infinity?

We will answer these questions by taking the BMI and filling it in in an appropriate way, with a fully comprehensible process that will produce a final result, notated by "∞," in which parallel lines meet at a point *at* infinity. Our
answer must fit an important constraint. The point at infinity must function
like any other point; for example, it must be able to function as the intersection
of lines and as the vertex of a triangle. Theorems about intersections of lines
and vertices of triangles must hold of "points at infinity."

To produce such a special case of the BMI, we have to specify certain
parameters:

- A *subject-matter frame,* indicating that the subject matter is geometry—
  specifically, that it is about lines that intersect at a point.
- The initial step of the process.
- The *iterated process,* or a *step-by-step condition* that links each state
  resulting from the process to the next state.
- The *resultant state* after each iteration.
- An *entailment* of the uniqueness of the final resultant state.

In the case of projective geometry, we take as the subject-matter frame an
isosceles triangle, for reasons that will shortly become clear (see Figure 8.2).

The Isosceles Triangle Frame

An isosceles triangle, ABC_n with
    Sides: AB, AC_n, BC_n
    Angles: α_n, β_n
    D_n: The distance from A to C_n
Where:
    $AC_n = BC_n$
    $\alpha_n = \beta_n$
Inference:
    If AC_n lies on L_{1n} and BC_n lies on L_{2n},
      then L_{1n} intersects L_{2n}.

The iterative process in this case is to move point C_n further and further away from points A and B. As the distance D_n between A and C_n gets larger, the angles α_n and β_n approach 90 degrees more and more closely. As a result, the intersecting lines L_{1n} and L_{2n} get closer and closer to being parallel. This is an unending, infinite process. At each stage n, the lines meet at point C_n. (For details, see Kline, 1962, ch. 11; Maor, 1987.)

The idea of "moving point C_n further and further" is captured by a sequence of moves, with each move extending over an arbitrary distance. By the condition "arbitrary distance" we mean to quantify over *all* the distances for which the condition holds.

Here are the details for filling in the parameters in the BMI in this special case.

Parallel Lines Meet at Infinity

Target Domain Iterative Processes That Go On and On		*Special Case* Projective Geometry
The beginning state (0)	$\Rightarrow$	The isosceles-triangle frame, with triangle ABC_0
State (1) resulting from the initial stage of the process	$\Rightarrow$	Triangle ABC_1, where the length of AC_1 is D_1
The process: From a prior intermediate state $(n-1)$, produce the next state (n).	$\Rightarrow$	Form AC_n from AC_{n-1} by making D_n arbitrarily larger than D_{n-1}.
The intermediate result after that iteration of the process (the relation between n and $n-1$)	$\Rightarrow$	$D_n > D_{n-1}$ and $(90° - \alpha_n) < (90° - \alpha_{n-1})$.

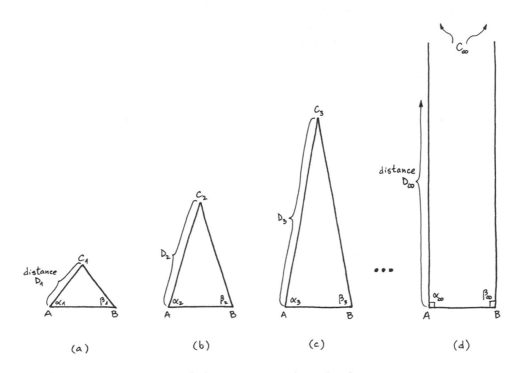

FIGURE 8.2 The application of the Basic Metaphor of Infinity to projective geometry. Drawing (a) shows the isosceles triangle ABC_1, the first member of the BMI sequence. Drawings (b) and (c) show isosceles triangles ABC_2 and ABC_3, in which the equal sides get progressively longer and the equal angles get closer and closer to 90°. Drawing (d) shows the final resultant state of the BMI: ABC_∞, where the angles *are* 90°. The equal sides are infinitely long, and they metaphorically "meet" at infinity—namely the unique point C_∞.

"The final resultant state" (actual infinity "∞") $\Rightarrow$	α_∞ = 90°. D_∞ is infinitely long. Sides AC_∞ and BC_∞ are infinitely long, parallel, and meet at C_∞—a point "at infinity."
Entailment *E*: The final resultant state ("∞") is unique and follows every nonfinal state. $\Rightarrow$	**Entailment *E*: There is a unique** AC_∞ (distance D_∞) that is longer than AC_n (distance D_n) for all finite *n*.

As a result of the BMI, lines L_1 and L_2 are parallel, meet at infinity, and are separated by the length of line segment AB. Since the length AB was left unspecified, as was the orientation of the triangle, this result will "fit" all lines parallel to L_1 and L_2 in the plane. Thus, this application of the BMI defines the same system of geometry as the basic axiom of projective geometry—namely, that all parallel lines in the plane meet at infinity. Thus, for each orientation there is an infinite family of parallel lines, all meeting "at infinity." Since there is such a family of parallel lines for each orientation in the plane, there is a "point at infinity" in every direction.

Thus, we have seen that there is a special case of BMI that defines the notion "point at infinity" in projective geometry, which is a special case of actual infinity as defined by the BMI. We should remind the reader here that this is a cognitive analysis of the concept "point at infinity" in projective geometry. It is not a mathematical analysis, is not meant to be one, and should not be confused with one. Our claim is a cognitive claim: The concept "point at infinity" in projective geometry is, from a cognitive perspective, a special case of the general notion of actual infinity.

We have at present no experimental evidence to back up this claim. In order to show that the claim is a plausible one, we will have to show that a wide variety of concepts of infinity in mathematics arise as special cases of the BMI. Even then, this will not prove empirically that they are; it will, however, make the claim highly plausible.

The significance of this claim is not only that there is a single general cognitive mechanism underlying all human conceptualizations of infinity in mathematics, but also that this single mechanism makes use of common elements of human cognition—aspect and conceptual metaphor.

The Point at Infinity in Inversive Geometry

Inversive geometry also has a concept of a "point at infinity," but it is a concept very different from the one found in projective geometry. Inversive geometry is defined by a certain transformation on the Cartesian plane. Consider the Cartesian plane described in polar coordinates, in which every point is represented by (r, θ) where θ is an angle and r is the distance from the origin. Consider the transformation that maps r onto $1/r$. This transformation maps the unit circle onto itself, the interior of the unit circle onto its exterior, and its exterior onto its interior. Let us consider what happens to zero under this transformation.

Consider a ray from zero extending outward indefinitely at some angle θ (see Figure 8.3). As r inside the circle gets closer to 0, $1/r$ gets further away. Thus, $1/1{,}000$ is mapped onto $1{,}000$, $1/1{,}000{,}000$ is mapped onto $1{,}000{,}000$, and so on. As r approaches 0, $1/r$ approaches ∞. What is the point at 0 mapped onto? It is tempting to map 0, line by line, into a point at ∞ on that line. But there would be many such points, one for each line. What inversive geometry does is define a single "point at infinity" for all lines, and it maps 0, which is unique, onto the unique "point at infinity."

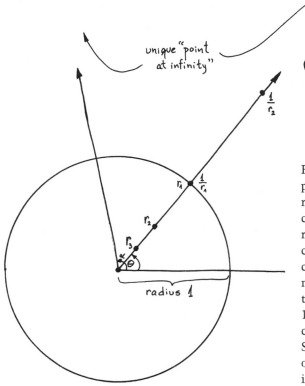

FIGURE 8.3 Inversive geometry: a plane with a circle of radius 1 and rays extending from the center of the circle. Each ray is a nonnegative real-number line, with zero at the center of the circle and 1 on the circle. There is a function $f(x) = 1/x$, mapping points r on each ray inside the circle to corresponding points $1/r$ on that ray outside the circle, and conversely, with zero mapped to ∞. Since there is only one zero, there is only one "point at infinity," which is in all the rays.

If we are correct that the BMI characterizes actual infinity in all of its many forms, then the point at infinity in inversive geometry should also be a special case of the BMI. But, as in most cases, the precise formulation takes some care. Whereas in projective geometry there is an infinity of points at infinity, in inversive geometry there is only one. This must emerge as an entailment of the BMI, given the appropriate parameters: the frame and the iterative process. Here is the frame, which fills in the "subject matter" parameter of the BMI. The frame is straightforward.

THE INVERSIVE GEOMETRY FRAME

The Cartesian plane, with polar coordinates.
The one-to-one function $f(x) = 1/x$, for x on a ray that
extends from the origin outward.

This frame picks a special case of the target domain of the BMI. To fill out this special case of the BMI, we need to characterize the iterative process.

The idea behind the process is simple. Pick a ray and pick a point x_1 on that ray smaller than 1. That point has an inverse, which is greater than 1. Keep picking points x_n on the same ray, closer and closer to zero. The inverse points $1/x_n$ get further and further from the origin. This is an infinite, unending process.

We take this process as the iterative process in the BMI. Here are the details:

THE POINT AT INFINITY IN INVERSIVE GEOMETRY

Target Domain ITERATIVE PROCESSES THAT GO ON AND ON		Special Case INVERSIVE GEOMETRY
The beginning state (0)	$\Rightarrow$	The origin, with rays projecting outward in every direction, and with an arbitrary ray r at angle θ designated
State (1) resulting from the initial stage of the process	$\Rightarrow$	A point x_1 on r at an arbitrary distance $d_1 < 1$ from the origin. There is an inverse point x_1', at a distance $1/d_1 > 1$ from the origin.
The process: From a prior intermediate state $(n–1)$, produce the next state (n).	$\Rightarrow$	Given point x_{n-1} at distance d_{n-1} from the origin, find point x_n at distance d_n from the origin, where d_n is arbitrarily smaller than d_{n-1}.
The intermediate result after that iteration of the process (the relation between n and $n–1$)	$\Rightarrow$	There is an inverse point, x_n', at a distance $1/d_n$ from the origin and further from the origin than x_{n-1}' is.
"The final resultant state" (actual infinity "∞")	$\Rightarrow$	**x_∞ is at the origin; that is, its distance from the origin is zero. x_∞' is a unique point infinitely far from the origin.**
Entailment E: The final resultant state ("∞") is unique and follows every nonfinal state.	$\Rightarrow$	**Entailment E: x_∞ is unique and is closer to the origin than any other point on ray r. There is a unique x_∞' that is further from the origin than any other point on ray r.**

The BMI then applies to this infinite, unending process and conceptualizes it metaphorically as having a unique final resultant state "∞." At this metaphorical final state, x_∞ is at the origin, and the distance from the origin to x_∞' is infinitely long. In inversive geometry, arithmetic is extended to include ∞ as a number with respect to division: $D_\infty = \infty$, $1/D_\infty = 1/\infty = 0$, and $1/0 = \infty$. This extension of arithmetic says nothing about the addition or subtraction of ∞. It is peculiar to inversive geometry because of the way this special case of the BMI is defined: All points must have inverses, including the point at the origin. Where dividing by zero is normally not possible, this metaphor extends ordinary division to give a metaphorical value to $1/0$ and $1/\infty$. In inversive geometry, ∞ does exist as a number and has limited specified possibilities for calculation.

Moreover, since there is only one zero point shared by all rays, and since $f(x) = 1/x$ is a one-to-one mapping that, via this metaphor, maps 0 onto ∞ and ∞ onto 0, there must therefore be only one ∞ point shared by all rays.

What is interesting about this case is that the same general metaphor, the BMI, produces a concept of the point at infinity very different from that in the case of projective geometry. In projective geometry, there is an infinity of points at infinity (for an image, think of the horizon line), while in inversive geometry there is only one. Moreover, projective geometry has no implicit associated arithmetic, while inversive geometry has an implicit arithmetic (differing from normal arithmetic in its treatment of zero and infinity).

Finally, though we have characterized inversive geometry in terms of a cognitive mechanism, the BMI, mathematicians of course do not. They simply define inversive geometry in normal mathematical terms. Our goal is to show how the same concept of infinity is involved in inversive geometry and other forms of mathematics, while respecting the differences in the concept of infinity across branches of mathematics.

The Infinite Set of Natural Numbers

Within formal arithmetic, the natural numbers are usually characterized by the successor operation: Start with 1. Add 1 to yield a result. Add 1 to the result to yield a new result. And so on. This is an unending, infinite process. It yields the natural numbers, one at a time. Not the *infinite set* containing *all* the natural numbers. Just the natural numbers themselves, each of which is finite. Since it is incapable of being used in calculation, ∞ is not a full-fledged member of the infinite set of natural numbers.

Here is the problem of characterizing the set of natural numbers. The set must be infinite since it contains *all* of the infinitely many numbers, but it cannot contain ∞ as a number.

To get the *set* of natural numbers, you have to collect up each number as it is formed. The set keeps growing without end. To get the *entire* set of natural

numbers—all of them, even though the set never stops growing—you need
something more. In axiomatic set theory, you add an axiom that simply stipu-
lates that the set exists. From a cognitive perspective, that set can be constructed conceptually via a version of the Basic Metaphor of Infinity. The BMI
imposes a metaphorical completion to the unending process of natural-number
collection. The result is the *entire collection*, the set of *all* natural numbers!

THE SET OF ALL NATURAL NUMBERS

Target Domain ITERATIVE PROCESSES THAT GO ON AND ON		*Special Case* THE SET OF NATURAL NUMBERS
The beginning state (0)	⇒	The natural number frame, with a set of existing numbers and a successor operation that adds 1 to the last number and forms a new set
State (1) resulting from the initial stage of the process	⇒	The empty set, the set of natural numbers smaller than 1.
The process: From a prior intermediate state (n–1), produce the next state (n).	⇒	Given S_{n-1}, the set of natural numbers smaller than n–1, form $S_{n-1} \cup \{n-1\} = S_n$.
The intermediate result after that iteration of the process (the relation between n and n–1)	⇒	At state n, we have S_n, the set of natural numbers smaller than n.
"The final resultant state" (actual infinity "∞")	⇒	**S_∞, the set of *all* natural numbers smaller than ∞—that is, the set of *all* natural numbers (which does not include ∞ as a number).**
Entailment *E*: The final resultant state ("∞") is unique and follows every nonfinal state.	⇒	**Entailment *E*: The set of all natural numbers is unique and includes every natural number (no more, no less).**

This special case of the BMI does the same work from a cognitive perspective
as the axiom of Infinity in set theory; that is, it ensures the existence of an infinite set whose members are all the natural numbers. This is an important
point. The BMI, as we shall see, is often the conceptual equivalent of some
axiom that guarantees the existence of some kind of infinite entity (e.g., a least
upper bound). And just as axioms do, the special cases of the BMI determines
the right set of inferences required.

This example teaches us something to be borne in mind throughout the remainder of this book. The meaning of "all" or "entire" is far from obvious when they are predicated of infinite sets. In general, the meaning of "all" involves completeness. One of the first uses of "all" learned in English is the child's use of "all gone" to indicate the completion of a process of consumption, removal, or destruction. *All twelve of the paintings were stolen* indicates completeness of the theft—the *entire* collection was stolen. But in the case of infinity, there is no such thing as the literal completeness of an infinite process. The BMI is necessary in some form—implicit or explicit—to characterize any "all" that ranges over an infinite process. Wherever there is infinite totality, the BMI is in use.

This BMI analysis shows us exactly how the same notion of infinity used to comprehend points at infinity can also be used to conceptualize the infinite set of natural numbers. In this case, we can see a somewhat different version of the concept of infinity. Points at infinity give us a concept of an *infinite extremity*. But the set of natural numbers is not an infinite extremity; rather, it is an *infinite totality*. Totalities always involve sets. Extremities only involve linear orderings.

Mathematical Induction

The process of mathematical induction is crucial in mathematical proofs. It works as follows: Prove that

1. The statement S is true for 1;
2. If the statement S is true for $n-1$, then it is true for n.

Literally, all this does is provide an unending, infinitely unfolding sequence of natural numbers for which the statement S is true. It does not prove that the statement is true for *all* natural numbers. Something is needed to go from the unending sequence of individual natural numbers, for which S is a single general truth, to *all* natural numbers.

In axiomatic arithmetic, there is a special axiom of Mathematical Induction needed to bridge the gap between the unending sequence of *specific* truths for each number, one at a time, and the single generalization to the infinite totality of *all* natural numbers. The axiom simply postulates that if (1) and (2) are proved, then the statement is true for every member of the set of *all* natural numbers.

The axiom of Mathematical Induction is the equivalent of the Basic Metaphor of Infinity applied to the subject matter of inductive proof. To see why, we can conceptualize the content of the axiom in terms of the BMI. Note that this instance of the BMI involves an infinite totality—the infinite set of all finite natural numbers—not an infinite extremity such as the "number" ∞. In this version of the BMI, a set is built up of all the finite natural numbers that the given statement is true of. At the final resultant state, there is an infinite set of finite numbers.

THE BMI FOR MATHEMATICAL INDUCTION

Target Domain ITERATIVE PROCESSES THAT GO ON AND ON		*Special Case* MATHEMATICAL INDUCTION
The beginning state (0)	⇒	A statement $S(x)$, where x varies over the set of natural numbers
State (1) resulting from the initial stage of the process	⇒	$S(1)$ is true for the members of $\{1\}$— the set containing the number 1.
The process: From a prior intermediate state $(n–1)$, produce the next state (n).	⇒	Given the truth of $S(n–1)$ for the members of the set $\{1, \ldots, n–1\}$, establish the truth of $S(n)$ for the members of $\{1, \ldots, n\}$.
The intermediate result after that iteration of the process (the relation between n and $n–1$).	⇒	$S(n)$ is true for the members of the set $\{1, \ldots, n\}$.
"The final resultant state" (actual infinity "∞")	⇒	**$S(\infty)$ is true for the members of the set of *all* natural numbers.**
Entailment *E:* The final resultant state ("∞") is unique and follows every nonfinal state.	⇒	**Entailment *E:* The set of natural numbers for which $S(\infty)$ is true is unique and includes *all* finite natural numbers.**

Thus, mathematical induction can also be seen as a special case of the BMI.

Generative Closure

The idea of generative closure for operations that generate infinite sets also makes implicit use of the BMI. For example, suppose we start with the set containing the integer 1 and the operation of addition. By adding 1 to itself at the first stage, we get 2, which is not in the original set. That means that the original set was not "closed" under the operation of addition. To move toward closure, we can then extend that set to include 2. At the next stage, we perform the binary operation of addition on 1 and 2, and on 2 and 2, to get new elements 3 and 4. We then extend the previous set by including 3 and 4. And so on. This defines an infinite sequence of set extensions. If we apply the BMI, we get "closure" under addition—the set of *all* resulting extensions.

This will work starting with any finite set of elements and any finite number of binary operations on those elements. At least in this simple case, the concept of closure can be seen as a special case of the BMI. We can also see from this case how to state the general case of closure in terms of the BMI. We let C_0 be any set of elements and O be any finite set of operations, each applying to some element or pair of elements in C_0. Here is the iterative process:

- At stage 1, apply every operation once in every way it can apply to the elements of C_0. Collect the results in a set S_0 and form the union of S_0 with C_0 to yield C_1.
- At stage 2, apply every operation once in every way it can apply to the elements of C_1. Collect the results in a set S_1 and form the union of S_1 with C_1 to yield C_2.
- And so on.

If this process fails to yield new elements at any stage—that is, if $C_{n-1} = C_n$—then the closure is finite and the process stops. Otherwise it goes on. Note that for each n, C_n contains as a subset all of the C_i, for $i < n$.

Letting this be the iterative process, we can characterize infinite closure using the Basic Metaphor of Infinity in the following way. At ∞, C_∞ contains every C_n, for $n < \infty$. That is, it contains every finite combination of operations.

THE BMI FOR GENERATIVE CLOSURE

Target Domain ITERATIVE PROCESSES THAT GO ON AND ON		Special Case GENERATIVE CLOSURE
The beginning state (0)	$\Rightarrow$	A finite set of elements C_0 and a finite collection O of binary operations on those elements
State (1) resulting from the initial stage of the process	$\Rightarrow$	C_1 = The union of C_0 and the set S_0 of elements resulting from a single application of each operator in O to each pair of elements of C_0
The process: From a prior intermediate state $(n-1)$, produce the next state (n).	$\Rightarrow$	Given C_{n-1}, form S_{n-1}, the set of elements resulting from a single application of each operator in O to each pair of elements of C_{n-1}. $C_n = C_{n-1} \cup S_{n-1}$.

The intermediate result after that iteration of the process (the relation between n and $n-1$).	$\Rightarrow$	At state n, we have C_n = the union of C_0 and the union of all S_k, for $k < n$. If $C_{n-1} = C_n$, then the closure is finite and the process ends. Otherwise it continues.
"The final resultant state" (actual infinity "∞")	$\Rightarrow$	C_∞ = **the infinite closure of C_0 and its extensions under the operations in O.**
Entailment E: The final resultant state ("∞") is unique and follows every nonfinal state.	$\Rightarrow$	**Entailment E: C_∞ is unique and includes all possible finite iterations of applications of the operations in O to the elements of C_0 and elements resulting from those operations.**

Notice that although the closure is infinite, there are no infinite sequences of operations. Each sequence of operations is finite, but there is no bound to their length.

"All"

It is commonplace in formal logic to use the symbol for the universal quantifier "$\forall$" in statements or axioms concerning such entities as the set of *all* natural numbers. This symbol is just a symbol and requires an interpretation to be meaningful. In formal logic, interpretations of such quantifiers are given in two ways: (1) via a metalanguage and (2) via a mapping onto a mathematical structure—for instance, the generative closure produced from the set {1} under the operation "+", written "Closure [{1}, +]".

1) "$(\forall x: x \in N)$ Integer (x)" is true if and only if x is an integer for *all* members of the set of natural numbers, N.
2) "$(\forall x: x \in N)$ Integer (x)" is true if and only if x is a member of Closure [{1}, +].

In statement (1) above, the word "all" occurs in the metalanguage expression on the right of "if and only if." In statement (2), the closure of {1} and its extensions under the operation of addition occurs on the right of "if and only if."

Now, in statement (1), the symbol $\forall$ is defined in terms of a prior understanding of the word "all." From a cognitive point of view, this just begs the question of what ∀ means, since it requires a cognitive account of what "all" means when applied to infinite sets. Cognitively, "all" is understood in terms of a linear scale of inclusion, with "none" at one endpoint, "all" at the other, and values like "some," "most," and "almost all" in between. To give a cognitive account of the meaning of "all" applied to an infinite set, we need both the scale of inclusion and a cognitive account of infinite sets, which is given by the BMI.

Statement (2) is an attempt to characterize the meaning of ∀ for a given infinite set without using the word "all" and instead using the concept of generative closure. From a cognitive perspective, this requires a cognitive account of the meaning of generative closure for infinite sets, which we have just given in terms of the BMI.

The moral is that, in either case, one version of the BMI or another is needed to give a cognitive characterization of the meaning of ∀ when it is applied to infinite sets. The use of the symbol ∀ in symbolic logic does not get us out of the problem of giving a cognitive characterization of infinite sets. As far as we can tell, the BMI does suffice in all cases, as we shall see in the chapters to come.

Up to this point, we have shown six diverse mathematical uses of actual infinity which can each be conceptualized as a special case of the Basic Metaphor of Infinity. Each case involves either *infinite extremity* (points at infinity and the number ∞) or *infinite totality* (the set of natural numbers and mathematical induction).

The Basic Conclusions

By comparing the special cases, one can see precisely how a wide variety of superficially different mathematical concepts have a similar structure from a cognitive perspective. Since much of mathematics is concerned with comparisons of kinds of mathematical structures, this sort of analysis falls within the tradition of structural comparison in mathematics. These analyses can be seen as being very much in the tradition of algebra or category theory, in that they seek to reveal deep similarities in cases that superficially look dissimilar.

The difference, of course, is that we are discussing conceptual structure from a *cognitive* perspective, taking into account cognitive constraints, rather than from a purely mathematical point of view, which has no cognitive constraints whatsoever. That is, mathematicians are under no obligation to try to understand how mathematical understanding is embodied and how it makes use of normal cognitive mechanisms, like image schemas, aspectual structure, conceptual metaphors, and so on.

We are hypothesizing that the analyses given above are correct from a cognitive perspective. This hypothesis is testable, at least in principle. (See Gibbs, 1994, for an overview of literature for testing claims about metaphorical thought.) The hypothesis involves certain subhypotheses.

- The BMI is part of our *unconscious* conceptual system. You are generally not aware of conceptual mechanisms while you are using them. This should also hold true of the BMI. It is correspondingly *not* claimed that the use of the BMI is conscious.
- The concept of actual infinity, as characterized by the BMI, makes use of a metaphor based on the concept of aspect—imperfective aspect (in ongoing events) and perfective aspect (in completive events).
- As with any cognitive model, it is hypothesized that the entailments of the model are true of human cognition.

What we have provided in this chapter is a mathematical idea analysis of various concepts of infinity. Mathematics itself does not characterize what is common among them. The common mathematical notion for infinity—"...," as in the sequence "1 + 1/2 +1/4 + ... "—does not even distinguish between potential and actual infinity. If it is potential infinity, the sum only gives an endless sequence of partial sums always less than 2; if it is actual infinity, the sum is exactly 2. Our idea analysis in terms of the BMI makes explicit what is implicit.

In addition, our mathematical idea analysis provides a clear and explicit answer to the question of how finite embodied beings can have a concept of infinity—namely, via conceptual metaphor, specifically the BMI! And it shows the relationships among the various ideas of infinity found in mathematics.

9

Real Numbers and Limits

I T IS NO SURPRISE THAT THE CONCEPT OF INFINITY is central to the concepts of real numbers and limits. The idea of a limit arose in response to the fact that there are infinite series whose sums are finite. Real numbers are fundamentally conceptualized using the concept of infinity: in terms of infinite decimals, infinite sums, infinite sequences, and infinite intersections. Since we are hypothesizing that there is a single general concept of actual infinity characterized via the BMI, we will be arguing that the conceptualization of the real numbers inherently uses the BMI.

There is no small irony here. The real numbers are taken as "real." Yet no one has ever seen a real number. The physical world, limited at the quantum level to sizes no smaller than the Planck length, has nothing as fine-grained as a real number. Computers, which do most of the mathematical calculations in the world, can use only "floating-point numbers" (see Chapter 15), which are limited in information to a small number of bits, usually 32. No computers use—or could use—real numbers in their infinite detail.

We will be hypothesizing that the real numbers, which require the concept of infinity, are conceptualized using the Basic Metaphor of Infinity. The irony is that what are called "real" numbers are fundamentally metaphorical in their conceptualization. This has important consequences for an understanding of what mathematics is (see Chapters 15 and 16). For now, we will simply show how the BMI can conceptualize the mathematical ideas required to characterize the real numbers: infinite decimals, infinite polynomials, limits of infinite sequences, least upper bounds, and infinite intersections of intervals. We will put off discussing Dedekind cuts until Chapter 13, where we will argue that they, too, are instances of the BMI.

Numerals for the Natural Numbers

In building up to the infinite decimals, we have to start with the numerals for the natural numbers in decimal notation—that is, using the digits 0, 1, . . . , 9. To get the set of *all* numerals for natural numbers, just stringing digits one after another without end is not enough. That will give us a perpetually growing class (a potential infinity) but not *all* the numerals for *all* the natural numbers (an actual infinity). To get the infinite set with *all* the numerals, we need to use a special case of the BMI.

Here's how the BMI process works to build up an infinity of indefinitely long finite decimal representations for the infinite set of indefinitely large finite natural numbers. We will start at the beginning state with N_1, the empty set. In the first stage of the process, we will add the one-place decimal representations to form the set $N_2 = \{1, . . . , 9\}$. In the second stage, we will form N_3, the set of 2-place strings of numerals from the members of N_2 as follows: If s is a member of N_2, it is a member of N_3. Also, s followed by one of the numerals 0, 1, . . . , 9 is also a member of N_3. Thus, for example, let s be 5. Then the strings of digits 50, 51, . . . , 59 are all members of N_3. In this way, we get $N_3 = \{1, . . . , 9, 10, 11, . . . , 99\}$, the set of all one- or two-place numerals. We then generalize this process and use it as the iterative process in the BMI, as follows:

NUMERALS FOR NATURAL NUMBERS

Target Domain ITERATIVE PROCESSES THAT GO ON AND ON		*Special Case* THE INFINITE SET OF NUMERALS FOR THE NATURAL NUMBERS
The beginning state (1)	$\Rightarrow$	N_1: The empty set
State (2) resulting from the initial stage of the process	$\Rightarrow$	The set of numerals $N_2 = \{1, . . . , 9\}$
The process: From a prior intermediate state $(n{-}1)$, produce the next state (n).	$\Rightarrow$	Let N_{n-1} be the set of numerals whose number of digits is less than $n{-}1$. Let s be a string of digits in N_{n-1}.
		The members of N_n take one of the following forms: s, or $s0$, or $s1$, . . . , or $s9$.
The intermediate result after that iteration of the process	$\Rightarrow$	The set of numerals N_n, each member of which has less than n digits
"The final resultant state" (actual infinity "∞")	$\Rightarrow$	N_∞: **the set containing all the numerals for all the natural numbers**

Entailment *E:* The final resultant state ("∞") is unique and follows every nonfinal state.	⇒	Entailment *E:* N_∞ is unique and contains all the numerals for the natural numbers.

At each stage *n*, we get all the numerals with *less than n* digits. At the "final state," where *n* is the metaphorical "number" ∞, we get the *infinite totality* of all the finitely long numerals whose number of digits is less than ∞—namely, all the finitely long numerals.

Although our cognitive model of this use of the BMI is complicated, what it describes is something we all find quite simple. This is normal with cognitive models, whose job is to characterize every minute detail of what we understand intuitively as "simple."

Infinite Decimals

Consider π = 3.14159265.... π is a precise number characterized by an infinitely long string of particular digits to the right of the decimal point. This is not just a sequence that gets longer and longer but an *infinitely long* fixed sequence—a thing. It is a particular infinite sequence, not an ongoing process. This is a case of actual infinity, not potential infinity.

The same is true of any infinite decimal:

$$147963.674039839275....$$

Though we cannot write it down, the "..." is taken as representing a fixed infinitely long sequence of particular digits, again a case of actual infinity.

Thus, each infinite decimal can be seen as composed of two parts:

a. A finite sequence of digits to the left of the decimal point, which is a numeral indicating a natural number.
b. An infinitely long sequence of digits to the right of the decimal point, indicating a real number between zero and one.

We have already shown how the BMI characterizes (a). We now need to show how a further use of a special case of the BMI characterizes (a) with (b) added to the right.

We start out with the numerals for the natural numbers—those to the left of the decimal point. That is the set N_∞, which was characterized in the previous section.

INFINITE DECIMALS

Target Domain ITERATIVE PROCESSES THAT GO ON AND ON		Special Case NUMERALS FOR REAL NUMBERS
The beginning state (0)	$\Rightarrow$	R_0: the set consisting of members of N_∞, each followed by a decimal point
State (1) resulting from the initial stage of the process	$\Rightarrow$	R_1: For each string of digits s in R_0, $s0, s1, s2, \ldots, s9$ is in R_1.
The process: From a prior intermediate state $(n-1)$, produce the next state (n).	$\Rightarrow$	The members of R_n are of one of the following forms: either $s0$, or $s1$, ..., or $s9$, where the string of digits s is a member of R_{n-1}.
The intermediate result after that iteration of the process	$\Rightarrow$	R_n: the set of numerals whose members have n digits after the decimal point
"The final resultant state" (actual infinity "∞")	$\Rightarrow$	**R_∞: the infinite set of numerals whose members have an infinite number of digits after the decimal point**
Entailment *E*: The final resultant state ("∞") is unique and follows every nonfinal state.	$\Rightarrow$	**Entailment *E*: R_∞ is unique and its members all have more digits after the decimal point than any members of any other R_n.**

Given the numerals for the natural numbers at the beginning, we get the resultant set R_1 by adding one digit (either 0, 1, ..., or 9) after the decimal point. From this, we get the next set R_2 by the specified process—namely, adding another digit to the members of R_1. This gives us the numbers to two decimal places by the end of state 2. By the end of state n, we have the set of all the numbers to n decimal places. The final state given by the BMI is a metaphorical ∞th resultant state in which the number of digits after the decimal point is infinite. The result of this special case of the BMI is the set of all infinite decimals.

This special case of the BMI creates two kinds of infinities at once: an *infinite totality* (the set consisting of the infinity of infinitely long decimals), and within each set, each member is an *infinite extremity* (a single numeral of infinite length).

Note, incidentally, that the number 4 in this notation is represented by the infinite decimal 4.0000. . . . Thus, all numerals characterized by this version of the BMI are infinite decimals, even numerals for natural numbers like 4.

These are the numerals for all the real numbers. They are not the real numbers themselves but only names for them. Moreover, we have not yet specified how to associate each name with a corresponding real number. This is a nontrivial job, since each name is infinitely long and there is an uncountable infinity of real numbers. The way this is accomplished by mathematicians is to associate each of these names with a unique infinite polynomial and to represent each real number as a unique infinite polynomial. This, too, requires a use of the BMI.

Infinite Polynomials

A polynomial is a sum—a sum of products of a certain form. A simple polynomial is: $(4 \cdot 10^3) + (7 \cdot 10^2) + (2 \cdot 10^1) + (9 \cdot 10^0)$. This polynomial corresponds to the natural number written in decimal notation as 4,729. Such simple polynomials are said to have a "base" 10, since they are sums of multiples of powers of 10. The preceding polynomial has the form:

$$(a_3 \cdot 10^3) + (a_2 \cdot 10^2) + (a_1 \cdot 10^1) + (a_0 \cdot 10^0),$$

where $a_3 = 4$, $a_2 = 7$, $a_1 = 2$, and $a_0 = 9$. Polynomials are numbers; decimals are numerals—names for polynomials. 83.76 is the numeral naming the polynomial: $(8 \cdot 10^1) + (3 \cdot 10^0) + (7 \cdot 10^{-1}) + (6 \cdot 10^{-2})$.

In general, a polynomial has the form:

$$a_n \cdot 10^n + \ldots + a_0 \cdot 10^0 + a_{-1} \cdot 10^{-1} + a_{-2} \cdot 10^{-2} + \ldots + a_{-k} \cdot 10^{-k},$$

where n is a non-negative integer and k is a natural number and a_n and a_{-k} are integers between 0 and 9. For example, in 459.6702, $a_2 = 4$ and $a_{-4} = 2$.

Since we have already shown how the infinite decimals can be characterized using special cases of the BMI, we can now map the infinite decimals conceptualized in this way onto corresponding infinite polynomials. Each infinite decimal has the form:

$$a_n \ldots a_0 . a_{-1} \ldots a_{-k} \ldots$$

It corresponds to an infinite polynomial:

$$a_n \cdot 10^n + \ldots + a_0 \cdot 10^0 + a_{-1} \cdot 10^{-1} + \ldots + a_{-k} \cdot 10^{-k} + \ldots$$

In other words, there is a symbolic map characterizing the one-to-one relationship between infinite decimals and infinite polynomials.

HOW NUMERALS SYMBOLIZE POLYNOMIALS

Conceptual Domain INFINITE POLYNOMIALS	*Symbolic Domain* INFINITE DECIMALS
$a_n \cdot 10^n + \ldots + a_0 \cdot 10^0 + a_{-1} \cdot 10^{-1} + \ldots$ $+ a_{-k} \cdot 10^{-k} + \ldots$ $\leftrightarrow$	$a_n \ldots a_0 . a_{-1} \ldots a_{-k} \ldots$

By virtue of this one-to-one mapping, the BMI structure of infinite decimals is mapped onto infinite polynomials.

Infinite polynomials are real numbers. Indeed, one of the *definitions* of real numbers is that they are infinite polynomials. The relationship between real numbers and infinite polynomials will become clearer as we proceed.

The irony should not be lost in the details: It takes a metaphor—the BMI—to conceptualize the "real" numbers.

Limits of Infinite Sequences

An infinite sequence of real numbers is commonly conceptualized as a function from the natural numbers to the real numbers. The natural numbers constitute an infinite set conceptualized via the BMI. Thus, the BMI is used implicitly in conceptualizing an infinite sequence. Each infinite sequence is usually conceptualized as an infinite set of ordered pairs (i, r_i), where the i's are the members of the set of natural numbers and each appears exactly once in the first place of some pair, while the second place in each pair is a real number. The use of the BMI in characterizing what is infinite about an infinite sequence is therefore straightforward: It is like the use of the BMI to characterize the infinite set of natural numbers, since there is one term of a sequence for each natural number.

The notion of a limit of an infinite sequence is more complex. A limit is a real number that the values of the sequence "get infinitely close to" as the number of terms increases "to infinity." We normally conceptualize the "convergence" of an infinite sequence to a limit by means of the concept of "approaching": The sequence "approaches" the limit as the number of terms "approaches infinity." That is, the value of x_n gets progressively closer to L (for limit) as n gets progressively "closer to infinity."

Lurking in the background here is the spatial metaphor that numbers are points on a line. Metaphorically, the limit is a fixed point L on the number line. As the number of terms n gets progressively "closer to infinity," the values x_n of the sequence get progressively closer to point L. As "n approaches ∞," the sequence comes "infinitely close to" L. It is this conception of the limit of a sequence that we shall attempt to characterize via the BMI. The difficulty will be to show just what it means for natural numbers to "approach infinity" and for the terms of the sequence to come "infinitely close" to the limit.

The metaphors involved here are fairly obvious. Nothing is literally moving or "approaching" anything. The value of each term of a sequence is fixed; since the values don't change, the terms can't literally "approach." And infinity is not something you can literally get "close to."

There are three reasons for undertaking this task. First, we want to show exactly how the BMI enters into the notion of a limit. Second, we want to show how limits are usually conceptualized. And third, we want to show that there is a precise arithmetic way to characterize limits using the BMI.

Before we begin, however, we should contrast the concept of a limit as we will describe it with the formal definition usually used. The formal definition of a limit of a sequence does not capture the idea of "approaching" a limit. Here is the formal definition.

> The sequence $\{x_n\}$ has L as a limit if, for each positive number ε, there is a positive integer n_0 with the property that $|x_n - L| < \varepsilon$ for all $n \geq n_0$.

This is simply a static condition using the quantifiers "for each" and "there is." This static condition happens to cover the case where $\{x_n\}$ "approaches L" as n "approaches infinity." It also happens to cover all sorts of other irrelevant cases where there are combinations of epsilons that have nothing to do with "approaching" anything.

Since such irrelevant cases are usually not discussed in textbooks, it is worthwhile for our purposes to give an example here. Consider the sequence: $\{x_n\} = \frac{n}{n+1}$. The terms are: 1/2, 2/3, 3/4, 4/5, ... The limit is 1. In the formal definition, epsilon ranges over all positive numbers. Take the positive number 43. No matter what n_0 one picks, the condition will be met. For example, let $n = 999$. Then $|x_n - L| = |999/1000 - 1| = 0.001$. This is certainly less than 43. So what? This choice of values is irrelevant to the question of whether the sequence converges to a limit. Our normal, everyday understanding of converging to a limit has a concept of a limit in which such irrelevant values don't occur. The point is not that the irrelevant values are mathematically harmful; they aren't. The definition works perfectly well from a mathematical perspective. And the fact that irrelevant cases fit the definition does not matter from a mathematical perspective.

But it does matter from a cognitive perspective. We normally conceptualize limits using an idea of a limit without such irrelevant cases. We hypothesize that a special case of the BMI is used, and that it can be formulated precisely.

Our formulation uses the idea of the "process" in the BMI to characterize the process of "approaching" in the prototypical case. The process is one of getting "progressively closer," so that at each stage n the "distance" between the limit L and x_n becomes smaller. "Distance" in the geometric metaphor Numbers Are Points on a Line (see Chapter 12) is characterized metaphorically in terms of arithmetic difference: the absolute value $|x_n - L|$, which must become smaller as n becomes larger. That "distance" should "approach zero" as n "approaches infinity."

This is achieved using the BMI in the following way.

- The concept "n gets progressively larger" is characterized by making the iterative process in the BMI the addition of 1 to n, which is the way we understand "n getting progressively larger" in a sequence.
- The concept "n approaches infinity" is characterized via the BMI, which creates a metaphorical final, ∞th resultant state of the process.
- The concept "approach" implicitly uses the metaphor Numbers Are Points on a Line, where the *distance* between points is metaphorically the *difference* between numbers. The magnitude of metaphorical "distance" between a term of the sequence x_n and the limit L is thus a positive real number r_n specifying the difference $|x_n - L|$.
- To go into greater detail, the "distance" between x_n and the limit L is the interval on the number line between x_n and L. According to the metaphor A Line Is a Set of Points, that interval is metaphorically a set of points. Since each such "point" is metaphorically a real number, the set of such points is, again metaphorically, the set R_n of real numbers r greater than zero and less than $|x_n - L|$.
- The "approach" is characterized by specifying the iterative process in the BMI to be the process by which the remaining metaphorical distance to the limit—the arithmetical difference $|x_n - L|$—gets smaller, and hence the set R_n excludes more and more real numbers. In other words, $R_n \subset R_{n-1}$.
- As the process continues, more and more terms of the sequence are generated. We collect them step by step in sets, so that at stage n we have set S_n, which contains the first n terms of the sequence. Thus, the set formed at step n contains all the terms of the sequence generated so far.
- By the final stage of the BMI, the ∞th stage, all of the terms of the sequence have been generated and collected in the set S_∞. Moreover, all the sets R_n of real numbers r, such that $0 < r < |x_n - L|$, have been generated. As x_n gets progressively closer to L, the set R_n comes to exclude more and more real numbers, until, at the ∞th stage of the process, R_∞ is completely empty. That is, x_n has gotten so close to L that there is no positive real number r such that for every finite n, $0 < r < |x_n - L|$.

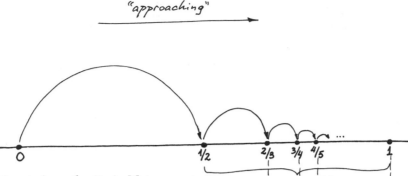

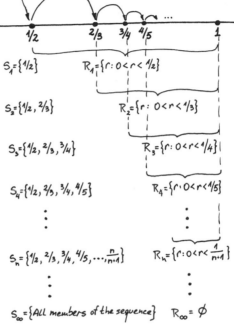

FIGURE 9.1   Here is how the Basic Metaphor of Infinity characterizes the idea of a sequence approaching a limit. The sequence here is $\{x_n\} = n/(n + 1)$. The limit is 1. In the special case of the BMI, two sets are formed at each stage. The S_n's progressively gather the terms of the sequence as they are generated. The corresponding sets R_n characterize the real numbers between 0 and $1 - x_n$—the portion of the real line (which remains at the nth stage) between the nth term of the sequence and the limit. At the final resultant state, S_∞ contains all the members of the sequence and R_∞ is empty. This will be true only for the number that is the limit—namely, 1. Note that, in this case, the limit does not occur in the sequence.

- This is what is implicitly meant when we say that the infinite sequence x_n "approaches L as a limit." Note, incidentally, that L can be entirely outside the sequence and still have terms of the sequence infinitely close to it.

Here is what this looks like as a special case of the BMI. First we characterize the "Sequence and Limit frame," and then we use it in the special case of the BMI (see Figure 9.1). Note that n is used to name the stages of the BMI and also to name the indexes of the terms of the sequence $\{x_n\}$.

THE SEQUENCE AND LIMIT FRAME (PROTOTYPICAL VERSION)

A statement defining a sequence $\{x_n\}$.
A set S_n containing the first n terms of $\{x_n\}$.
A finite number L.
A set R_n of real numbers r such that $0 < r < |x_n - L|$.

THE BMI FOR INFINITE SEQUENCES (PROTOTYPICAL VERSION)

Target Domain ITERATIVE PROCESSES THAT GO ON AND ON		*Special Case* INFINITE SEQUENCES WITH A LIMIT L
The beginning state (0)	$\Rightarrow$	The Sequence and Limit frame
State (1) resulting from the initial stage of the process	$\Rightarrow$	S_1 = the set containing the first term of the sequence
The process: From a prior intermediate state $(n–1)$, produce the next state (n).	$\Rightarrow$	From S_{n-1} containing the first $n–1$ terms of the sequence, form S_n containing the first n terms of the sequence.
The intermediate result after that iteration of the process	$\Rightarrow$	The set S_n. The set R_n containing all positive real numbers r such that $0 < r < |x_n - L|$. $R_n \subset R_{n-1}$.
"The final resultant state" (actual infinity "∞")	$\Rightarrow$	**The set S_∞ containing all the terms of the sequence**
		There is no positive real number r such that $0 < r < |x_i - L|$ for all x_i in S_∞. Hence, $R_\infty = \varnothing$. L is the limit of the sequence.
Entailment E: The final resultant state ("∞") is unique and follows every nonfinal state.	$\Rightarrow$	**Entailment E: L is the unique limit of the sequence.**

To clarify this further, here are all the stages and the sets S_n at each stage.

Stage $n = 1$	Stage $n = 2$	Stage $n = 3$	Stage n	Stage $n = \infty$
S_1	S_2	S_3	S_n	S_∞
The set consisting of the first term of the sequence $\{x_n\}$:	The set consisting of the first two terms of the sequence $\{x_n\}$:	The set consisting of the first three terms of the sequence $\{x_n\}$:	The set consisting of the first n terms of the sequence $\{x_n\}$:	The set consisting of all the terms of the sequence $\{x_n\}$:
$\{x_1\}$	$\{x_1, x_2\}$	$\{x_1, x_2, x_3\}$	$\{x_1, x_2, \ldots, x_n\}$	$\{x_1, x_2, \ldots\}$

A similar table could be built for the R_n's. The bottom row of such a table would show sets that exclude more and more elements. The set in the lower right cell of such a table would be the empty set.

Let us look again at our example of how this works.

- Consider the sequence $\{x_n\} = \frac{n}{n+1}$.
- The terms are: $x_1 = 1/2$, $x_2 = 2/3$, $x_3 = 3/4$, $x_4 = 4/5$, ...
- The set $S_1 = \{1/2\}$; $S_2 = \{1/2, 2/3\}$; $S_3 = \{1/2, 2/3, 3/4\}$; $S_4 = \{1/2, 2/3, 3/4, 4/5\}$; . . .
- The limit $L = 1$.
- The set $R_1 = \{r: 0 < r < 1/2\}$. $R_2 = \{r: 0 < r < 1/3\}$. $R_3 = \{r: 0 < r < 1/4\}$. $R_4 = \{r: 0 < r < 1/5\}$.

As n gets larger, the sets S_n come to have progressively more terms that get closer and closer to 1. The sets R_n come to exclude more and more positive real numbers. At stage $n = \infty$, S_∞ contains all of the infinite number of terms of the sequence $\{x_n\} = \frac{n}{n+1}$. It does not contain the number 1. At stage $n = \infty$, there is no positive real number r such that $r < |\frac{n}{n+1} - 1|$ for every finite n. Thus, the set R_∞ is empty. In this special case of the BMI, this is what it means for the infinite sequence $\frac{n}{n+1}$ to "approach" 1 as a limit as n "approaches ∞."

Notice that there are no epsilons and no quantifier statement "For every number epsilon there is an integer n_0." The irrelevant cases are not here at all. If we put this special case of the BMI together with the Numbers Are Points on a Line metaphor and Talmy's fictive-motion schema (see Chapter 2), we get the metaphor of "approaching" a limit as "n approaches infinity."

The values of the sequence are metaphorically conceptualized as locations along a line. Visualizing the process via these metaphors, there are two coordinated trajectors in motion: As the first moves from integer to integer starting with 1, the second moves correspondingly from point-location to point-location on the number line, starting with 1/2. As the first trajector moves from 1 to 2, the second moves from 1/2 to 2/3. Via the BMI, ∞ is the endpoint of the line, the final point-location that the first trajector can metaphorically "approach" infinitely close to. When the first trajector "reaches infinity," the second trajector "approaches the limit"; that is, it gets to a sequence of point-locations infinitely close to the limit, so close that there is no positive real number that can measure any distance between such point-locations and the limit.

This is how the prototypical idea of "approaching a limit as n approaches infinity" can be precisely conceptualized via the BMI and other conceptual metaphors.

The General Notion of a Limit Using the BMI

So far we have analyzed the prototypical version of a sequence that approaches a limit—namely, the case in which a sequence converges directly toward the limit. But many sequences converge indirectly—winding around and going back and forth as they ultimately converge to a limit. Because such cases exist, we will have to construct a fully general version of the BMI for convergent sequences.

To get an idea of the problem, let us consider what we will call "teaser sequences." Here is an example:

$$3/6,\ 4/6,\ 5/6,\ 9/12,\ 10/12,\ 11/12,\ 15/18,\ 16/18,\ 17/18,\ 21/24,\ \ldots$$

The sequence grows from 3/6, to 5/6, then drops back to 9/12, which is between 4/6 and 5/6 in size. Then it grows again to 11/12, which is more than 5/6. Then it drops back again—not all the way to 9/12, but only to 15/18, which is equivalent to 5/6 in size. Then it grows to 17/18, which is more than 11/12. Then it drops back to 21/24, which is between 15/18 and 16/18 in size. Then it grows again and repeats the pattern (see Figure 9.2).

From the figure we can see that the *teaser elements*—those that drop back in size—are 9/12, 15/18, 21/24, ... The endpoints of each subsequence are 5/6, 11/12, 17/18, 23/24, ... The teaser sequence has two important properties:

1. The teaser elements form a directly convergent sequence.
2. Each teaser element is smaller than all the elements of the sequence that follow it.

From these properties it follows that the sequence as a whole converges.

How can we reformulate the special case of the BMI for limits of sequences to accommodate all such cases of indirect convergence? To answer this question, we have to see where the limitations arise in our previous formulation.

In the case of directly convergent sequences, the n in the definition of the sequence $\{x_n\}$ corresponds to the n characterizing the stages of the BMI. This does not work for teaser sequences. To generalize the notion of the limit, we need to distinguish between the positive integers characterizing the stages of the BMI and the positive integers indexing the elements of the sequence. We will continue to use the variable n for the integers characterizing the stages of the BMI, but we shall use the variable m for the positive integers characterizing the teaser sequence, which we will now call $\{x_m\}$. The following table illustrates the

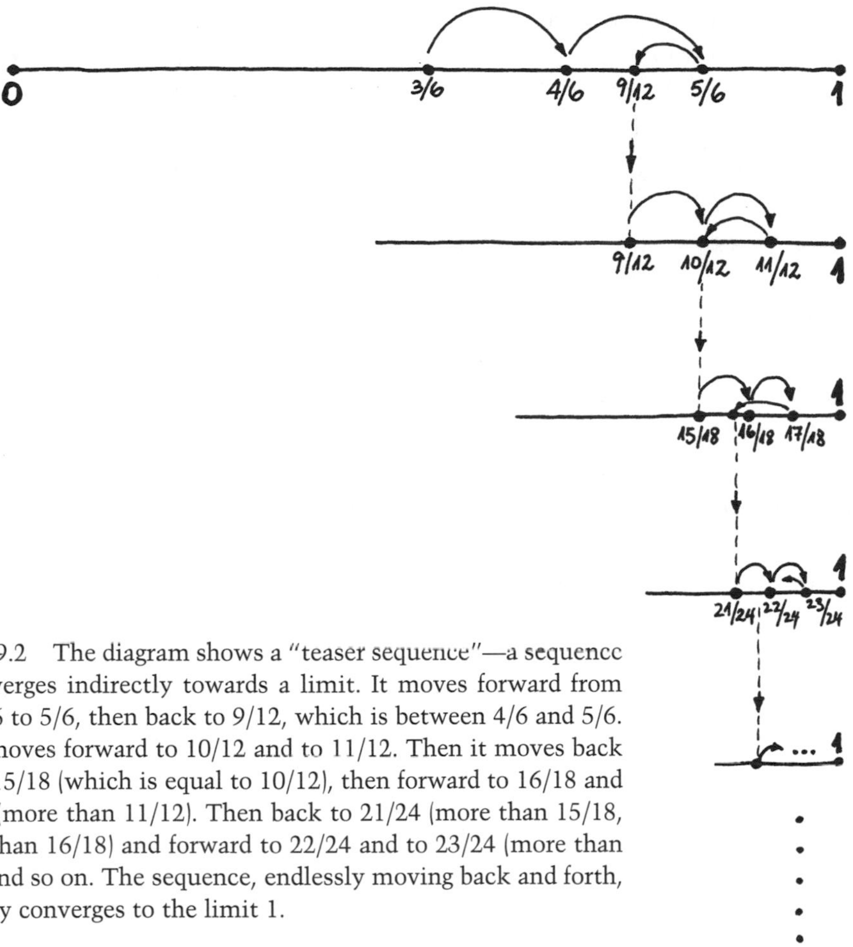

FIGURE 9.2 The diagram shows a "teaser sequence"—a sequence that converges indirectly towards a limit. It moves forward from 3/6 to 4/6 to 5/6, then back to 9/12, which is between 4/6 and 5/6. Then it moves forward to 10/12 and to 11/12. Then it moves back again to 15/18 (which is equal to 10/12), then forward to 16/18 and to 17/18 (more than 11/12). Then back to 21/24 (more than 15/18, but less than 16/18) and forward to 22/24 and to 23/24 (more than 17/18). And so on. The sequence, endlessly moving back and forth, ultimately converges to the limit 1.

kind of relationship between n and m that would be needed to handle the convergence of teaser sequences.

CONVERGENT SEQUENCE OF TEASER ELEMENTS

BMI stage n	Sequence term m	Value of the nth term of $\{x_t\}$, the sequence of teaser elements of $\{x_m\}$
1	1	3/6
2	4	9/12
3	7	15/18
4	10	21/24
5	13	27/30
. . .	. . .	. . .
n	$3n - 2$	$3(2n - 1)/6n$
. . .	. . .	. . .

We can characterize the relationship between the stage n of the BMI with the term m of the sequence $\{x_m\}$ with a statement. In this case, the statement is $m = 3n - 2$. Thus the second stage of the BMI maps onto the fourth term of the sequence, 9/12, which is the second teaser term. The third stage of the BMI maps onto the seventh term of the sequence, 15/18, which is the third teaser term; and so on. This generates a new sequence, the *sequence of teaser terms*, which we denote as $\{x_t\}$, for teasers.

Such statements allow us to adapt the earlier version of the BMI to characterize the direct convergence of this sequence of teaser terms $\{x_t\}$. Since the teaser elements x_t converge directly and since each is less than all the following elements in the original sequence $\{x_m\}$, the convergence of the teaser elements guarantees the convergence of the sequence $\{x_m\}$ as a whole. What we need to do now is characterize convergence for $\{x_t\}$, the sequence of teaser elements. To do so, we need to look at the differences between the limit and each teaser element.

Given that the limit L is equal to 1 in this case, we can form the sequence of differences between the limit and each of the teaser terms: $1 - 3/6 = 3/6$, $1 - 9/12 = 3/12$, $1 - 15/18 = 3/18$, and so on. This yields the following sequence of differences: 3/6, 3/12, 3/18, ... We will refer to the tth difference as ε_t. Thus, $\varepsilon_1 = 3/6$, $\varepsilon_2 = 3/12$, and so on, as we can see in the following table.

BMI stage n	$1 - x_t$	ε_t
1	3/6	ε_1
2	3/12	ε_2
3	3/18	ε_3
4	3/24	ε_4
5	3/30	ε_5
. . .	. . .	. . .
n	$3/6n$	ε_n
. . .	. . .	. . .

Thus, corresponding to each finite stage n of the BMI, there is an ε_t. As the BMI proceeds, a decreasing sequence of epsilons is produced, which characterizes the "approach of the sequence to the limit." As n approaches infinity, the epsilons approach zero, and the sequence approaches the limit.

When $n = m$, we have the prototypical version of approaching limits described earlier. When $n \leq m$, the BMI will map (via the statement $m = 3n - 2$) the nth stage onto the teaser elements of $\{x_m\}$. At the final stage where $n = \infty$, all the teaser elements have been chosen for every $n < \infty$. In other words, the entire infinite sequence $\{x_t\}$ of teaser terms has been chosen from the sequence $\{x_m\}$ as a whole.

To state this special case of the BMI, we need to revise the Sequence and Limit frame as follows. In describing this frame, we use the phrase "critical elements" for those terms of the sequence that must converge in order for the sequence as a whole to converge. Teaser elements are thus special cases of critical elements.

THE SEQUENCE AND LIMIT FRAME (GENERAL VERSION)

A statement defining a sequence $\{x_m\}$, where m is a positive integer.

A statement defining a sequence $\{x_t\}$ of *critical elements*, of $\{x_m\}$.

A set S_n containing the first n elements of $\{x_t\}$.

A finite number L.

A set R_n of real numbers r such that $0 < r < |L - x_n|$, where x_n is the nth term of $\{x_t\}$.

A constraint: Let x_n be the nth term of $\{x_t\}$; Let x_n' be the critical element in $\{x_m\}$ corresponding to the x_n; and let x_i be a term in $\{x_m\}$ with $i > n'$. $|L - x_n'| > |L - x_i|$.

The constraint in the frame sets a bound for each teaser element: no subsequent term of $\{x_m\}$ can be further from the limit L than a given critical element.

The special case of the BMI about to be stated will characterize two infinite sequences of sets, the S_n's and the R_n's. As n increases, the S_n's will come to contain more and more of the critical elements of the sequence and the R_n's will come to contain fewer and fewer real numbers between the nth term of $\{x_t\}$ and the limit. At ∞, S_∞ will contain all the elements of the sequence $\{x_t\}$ and R_∞ will be empty, which means that there will be no fixed real number that characterizes the difference between the limit and the terms of the sequence $\{x_t\}$, as n approaches ∞. The constraint in the frame guarantees that the sequence $\{x_m\}$ will converge to the limit L as n approaches ∞. Here is that special case of the BMI.

THE BMI FOR INFINITE SEQUENCES (GENERAL VERSION)

Target Domain ITERATIVE PROCESSES THAT GO ON AND ON		*Special Case* INFINITE SEQUENCES WITH A LIMIT L
The beginning state (0)	$\Rightarrow$	The Sequence and Limit frame (general version).
State (1) resulting from the initial stage of the process	$\Rightarrow$	$S_1 = $ the set containing the first term of the sequence $\{x_t\}$.

The process: From a prior intermediate state $(n-1)$, produce the next state (n).	⇒	From S_{n-1} containing the first $n-1$ terms of the sequence $\{x_t\}$, form S_n containing the first n terms of the sequence $\{x_t\}$.
The intermediate result after that iteration of the process	⇒	The set S_n. The set R_n containing all positive real numbers r such that $0 < r < |x_n - L|$. $R_n \subset R_{n-1}$.
"The final resultant state" **(actual infinity "∞")**	⇒	**The set S_∞ contains all the terms of the sequence $\{x_t\}$. There is no positive real number r such that $0 < r < |x_n - L|$ for all x_n in S_∞. Hence, $R_\infty = \varnothing$. L is the limit of the sequence $\{x_t\}$ of critical elements of $\{x_m\}$. L is also the limit of the entire sequence $\{x_m\}$.**
Entailment E: The final resultant state ("∞") is unique and follows every nonfinal state.	⇒	**Entailment E: L is the unique limit of the sequences $\{x_t\}$ and $\{x_m\}$.**

To clarify this further, here are all the stages and the sets S_n at each stage of the BMI for the sequence of teaser elements given earlier.

BMI stage $n = 1$	BMI stage $n = 2$	BMI stage $n = 3$	BMI stage n	BMI stage $n = \infty$
S_1	S_2	S_3	S_n	S_∞
The set consisting of the first term of the sequence $\{x_t\}$:	The set consisting of the first two terms of the sequence $\{x_t\}$:	The set consisting of the first three terms of the sequence $\{x_t\}$:	The set consisting of the first n terms of the sequence $\{x_t\}$:	The set consisting of all the terms of the sequence $\{x_t\}$:
$\{3/6\}$	$\{3/6, 9/12\}$	$\{3/6, 9/12, 15/18\}$	$\{3/6, \ldots, 3(2n-1)/6n\}$	$\{3/6, 9/12, \ldots\}$

Here we see the gradual formation of the sequence of teaser terms. The bottom right cell has the set consisting of *all* the terms of the sequence $\{x_t\}$. A similar table could be built for the R_n's. The bottom row of such a table would show sets that exclude more and more elements. The set in the lower right cell of such a table would be the empty set.

We have now shown how the BMI can characterize the general case of the convergence of a sequence to a limit.

Infinite Sums

Before we proceed, there is an additional metaphor that should be discussed. Consider the sequence of partial sums given by $\sum_{k=1}^{n} \frac{9}{10^k}$. The first three terms are 0.9, 0.99, and 0.999.

When k grows indefinitely, the result is an infinite sum $\sum_{k=1}^{\infty} \frac{9}{10^k}$. Mathematicians take this infinite sum as being *equal* to 1. This requires an extra metaphor:

INFINITE SUMS ARE LIMITS OF
INFINITE SEQUENCES OF PARTIAL SUMS

Source Domain	*Target Domain*
LIMITS OF INFINITE SEQUENCES	INFINITE SUMS

The limit of an infinite sequence of partial sums	$\rightarrow$	An infinite sum
$\lim_{n \to \infty} \sum_{k=1}^{n} a_k$	$\rightarrow$	$\sum_{k=1}^{\infty} a_k$

Via this metaphor, one can "define" an infinite sum as being the limit of an infinite sequence of partial sums. This metaphor piggybacks on the version of the BMI defining the limit.

We can see how this works in the classic example of the sum:

$$0.9 + 0.09 + 0.009 + 0.0009 + \dots .$$

In the special case of the BMI for infinite sequences, we plug in the following values:

- $s_n = \sum_{k=1}^{n} \frac{9}{10^k}$
- $L = 1$
- $s_1 = 0.9;\ s_2 = 0.99;\ s_3 = 0.999;\ \dots$
- $\varepsilon_1 = |1 - 0.9| = 0.1;\ \varepsilon_2 = |1 - 0.99| = 0.01;\ \varepsilon_3 = |1 - 0.999| = 0.001;\ \dots$
- s_∞ is the infinite sum, $0.99999\dots$
- Since there is no positive real number r that is less than $|L - s_\infty|$ for every n, it follows that $|L - s_\infty|$, which equals $|1 - 0.9999\dots| = 0$. Thus, $1 = 0.999\dots .$

In other words, the sequence of partial sums—0.9, 0.99, 0.999, . . . —has 1 as its limit. Via the infinite sum metaphor, An Infinite Sum Is the Limit of An Infinite Sequence of Partial Sums, the infinite sum 0.9999. . . equals 1. This is an entailment of two metaphors: the special case of the BMI for limits and the infinite sum metaphor.

Limits of Functions

One of the most basic ideas in calculus is that of the limit of a function. We can now straightforwardly extend the account we have just given of limits of infinite sequences using the BMI to the notion of the limit L of a function $f(x)$ as x "approaches" a real value a.

As we saw in the previous section, the notion of a function "approaching" a value can be characterized in terms of limits of sequences plus the metaphors:

- Numbers Are Points on a Line,
- Talmy's fictive-motion schema, and
- the Change of a Function Is the Coordinated Motion of Two Trajectors, one in the domain and one in the range of the function.

Given these conceptual metaphors, we can now extend the BMI account of the limit of a sequence to the limit of a function $f(x)$ as x approaches a.

LIMIT OF A FUNCTION $f(x)$ AS x APPROACHES a
(USING THE BMI DEFINED FOR SEQUENCES)

Let $f(x)$ be a function.
Suppose that for every infinite sequence
$$\{r_i\} = r_1, \ldots, r_i, \ldots$$
such that $\{r_i\}$ converges to a,
there is a corresponding infinite sequence
$$\{f(r_i)\} = (f(r_1), \ldots, f(r_i), \ldots)$$
that converges to L.
We define L to be the limit of $f(x)$ as x approaches a.
Moreover, if $f(x)$ is defined at a, then $f(a) = L$ (which assumes the continuity implicit in the above fictive-motion schema).

Applying the metaphors mentioned above, this will yield the concept of the limit L of a function $f(x)$ as x approaches a.

It is straightforward to show that whenever this condition holds, the traditional epsilon-delta condition holds:

LIMIT OF A FUNCTION
(USING EPSILONS AND DELTAS)

$\lim\limits_{x \to a} f(x) = L$ iff

for all $\varepsilon > 0$, there is a $\delta > 0$,

such that if $0 < |a - x| < \delta$,

then $|L - f(x)| < \varepsilon$.

If we let each choice of epsilon be some $f(r_i)$ in a sequence $\{f(r_i)\}$, then r_i will be a suitable delta, and the condition will be met.

The point of this is not to eliminate the epsilon-delta condition from mathematics but, rather, to comprehend how we understand it. We understand it first in geometric terms using the notion of "approaching a limit." And we understand "approaching a limit" in arithmetic terms via the BMI, as described in the previous section. Given that, we can understand the idea of "approaching a limit" for a function in arithmetic terms using the characterization above, which presupposes the BMI. In that characterization, each sequence $\{f(r_i)\}$ metaphorically defines a sequence of "steps" in some "movement" toward L, while each corresponding sequence $\{r_i\}$ defines a sequence of "steps" in some "movement" toward a.

So far, so good. We have depended upon the use of the BMI in the previous section, which characterized limits of sequences. This works fine for cases where $f(x)$ is defined at a. Then it is clear that $f(a) = L$. But what if $f(x)$ is not defined at a? Does our characterization of the limit still hold? And if it does, can we meaningfully add the value $f(a) = L$ without contradiction?

Let us take an example.

$$f(x) = \frac{x^2 - 1}{x - 1}$$

This function is not defined for $x = 1$, since that would make the denominator equal to zero. For values other than $x = 1$, the function acts the same as $f(x) = x + 1$, since the numerator $x^2 - 1 = (x + 1)(x - 1)$. Now consider the values $x = 0.9, 0.99, 0.999$, and so on. The corresponding values of $f(x)$ are 1.9, 1.99, 1.999, and so on. On our characterization of the limit of $f(x)$ as x approaches 1, there would be a limit—namely, 2. Even though strictly speaking it is not the case that $f(1) = 2$!

There are several morals to this discussion:

- The BMI as used for limits of sequences can provide a mechanism for characterizing limits of functions in arithmetic terms.
- With the appropriate additional metaphors, the analysis given also characterizes in precise terms the intuitive notion of "approaching a limit." It is thus false that such a mathematical idea cannot be made precise in intuitive terms. Indeed, the fact that conceptual metaphors can be precisely formulated allows us to make such intuitive ideas precise.
- The epsilon-delta definition of a limit does not characterize the mathematical idea of approaching a limit, since it does not involve any "approach" to anything. It is just a different idea that happens to cover the same cases, as well as countless irrelevant but mathematically harmless cases.

We can now see why students have problems learning the epsilon-delta definition of a limit. They are taught an intuitive idea of what a "limit" is in terms of a motion metaphor, and then told, incorrectly, that the epsilon-delta condition expresses the same idea (see Núñez, Edwards, & Matos, 1999).

Least Upper Bounds

We have seen an example of how the BMI can characterize real numbers—both the symbols for real numbers (infinite decimals) and the real numbers themselves (infinite polynomials). However, the common axiomatization of the real numbers does not use this characterization directly. The real numbers are taken to be whatever objects fit the axioms. There are ten axioms. The first six characterize fields, and the next three add ordering constraints, so that the first nine axioms characterize ordered fields. And the tenth, the *Least Upper Bound axiom*, characterizes complete ordered fields. The real numbers are the only complete ordered field. Here are the axioms.

1. Commutative laws for addition and multiplication.
2. Associative laws for addition and multiplication.
3. The distributive law.
4. The existence of identity elements for both addition and multiplication.
5. The existence of additive inverses (i.e., negatives).
6. The existence of multiplicative inverses (i.e., reciprocals).
7. Total ordering.
8. If x and y are positive, so is $x + y$.
9. If x and y are positive, so is $x \cdot y$.

10. The Least Upper Bound axiom: Every nonempty set that has an upper
    bound has a least upper bound.

Upper bounds and least upper bounds are defined in the following way:

Upper Bound
b is an *upper bound* for S if
$x \leq b$, for every x in S.

Least Upper Bound
b_0 is a *least upper bound* for S if
- b_0 is an upper bound for S, and
- $b_0 \leq b$ for every upper bound b of S.

Axioms 1 through 9 are not sufficient to distinguish the reals from the rationals, since the rational numbers also fit them. Add axiom 10 and you get the reals.

The question to be asked is, Why? What do infinite decimals and infinite polynomial sums have to do with least upper bounds? What does the BMI have to do with least upper bounds? To make the question clear, let us take an example. Consider $\pi = 3.14159265358979323846264\ldots$. The irrational number π has an infinite number of decimal places. We cannot write all of them down. We can write down only approximations to some number of decimal places. But each such approximation is a rational number, not an irrational number. For any approximation of π to n decimal places, we can write down a smallest upper bound for that approximation. The following table gives a sequence of rational approximations to π to a given number of decimal places, the corresponding smallest upper bound to that number of decimal places, and the difference between the two.

π to n Decimal Places	Least Upper Bounds of π to n Decimal Places	The Difference to n Decimal Places	The Number of Decimal Places
3.1	3.2	0.1	1
3.14	3.15	0.01	2
3.141	3.142	0.001	3
3.1415	3.1416	0.0001	4
3.14159	3.14160	0.00001	5
3.141592	3.141593	0.000001	6
3.1415926	3.1415927	0.0000001	7
3.14159265	3.14159266	0.00000001	8
3.141592653	3.141592654	0.000000001	9

Each of the first three columns provides the first nine terms of an infinite sequence. That sequence can be thought of from the point of view of potential infinity as ongoing, or from the perspective of actual infinity as *being* infinite. In the case of actual infinity, we can conceptualize these infinite sequences via the BMI. The first two sequences—the approximation of π to n places and its least upper bound to n places—become identical in more and more places as n increases. Moreover, the differences between them, as column three shows, become smaller and smaller as n increases. If we apply the BMI to the sequence of differences, then "at" $n = \infty$, the value of the difference "is" zero. That means that, "at" $n = \infty$, the value of the real number π is identical to the least upper bound of all rational approximations to π. π is also its *own* least upper bound!

This is extremely important in understanding the nature of a least upper bound. Recall the definition:

Upper Bound
b is an *upper bound* for S if
$x \le b$, for every x in S.

Least Upper Bound
b_0 is a *least upper bound* for S if
- b_0 is an upper bound for S, and
- $b_0 \le b$ for every upper bound b of S.

Each of the finite terms in the second column—a least upper bound to n places—is an upper bound for π. This is true for every finite term in the sequence defined by the second column. Thus, there is an infinite number of upper bounds for π descending sequentially in value; that is, the number of finite upper bounds getting smaller and smaller is endless. From the perspective of potential infinity, there *is no least upper bound!* It is only from the perspective of actual infinity, characterized via the BMI, that a least upper bound comes into existence. In other words, that least upper bound is "created," from a cognitive point of view, by the Basic Metaphor of Infinity. Moreover, by virtue of the BMI, that least upper bound is unique!

Here we can see clearly that the existence of least upper bounds is not guaranteed by the first nine axioms, which characterize the rational numbers. That is why axiomatically oriented mathematicians have added the Least Upper Bound axiom. It was brought into mathematics for the purpose of "creating" the real numbers from the rationals. It doesn't follow from anything else, and *it is not a special case of anything else!*

What is the difference between using the BMI to characterize real numbers and using the Least Upper Bound axiom? The question is a bit strange, something like comparing apples and oranges. The BMI is an unconscious cognitive mechanism that makes use of ordinary elements of human cognition: aspect, conceptual metaphor, and so on. The BMI has arisen spontaneously, outside mathematics, and has been applied within mathematics to characterize cases of actual infinity. The Least Upper Bound axiom is a product of formal mathematics, consciously and intentionally constructed to characterize the real numbers, given the first nine axioms. It is also used in other branches of mathematics, especially in set theory.

But the Least Upper Bound axiom is anything but intuitively clear on its own terms. Indeed, to understand it, most students of mathematics have to conceptualize it in terms of the BMI. From a cognitive perspective, the BMI is conceptually more basic.

Here is a characterization of the concept of a least upper bound from the perspective of the Basic Metaphor of Infinity.

THE BMI VERSION OF LEAST UPPER BOUNDS

Target Domain ITERATIVE PROCESSES THAT GO ON AND ON		*Special Case* LEAST UPPER BOUNDS
The beginning state (0)	$\Rightarrow$	S is a set of real numbers. A real number b is an upper bound for S if $b \geq$ every member of S. b_0 is a least upper bound of S if it is smaller than every upper bound b. Let B be an infinite set of upper bounds b_n for S.
State (1) resulting from the initial stage of the process.	$\Rightarrow$	Choose an upper bound b_1.
The process: From a prior intermediate state (n–1), produce the next state (n).	$\Rightarrow$	Given an upper bound b_{n-1}, choose an upper bound b_n arbitrarily less than b_{n-1}.
The intermediate result after that iteration of the process (the relation between n and n–1)	$\Rightarrow$	A finite sequence $\{b_i\}$ of upper bounds, with $b_n < b_{n-1}$
"The final resultant state" (actual infinity "∞")	$\Rightarrow$	**A least upper bound, b_∞**
Entailment *E*: The final resultant state ("∞") is unique and follows every nonfinal state.	$\Rightarrow$	**Entailment *E*: The least upper bound, b_∞, is unique and less than all other upper bounds.**

Of course, we can correspondingly characterize greatest lower bounds in a symmetrical fashion, substituting lower bounds for upper bounds and interchanging "<" and ">."

Is 0.9999. . . = 1.000. . . ?

Consider 0.9999999. . . . Its least upper bound is 1.0000000. . . . Within the real number system, these two infinite decimals will symbolize the same number (their corresponding polynomials would give the same infinite sum). The reasoning is the following: Within the real numbers, two numbers are identical if they are not distinct. Two numbers are distinct if there is a nonzero difference between them; that is, x and y are distinct if and only if there is a positive number $d > 0$, such that $|x - y| = d$.

In the case of 0.999999. . . and 1.000000. . . , there can be no such d. No matter how small a number d you pick, you can always take 0.999999. . . and 1.00000. . . out to enough decimal places so that the difference between them is less that than d.

We can see this clearly using the BMI. Suppose again we consider the least upper bounds to n places.

0.9. . . 9 to n places	1.0. . . 0 to n places	The difference to n places
0.9	1.0	0.1
0.99	1.00	0.01
0.999	1.000	0.001
0.9999	1.0000	0.0001
0.99999	1.00000	0.00001
0.999999	1.000000	0.000001

As in the case of π, the sequence of differences gets smaller and smaller. Applying the BMI to this sequence, the difference is zero "at infinity." Therefore, "at" metaphorical infinity, 0.9999. . . = 1.0000. . . . That is, there are no real numbers between the infinite decimals 0.9999. . . and 1.0000. . . . They are, then, just different symbols for the same number.

As we shall see in Chapter 11, this result does *not* hold for every number system. There are number systems beyond the reals (namely, the hyperreals) in which 0.9999. . . and 1.0000. . . are *distinct* numbers—not just distinct numerals but actually different numbers, with a hyperreal number characterizing the difference between 0.9999. . . and 1.000. . . . But we will get to that later. In the system of real numbers, 0.9999. . . does equal 1.000. . . .

At this point, we can see exactly why the Least Upper Bound axiom, when added to the first nine, characterizes the real numbers and not the rationals. The sequence of finite rational approximations of π to n places consists of rational numbers alone. π, the infinite sequence, is not in the set of finite rational sequences. As an irrational number, it stands outside the set of rational numbers altogether. But π is the least upper bound of that sequence. Thus, insisting that an ordered field (like the rationals) contain all its least upper bounds adds the irrational numbers—the numbers written with *infinite decimals*—to the rationals.

One of the beauties of the BMI is that it allows us to see clearly just *why* adding the Least Upper Bound axiom to the axioms for ordered fields takes us beyond the rationals to the reals.

Infinite Intersections of Nested Intervals

We have just seen how least upper bounds take us from an axiom system that can fit the rationals without the reals to one that necessarily includes the reals. That is, if we add least upper bounds to all convergent sequences of rationals, we get the reals. The BMI, as we have just seen, can be used to show why this is so.

We are now in a position to use the BMI to shed light on another way in which mathematicians commonly characterize the real numbers—as infinite intersections of nested closed intervals with rational endpoints on the number line. In the process, we can show why the infinite-intersection characterization of the reals is equivalent to the least-upper-bound characterization.

Let us once more take as an example π = 3.141592653589793238462643. . . . Let us take the number line containing the rational numbers. Consider the following set of nested intervals, as given in the following table. The first interval is [3.1, 3.2]. The second is [3.14, 3.15]. The third is [3.141, 3.142]. And so on.

Left-Hand Side of the Closed Interval	Right-Hand Side of the Closed Interval
3.1	3.2
3.14	3.15
3.141	3.142
3.1415	3.1416
3.14159	3.14160
3.141592	3.141593
3.1415926	3.1415927
3.14159265	3.14159266
3.141592653	3.141592654

This is exactly the same table that we used in the last section, but with approximations to n places conceptualized as left ends of the intervals, while upper bounds to n places are conceptualized as right ends of the intervals. We are now in a position to characterize infinite sequences of nested intervals as special cases of the BMI.

Nested Intervals

Target Domain Iterative Processes That Go On and On		Special Case Infinite Sequences of Nested Intervals
The beginning state (0)	$\Rightarrow$	An unending sequence of nested closed intervals, each defined by the pair of rational numbers $[l_n, u_n]$, where $l_n > l_{n-1}$ and $u_n < u_{n-1}$
State (1) resulting from the initial stage of the process	$\Rightarrow$	The first term of the sequence: the interval $[l_1, u_1]$
The process: From a prior intermediate state $(n-1)$, produce the next state (n).	$\Rightarrow$	Form the intersection of $[l_{n-1}, u_{n-1}]$ and $[l_n, u_n]$.
The intermediate result after the nth finite iteration of the process	$\Rightarrow$	The interval $[l_n, u_n]$
"The final resultant state" (actual infinity "∞")	$\Rightarrow$	**The "interval" $[l_\infty, u_\infty]$, where $l_\infty > l_n$ and $u_\infty < u_n$ for every finite n. The distance between l_∞ and u_∞ is zero. That is, the "interval" is a point,** $p = l_\infty = u_\infty.$
Entailment E: The final resultant state ("∞") is unique and follows every nonfinal state.	$\Rightarrow$	**Entailment E: p is a unique point and is greater than every l_n and less than every u_n.**

This is a general way of characterizing the reals, given the rationals: p need not be a rational number, depending on how the intervals are chosen at each of the infinitely many finite stages. Now we can see the equivalence of various ways of using the BMI to get the real numbers from the rationals: infinite decimals (which symbolize infinite polynomials), least upper bounds, and infinite intersections of nested intervals.

Conclusion

The concepts discussed in this section—infinite decimals, infinite polynomials, limits of infinite sequences (and of functions), infinite sums, least upper bounds, and infinite intersections of nested intervals—all involve the notion of actual infinity. We have argued that actual infinity is conceptualized metaphorically using the Basic Metaphor of Infinity. What we have shown in this chapter is how a single cognitive mechanism—the BMI—with different special cases, can characterize each of these concepts.

Since real numbers are characterized only via such concepts (and the Dedekind cut; see Chapter 13), we have effectively shown how the real numbers are conceptualized via the Basic Metaphor of Infinity. In short, the "real" numbers are, from a cognitive point of view, a metaphorical construct.

The various "definitions" of the real numbers are implicitly metaphorical definitions, making implicit conceptual use of the Basic Metaphor of Infinity. These metaphorical definitions are equivalent; that is, they have the same (or equivalent) entailments. Given the equivalence of their entailments, we can think of them as characterizing the same "entities" with the same properties; that is, the entities and the properties these different special cases define can be put in one-to-one correspondence. For this reason, many mathematicians speak of *the* real numbers, as if they were objectively existing entities in the universe. Because of their metaphorical equivalence, this creates no practical problems: If you happen to think of these metaphorical objects as if they exist objectively, you will not get into any mathematical difficulties.

It is by no means obvious that the real numbers are necessarily conceptualized only via one version or another of the Basic Metaphor of Infinity. It requires *mathematical idea analysis* to show this. One of the benefits of mathematical idea analysis is that it shows explicitly how the various characterizations of the real numbers are related to one another. It also makes explicit the distinction between the symbolization of real numbers (using infinite decimal representations) and the numbers themselves.

10

Transfinite Numbers

GEORG CANTOR IS PERHAPS MOST FAMOUS for having created a mathematical system for precisely characterizing numbers of infinite size with a precise arithmetic. In Cantor's system, the "transfinite" numbers are not merely infinite but have different degrees of infinity, with one infinite number being "larger" than another.

The concepts Same Size As and Larger Than for infinite numbers came out of Cantor's metaphor, which conceptualized two sets as having the "same size"—the same *number* of members—if they could be put in one-to-one correspondence (see Chapter 7). We will designate the size of a set determined via Cantor's metaphor as the *Cantor size,* or cardinality, of the set.

As we saw in Chapter 8, the BMI is used implicitly to conceptualize the infinite set of all natural numbers. Cantor, by his famous rational array diagram, proved that the set of rational numbers (fractions) has the same Cantor size as the set of natural numbers. He achieved this by constructing a sequence of the rationals, thus implicitly showing that the rationals could be put in one-to-one correspondence with the natural numbers. Figure 10.1 shows Cantor's display, indicating how the sequence can be constructed.

As we trace along the line in the diagram, we form a series of the rational numbers we encounter:

$$1, 2, 1/2, 1/3, 3, 4, 3/2, 2/3, 1/4, \ldots$$

Note that the diagram is not, and cannot be, complete, since the line in itself is intended to cover infinitely many steps. What is implicit in how the diagram is intended to be understood is the BMI. The BMI is implicitly and unconsciously used in the following ways in comprehending this diagram:

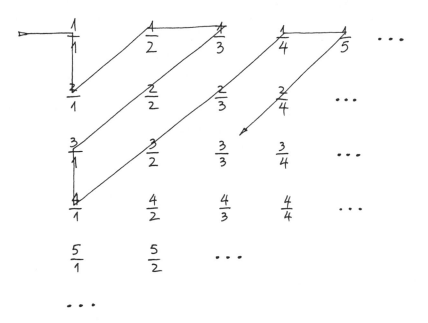

FIGURE 10.1   Cantor's one-to-one correspondence between the natural numbers and the rational numbers. This is an array in which all possible fractions (rational numbers) occur. All the fractions with 1 as numerator are in the first row, all those with 2 as numerator are in the second row, and so on. All those with 1 as denominator are in the first column, all those with 2 as denominator are in the second column, and so on. The zigzag line systematically goes through all the rational numbers. Cantor mapped the first rational number the line hits (1/1) to 1, the second (2/1) to 2, the third (1/2) to 3, and so on. This established a one-to-one correspondence between the natural numbers and the rationals.

- First, the special case of the BMI for the infinite set of natural numbers is used to guarantee that the first column of the matrix contains *all* the natural numbers—or equivalently, all the natural numbers as numerators with 1 as a denominator.
- Next, a similar version of the BMI is used to guarantee that the second column contains *all* the fractions with denominator 2.
- So far we have the beginning of a sequence of sequences: (the sequence of all fractions with denominator 1, the sequence of all the fractions with denominator 2, . . .). This sequence of sequences (extending downward) is also unending, and another use of the BMI is implicit to guarantee that this infinite sequence of sequences is complete. This gives us an understanding of the infinite array of rational numbers, *without any path through it.*

- Finally, there is the linear path that systematically covers every member of the array, creating a single sequence 1/1, 2/1, 1/2, 1/3, 2/2, 3/1, 4/1, 3/2, 2/3, 1/4, . . . which exhausts *all* the rational numbers. Here, too, a version of the BMI is used to guarantee that *all* rational numbers are included in the infinite sequence.
- Since a sequence, technically, is a mapping from the natural numbers to the terms of the sequence, the sequence itself constitutes a one-to-one correspondence between the natural numbers and the rational numbers. By Cantor's metaphor, the set of rationals is thereby assigned the same Cantor size as the set of natural numbers.

In short, the BMI is used over and over, implicitly and unconsciously, in comprehending this diagram. The diagram, via Cantor's metaphor, is taken as a *proof* that the natural numbers and the rational numbers have the same Cantor size—that is, the same cardinality.

Of course, according to our normal, everyday notion of size, there are more rational numbers than natural numbers, since all the natural numbers are contained in the rationals, and if they are taken away there are lots of rationals left over. Moreover, between any two natural numbers, there are an infinite number of rationals that are not natural numbers. In short, according to our normal non-Cantorian notions of number and size, there are zillions more rational numbers than there are natural numbers.

However, this chapter is not about our normal notion of number and size but about Cantor's. If one accepts Cantor's metaphor and the BMI, remarkable things follow.

Cantor's Diagonalization Proof

Cantor's most celebrated proof is his diagonalization argument, which demonstrated that the real numbers cannot be put in a one-to-one correspondence with the rational numbers. The proof works by contradiction. If the reals can be put in one-to-one correspondence with the rationals, then they can also be put into one-to-one correspondence with the natural numbers, which means they can be put in a list. Assume that each real number between zero and one is represented by an infinite decimal. Then there should be a list of the sort given in Figure 10.2, where line 1 would contain the first infinite decimal, line 2 the second infinite decimal, and so on.

Cantor showed that no matter what such a list looks like, there must be at least one real number that is not on the list. He drew a diagonal line that picks

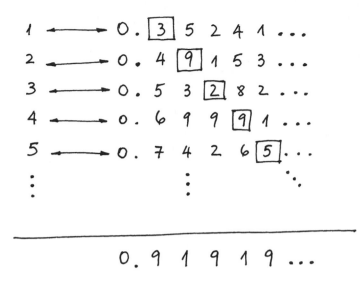

FIGURE 10.2   Cantor's diagonal argument shows that there can be no one-to-one correspondence between the natural numbers and the real numbers. It suffices to prove the result for the reals between 0 and 1, since if *they* cannot be put in one-to-one correspondence with the natural numbers, all the reals certainly cannot. Suppose there were such a correspondence, as given in the diagram, where the natural numbers are on the left and the corresponding real numbers on the right. Cantor showed how to construct a real number that, contrary to assumption, cannot be on the list. He constructed an infinite decimal as follows: For the nth number on the list, he replaced the digit in the nth decimal place by another digit; for instance, if that digit was 9, he replaced it by 1, and if it was not 9, he replaced it by 9. The resulting number differed from every infinite decimal on the list in at least one decimal place. Therefore, any assumed one-to-one correspondence between natural numbers and real numbers cannot exist.

out the first digit of the first number, the second digit of the second number, and so on. Then he constructed a new infinite decimal by using a rule for replacing the digits of the diagonal (e.g., where there is a 9 he puts a 1, and where there is a digit other than 9 he puts a 9). He replaced the first digit, then the second, and so on infinitely. The result is an infinite decimal that is not on the list, since it must differ from *every* infinite decimal on the list in at least one decimal place.

There is one minor technicality in the proof. Each number must be represented uniquely. Since the two numerals .499999. . . and .50000. . . represent the same number, one representation must be chosen. Cantor chose the first in this case and all similar cases.

As before, this proof of Cantor's uses an array. Each place in the array is indexed by two integers (n, m): the row of the array (the nth infinite decimal) and the column of the array (the mth digit). The diagonal line goes through all the digits where $n = m$.

There are many uses of the BMI implicit in Cantor's proof.

- First, there is the use of the special case of the BMI for infinite decimals. Each line is unending, yet complete.
- Then there is the use of the special case of the BMI for the set of all natural numbers. Each row corresponds to a natural number, and *all of them* must be there.
- Third, there is the sequence along the diagonal $(a_{11}, a_{22}, a_{33}, \ldots)$. It, too, is assumed to include *all* the digits on the diagonal. This is another implicit use of the BMI.
- And finally, there is the process of replacing each digit on the diagonal with another digit. The process is unending, but must cover the *whole* diagonal. Another implicit special case of the BMI.

These implicit uses of the BMI in Cantor's proof are of great theoretical importance. The proof is usually taken to be a formal proof, and the result is taken to be objective—independent of any minds. But the proof has not been, and cannot be, written down formally using actual infinity, with an array that is *actually* infinitely long and wide. The actual infinities involved in the proof are all mental entities, conceptualized via the BMI. In other words, the proof is not literal; it is inherently metaphorical, using the Basic Metaphor of Infinity in four places. Since it uses metaphor, it makes use of a cognitive process that exists not in the external, objective world but only in minds. Thus, the result that there are more real numbers than rational numbers is an *inherently* metaphorical result, not a result that transcends human minds.

This, of course, does not make it a less valid mathematical result, any more than the fact that the real numbers require the BMI makes them any less "real."

Cantor's proof that the reals cannot be put into one-to-one correspondence with the natural numbers and, hence, with the rationals had a dramatic import. It showed that the set of reals has a larger Cantor size than the rationals, that there are "more" reals than rationals (if one accepts Cantor's metaphor). Since the set of natural numbers is infinite, this meant that the set of real numbers is "larger than" that infinity, that there are *degrees of infinity*, that some infinities are "larger than" others (in a mathematics accepting Cantor's metaphor). Cantor called the number of elements in the set of natural numbers "$\aleph_0$" and the number of elements in the set of real numbers "C" for "continuum" (see Chapter 12).

Cantor further proved a remarkable result about power sets, the set of all subsets of a given set. He proved that there can be no one-to-one correspondence between the members of a set and its power set—even for infinite sets! Recall (from Chapter 7) that he had previously shown that an infinite set (the natural numbers) could be put in one-to-one correspondence with one of its subsets (the even positive integers). The power-set proof showed that there was at least one general principle about the impossibility of one-to-one correspondences between infinite sets:

Number of Elements in a Set	Number of Elements in the Corresponding Power Set
1	2
2	4
3	8
.	.
.	.
.	.
.	.
.	.
n	2^n
.	.
.	.
.	.
.	.
.	
$\aleph_0$	$2^{\aleph_0}$

From this table we can see that for finite sets with n members, the number of elements in a power set is 2^n. Cantor called the number of elements in the power set of the natural numbers "$2^{\aleph_0}$". Whereas 2^3 equals 2 times 2 times 2, $2^{\aleph_0}$ is not $2 \cdot 2 \cdot 2 \cdot \ldots \cdot 2$ an $\aleph_0$ number of times. Such a concept is not defined. Rather, the symbol $2^{\aleph_0}$ is a name for the number of elements in the power set of natural numbers—whatever that number is. What Cantor proved was that that number is "greater than" $\aleph_0$, assuming Cantor's metaphor. Of course, the symbol $2^{\aleph_0}$ was not chosen randomly. It was a choice based on the application of the BMI to *numerals*, not numbers. For each finite set, the numeral naming the number of elements in the power set of a set with n elements is 2^n. Applying the BMI to the numerals, we get a new symbol—a new numeral naming the number of elements in the power set of the natural numbers.

Cantor was able to put his two remarkable proofs together. He had proved, using his central metaphor, that

- The number of real numbers is "greater than" the number of natural numbers.
- The number of elements in a power set is "greater than" the number of elements in the set it is based on.

He was able to prove a further remarkable result:

- The number of elements in the set of real numbers is the "same" as the number of elements in the power set of the natural numbers. Or, as written in symbols: $C = 2^{\aleph_0}$.

Transfinite Arithmetic

Cantor's metaphor, together with the BMI, does something that the BMI alone does not do. It creates for infinite sets unique, precise cardinal numbers, which have a well-defined arithmetic. Recall that without Cantor's metaphor, the BMI alone can create the "number" ∞, the largest natural number. But ∞, not being one of the natural numbers, does not combine in a meaningful way with other natural numbers. As we saw, $0 \cdot \infty$, $\infty - \infty$, ∞/∞, and ∞/n are not well defined as operations on natural numbers.

Cantor's metaphor, together with the Basic Metaphor of Infinity, defines the transfinite cardinal number $\aleph_0$ differently from the way the BMI alone characterizes ∞ as the endpoint of the natural-number sequence. Transfinite cardinals were carefully characterized by Cantor so that they would have a meaningful arithmetic.

Consider $\aleph_0 + 1$. This is the cardinality of all sets in one-to-one correspondence with a set formed as follows: Take the set of natural numbers and include in it one additional element. The resulting set can be put in one-to-one correspondence with the set of natural numbers in the following way:

Start with the set of natural numbers: $\{1, 2, 3, \ldots\}$. Call the additional element "*." Adding *, we get the set $\{*, 1, 2, 3, \ldots\}$. We can now map the set of natural numbers one-to-one onto this set in the following way: Map 1 to *, 2 to 1, 3 to 2, and so on. Thus the set with cardinal number $\aleph_0 + 1$ is in one-to-one correspondence with the set of natural numbers which has the cardinal number $\aleph_0$. Thus, $\aleph_0 + 1 = \aleph_0$.

Note, incidentally, that including an additional element * in the set of natural numbers yields the same set as that including all the natural numbers in a set with the element *. For this reason, addition is commutative for the transfinite cardinal numbers: $1 + \aleph_0 = \aleph_0 + 1$.

By arguments of the above form, we can arrive at some basic principles of cardinal arithmetic.

$k + \aleph_0 = \aleph_0 + k = \aleph_0$, for any natural number k.

$\aleph_0 + \aleph_0 = \aleph_0$

$2 \cdot \aleph_0 = \aleph_0$

$k \cdot \aleph_0 = \aleph_0$, for any natural number k.

$\aleph_0 + \aleph_0 + \aleph_0 + \ldots = \aleph_0$

$\aleph_0 \cdot \aleph_0 = \aleph_0$

$(\aleph_0)^2 = \aleph_0$

$(\aleph_0)^k = \aleph_0$, for any natural number k.

Beyond $\aleph_0$

Cantor, as we have seen, proved that there are sets with a number of elements greater than $\aleph_0$. He used the symbol $\aleph_1$ to indicate the next largest cardinal number greater than $\aleph_0$. He hypothesized that that number was the number of members of the set of real numbers, which he had called "C" and had proved equal to $2^{\aleph_0}$. This is called the *Continuum hypothesis*, written as: $2^{\aleph_0} = \aleph_1$. It says that if "larger" and "smaller" are defined by Cantor's metaphor, then the smallest set larger than $\aleph_0$ has the cardinal number (i.e., the Cantor size) $2^{\aleph_0}$.

Cantor stated the Continuum hypothesis as a conjecture, which he was never able to prove. Later, Kurt Gödel (in 1938) and Paul Cohen (in 1963) proved theorems that together showed that the conjecture is neither absolutely true nor absolutely false relative to generally accepted axioms for set theory. Its truth depends on what further axioms one takes "set" to be defined by. From the perspective of our mathematical idea analysis, this means that whether or not the Continuum hypothesis is "true" depends on the underlying conceptual metaphors characterizing the concept "set."

The Hierarchy of Transfinite Cardinals

A power set is always larger than the set it was based on. Moreover, given a power set, we can always form the power set of that power set, and so on. This is true for infinite as well as finite sets within Cantor's set theory.

Given a set of cardinality $\aleph_0$, we can form the power set of the power set of ... of the power set of the natural numbers. This is a hierarchy of sets of ever-increasing Cantor size:

$$\aleph_0 < 2^{\aleph_0} < 2^{2^{\aleph_0}} < \ldots$$

To summarize, the Basic Metaphor of Infinity allows us to form the entire set of natural numbers. Cantor's metaphor, applied with the BMI, allows us to derive the cardinal number, $\aleph_0$, for the entire infinite set of natural numbers. Cantor's metaphor and the BMI jointly allow us to "prove" that there are "more" reals than natural numbers and rational numbers. Cantor's metaphor allowed Cantor to prove that there are always more elements in a power set than in its base set, even for infinite sets. And this result provides a hierarchy of transfinite cardinal numbers in Cantor's set theory.

Ordinal Numbers

Take an ordinary natural number—say, 17. This number can be used in at least two ways. It can have a *cardinal* use; that is, it can be used to indicate how many elements there are in some collection. For example, there might be 17 pens in your desk drawer. It can also have an *ordinal* use; that is, it can be used to indicate a position in a sequence. For example, today might be the 17th day since it last rained. These are two very different uses of numbers.

Natural numbers are neutral between these two uses. They have exactly the same properties, independently of how they are being used. For example, 17 is prime, whether you use it to estimate size or position in a sequence. 17 + 1 = 1 + 17, no matter which of those two uses you are making of numbers.

Most important, there is a simple relation between the cardinal and ordinal uses of natural numbers. Suppose you are forming a collection. You lay out 17 items in sequence to go into the collection. The last item is the 17th item. This is an ordinal use of 17. For convenience we will say that such a sequence is 17 elements long—or equivalently, that it has a length of 17. When we put this sequence of elements into a collection, the number of the items in the collection is 17. This is a cardinal use of 17. In this case, the (ordinal) number for the last item counted corresponds to the (cardinal) number for the size of the collection. Ordinality and cardinality always yield the same number, no matter how you count. For this reason, the arithmetic of the natural numbers is the same for cardinal and ordinal uses.

But this is not true for transfinite numbers. Cantor's metaphor determines "size" for an infinite collection by pairing, not counting in a sequence. Cantor's metaphor, therefore, is only about cardinality (i.e., "size") not about ordinality (i.e., sequence). When Cantor's metaphor is used to form transfinite numbers, those metaphorically constituted "numbers" are defined in terms of sets, with no sequences. Such sets have no internal ordering and so can have only cardinal uses, not ordinal uses.

What this means is that in the realm of the infinite there are no transfinite numbers that can have both cardinal and ordinal uses. Rather, two different types of numbers are needed, each with its own properties and its own arithmetic. And as we will see shortly, there is another strange difference between the cardinal and the ordinal in the domain of the infinite: You can get different results by "counting" the members of a fixed collection in different orders! The reason for this has to do with the different metaphors needed to extend the concept of number to the transfinite domain for ordinal as opposed to cardinal uses.

The Transfinite Ordinals

In the Basic Metaphor of Infinity, let the unending process be the process of generating the natural numbers so as to form a sequence, with number 1 first, number 2 second, and so on. The sequence—not the set!—formed is $(1, 2, 3, \ldots)$. This is the sequence of the natural numbers *in their natural order*. Here 2 is the second ordinal number, 3 is the third ordinal number, and so on.

Given this process of sequence formation, a special case of the BMI will create a final resultant state of the process: the infinite sequence of all the natural numbers in their natural order. Such a special case of the BMI will also assign to this entire infinite sequence an ordinal number, which Cantor called "ω." This is the first transfinite ordinal number. Every sequence of length ω is an infinite sequence that has been "completed" by virtue of the BMI. It has an endpoint—the number ω. The last position in this sequence is thus called the ω position.

TRANSFINITE ORDINALS

Target Domain ITERATIVE PROCESSES THAT GO ON AND ON		Special Case THE SEQUENCE OF THE NATURAL NUMBERS IN THEIR NATURAL ORDER
The beginning state (0)	⇒	No sequence
State (1) resulting from the initial stage of the process	⇒	The 1-place sequence (1), whose length is the ordinal number 1
The process: From a prior intermediate state $(n{-}1)$, produce the next state (n).	⇒	Given sequence $(1, \ldots, n{-}1)$, form the next longest sequence $(1, \ldots, n)$.
The intermediate result after that iteration of the process	⇒	The sequence $(1, \ldots, n)$ whose length is the ordinal number n

"The final resultant state" (actual infinity "∞")	⇒	The infinite sequence $(1, \ldots n, \ldots)$, whose length is the transfinite ordinal number ω.
Entailment E: The final resultant state ("∞") is unique and follows every nonfinal state.	⇒	Entailment E: ω is the unique smallest transfinite ordinal number, and it is larger than any finite ordinal number.

What distinguishes this special case of the BMI from others—for example, the set of natural numbers, the "number" ∞, and so on? Here we are building up *sequences* as *results* of stages of processes. In the case of ∞, the results were numbers. In the case of the set of all natural numbers, the results were sets of natural numbers. Infinite sequences, counted by ordinal numbers, just work differently than individual numbers or sets of numbers—as Cantor recognized clearly.

Other sequential arrangements of the same numbers in different orders will also have a length given by the ordinal number ω. For example, the following sequences all have length ω:

$$2, 1, 3, 4, 5, \ldots$$
$$3, 2, 1, 4, 5, \ldots$$
$$4, 1, 2, 3, 5, 6, \ldots$$

Cantor's Ordinal Metaphor

To make arithmetic sense of infinite ordinal numbers for infinite sequences of all sorts, Cantor assumed another metaphor:

CANTOR'S ORDINAL METAPHOR	
Source Domain ONE-TO-ONE MAPPINGS FOR SEQUENCES	*Target Domain* ORDINAL NUMBERS
Sequences A and B can be put in a one-to-one correspondence →	Sequences A and B have the same ordinal Number

The problem here is: What is to count as a one-to-one correspondence for an infinite sequence? Cantor assumed an account of one-to-one correspondences for sequences that made implicit use of the BMI. He separated the finite terms of sequences using the comma symbol, as in $(1, 2, 3, \ldots)$. However, he used the semicolon to indicate the end of an implicit use of the BMI.

Consider the following sequences:

$$1, 3, 5, 7, \ldots$$
$$2, 4, 6, 8, \ldots$$

Just as the set of odd numbers can be put in one-to-one correspondence with the natural numbers, so the *sequence* of odd numbers can be put in one-to-one correspondence with the sequence of the natural numbers. Therefore, they both have the same ordinal number, ω. The same is true of the even numbers. Each of these has one use of "..." at the end, indicating one implicit use of the BMI to complete the sequence.

Thus, Cantor implicitly used the BMI as part of his understanding of what a "one-to-one correspondence for an infinite sequence" is to mean. The "..." indicates an implicit use of the BMI, and it counts as part of the definition of one-to-one correspondence for a sequence.

Now consider the sequence of all the odd numbers "followed by" the sequence of all the even numbers:

$$1, 3, 5, 7, \ldots ; 2, 4, 6, 8, \ldots$$

The sequence of odd numbers $(1, 3, 5, 7, \ldots)$ contains a "...," indicating the implicit use of the BMI to complete the sequence. The sequence of all the odd numbers "followed by" the sequence of all the even numbers contains two uses of "...," indicating two implicit uses of the BMI to complete the sequences. Since the complete sequence of the odd numbers has length ω, and the complete sequence of even numbers following it also has length ω, the entire sequence has length $\omega + \omega = 2\omega$.

Defining one-to-one correspondence for an infinite sequence in this fashion has important consequences for an arithmetic of transfinite ordinal numbers. It is quite different from the arithmetic of transfinite cardinal numbers. Recall that in transfinite cardinal arithmetic, $\aleph_0 + \aleph_0 = \aleph_0$, whereas in transfinite ordinal arithmetic, $\omega + \omega = 2\omega$.

Recall that "addition" for transfinite cardinals is defined in terms of set union. Recall also that the set of positive odd integers can be put in one-to-one correspondence with the natural numbers, and so it has the cardinal number $\aleph_0$. The same is true of the set of positive even integers. Thus, the union of the set of positive odd integers and the set of positive even integers has the cardinality $\aleph_0 + \aleph_0$. Since this set union is just the set of all natural numbers $\aleph_0$, it follows that $\aleph_0 + \aleph_0 = \aleph_0$. Similarly, $\aleph_0 + 1 = \aleph_0$, since the union of a set the size of the natural numbers with a set of one element can be put in one-to-one correspondence with the set of natural numbers, as we have seen.

But the situation is very different with the ordinal numbers. Once we have established the ω position in an infinite sequence—the "last" position in that sequence—we can go on forming longer sequences by appending a further sequence "after" the ω position. If we add another element to the sequence already "counted" up to ω, the element added will be the $(\omega + 1)$ member of the sequence. Add another element and we have the $(\omega + 2)$ member of the sequence. Add ω more elements (via another application of the BMI) and we have the $(\omega + \omega)$, or (2ω), member of the sequence. If we keep adding ω elements to the sequence and you do it ω times, then the BMI will apply ω times, and the resulting sequence will be ω^ω elements long. We can keep on going, generating an unending hierarchy of transfinite ordinal numbers—namely:

$$\omega^{\omega^{\omega^{\cdots}}}$$

In this way, an ordinal arithmetic is built up.

$$\omega \neq \omega + 1$$
$$\omega + \omega = 2\omega$$
$$\omega \cdot \omega = \omega^2$$
$$\omega^a \cdot \omega^b = \omega^{a+b}$$

Incidentally, not all the basic laws of arithmetic apply to transfinite ordinals. For example, the commutative law does not hold: $1 + \omega \neq \omega + 1$. As we saw, $\omega \neq \omega + 1$. However, $1 + \omega = \omega$. Here is the reason: $1 + \omega$ is the length of a sequence. That sequence is made up of a sequence of length 1 followed by a sequence of length ω. For example, consider the sequence of the number 1 followed by the infinite sequence of the numbers 2, 3, 4, ... This is a sequence of length 1 followed by a sequence of length ω. But this sequence is the same as the sequence 1, 2, 3, 4, . . . , which is of length ω. For this reason, $1 + \omega = \omega$. But $\omega \neq \omega + 1$. Therefore, $1 + \omega \neq \omega + 1$, and the commutative law does not hold.

The general reason for this is that "+" is metaphorically defined in sequential terms in ordinal arithmetic, where the BMI applies to define infinite sequences. "$A + B$" in ordinal arithmetic refers to the length of a sequence. That sequence is made up of a sequence of length A followed by a sequence of length B. Addition is therefore asymmetrical when A is transfinite and B is not. If A is ω, which is the smallest transfinite ordinal number, adding a sequence with a finite length B after A will therefore always yield a sequence whose length is greater than ω. The reason is that the BMI produces the infinite sequence A with a "last" term. Putting a finite sequence before that "last" term does not result in a higher order of infinity. But putting it after that "last" infinite term produces a "longer" infinite sequence.

Suppose the sequence of finite length B comes first, followed by a sequence of length ω. We will then get a sequence: $1, \ldots, B$ followed by a sequence of length ω, the shortest infinite sequence. Adding a finite number of terms *at the beginning* of the shortest infinite sequence does not make that infinite sequence any "longer."

Summary

What is the difference between $\aleph_0$ and ω? The first is a transfinite cardinal that tells us "how many" elements are in an infinite set. The second is a transfinite ordinal that indicates a position in an infinite sequence.

The transfinite cardinal number $\aleph_0$ is, via the cardinal number metaphor, the set of all sets that are in one-to-one correspondence with the set of all natural numbers. The set of all natural numbers is generated by the Basic Metaphor of Infinity. $\aleph_0$ indicates the "size" of an infinite set by virtue of Cantor's metaphor (Cantor size). In this way, $\aleph_0$ is characterized by these three metaphors.

The transfinite ordinal number ω is the ordinal number corresponding to the well-ordered set of all natural numbers. It is generated by the Basic Metaphor of Infinity and Cantor's ordinal metaphor. This difference between transfinite cardinals and transfinite ordinals is the root of the difference between their arithmetic.

$\aleph_0$ and ω are both infinite numbers of very different sorts, with different arithmetics. Both are distinct from the infinite number ∞. As we saw in Chapter 8, ∞ is generated by the Basic Metaphor of Infinity when applied to the process of forming natural numbers via the Larger Than relation. Via this application of the BMI, ∞ is the largest of the natural numbers, which, as we saw, is the only numerical property it has. Otherwise, it is arithmetically defective. $\infty + 1$, for example, is not defined, while $\aleph_0 + 1$ and $\omega + 1$ are well defined and part of precise arithmetics—different precise arithmetics.

We should point out that the symbol + in "$\aleph_0 + 1$" and "$\omega + 1$" stands for different concepts than the + we find in ordinary arithmetic (e.g., in "$2 + 7$"). Types of transfinite addition are not mere generalizations over the usual forms of addition for finite numbers; rather, they are inherently metaphorical. They operate on metaphorical entities like $\aleph_0$ and ω, which are conceptualized via the BMI. Transfinite cardinal and ordinal addition not only work by different laws (different from ordinary addition and from each other) but are defined by infinite operations on sets and sequences, unlike ordinary addition.

What does mathematical idea analysis tell us in these cases? First, it shows explicitly how the BMI is involved in characterizing transfinite cardinals and ordinals. Second, it tells us explicitly what metaphors distinguish the transfinite cardinals from the transfinite ordinals. And, third, it shows not only how transfinite cardinal and ordinal arithmetics arise as consequences of these metaphor combinations but exactly *why* and *how* their arithmetics are different.

Finally, such an analysis shows how transfinite numbers are embodied. They are not transcendental abstractions but arise via such general, *nonmathematical* embodied mechanisms as aspect and conceptual metaphor.

11

Infinitesimals

Specks

When we look into the distance, normal-size objects that are very far away appear to us as specks, barely discernible entities so small as to have no internal structure. A large object in your field of vision—say, a mountain or a stadium—will have a discernible width. But the width of the stadium plus the width of the speck is not discernible at all. Nor is the width of one speck discernible from the widths of three specks or five specks. But you know that if you could zoom in on a speck sufficiently closely, it would look like a normal-size object.

Our conception of a speck, if arithmetized, might go like this: The widths of discernible objects correspond to real numbers. A speck-number ∂ is greater than zero but less than any real number. A real number plus a speck number is not discernibly different from the real number alone: $r + \partial \approx r$. A small number of adjacent specks are not discernibly different from one speck: $r \cdot \partial \approx \partial$, where r is small.

Infinitesimals

Our everyday experience with specks corresponds in these basic ways to mathematical entities called *infinitesimals*. Within mathematics, the idea of an infinitesimal became popular with Leibniz's development of calculus. Both Leibniz and Newton, in developing independently the calculus of instantaneous change, used the metaphor that Instantaneous Change Is Average Change over an Infinitely Small Interval. Thus, for a function $f(x)$ and an interval of length Δx, instantaneous change is formulated as $\frac{f(x + \Delta x) - f(x)}{\Delta x}$. The instantaneous change in

$f(x)$ at x is arrived at when Δx is "infinitely small." The concept of "infinitely small," of course, has to be arithmetized to get a numerical result. This requires an arithmetization metaphor.

Leibniz and Newton had different metaphors for arithmetizing "an infinitely small interval." For Newton, the derivative of a function for a given number was (metaphorically) the tangent of the curve of the function at the point corresponding to that number. But, as we have seen, that tangent cannot be used with the metaphor that Instantaneous Change Is Average Change over an Infinitely Small Interval, because *at* the tangent the infinitely small interval is zero and you can't divide by zero. So Newton came up with the limit metaphor, a version of the Basic Metaphor of Infinity (Chapters 8 and 9). The tangent line was conceptualized metaphorically as the limit of a sequence of secant lines, with the secant lines becoming progressively smaller but always having a real length. As Δx approached zero (without reaching it), the secant lines approached the tangent (see Figure 11.1). When Δx got arbitrarily small (but remained a real number), the difference between the secant line and the tangent was some arithmetic function of the Δx's. If that arithmetic function of the Δx's also got arbitrarily small—insignificant, for any practical purpose—*one could ignore it.* By ignoring this secant-tangent difference, the calculation based on a sequence of secants could be used for determining the tangent closely enough. Newton's solution is the one that has come down to us today as the "definition" of the derivative.

Leibniz solved the problem of arithmetizing "an infinitely small interval" in a very different way. He hypothesized infinitely small numbers—infinitesimals—to designate the size of infinitely small intervals. For Leibniz, dx was an infinitesimal number, a number greater than zero but less than all real numbers. For Leibniz, "$df(x)/dx$" was a ratio of two infinitesimal numbers (see Figure 11.2). Since neither number was zero, division was possible. The quotient—the tangent to the curve—was a real-valued function plus some arithmetic function of infinitesimals. Any arithmetic function of infinitesimals was so small it could be ignored.

As far as calculation was concerned, there was no difference between the two approaches. Both yielded the same results. For two hundred years after Leibniz, mathematicians thought in terms of infinitesimals, used them in calculations, and always got accurate results. But ontologically there is a world of difference between what Newton did and what Leibniz did. Newton did not need infinitesimal *numbers as existing mathematical entities.* His secants always had a real length, while Leibniz's had an infinitesimal length. Leibniz hypothesized a new kind of number. Newton did not.

On the other hand, Newtonians had to use geometry, with tangents and secants, to do calculus. Leibnizians could do calculus using arithmetic without

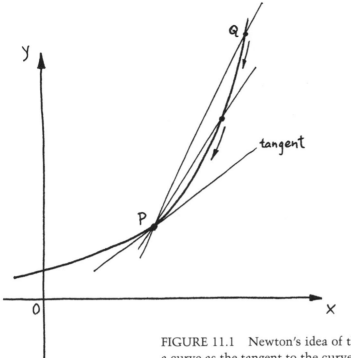

FIGURE 11.1 Newton's idea of the derivative at a point P on a curve as the tangent to the curve at P. He saw this tangent as the limit of a sequence of secants, generated by the movement of point Q on the curve toward point P.

geometry—by using infinitesimal *numbers*. Until the work of Karl Weierstrass in the late nineteenth century, the Newtonian approach to calculus relied on geometry and lacked a pure nongeometric arithmetization. Many mathematicians saw geometry as relying on "intuition," whereas arithmetic was seen as more precise and "rigorous." This perceived imprecision was considered a disadvantage of the Newtonian approach to calculus; its advantage was that it did not rely on infinitesimal numbers.

The Archimedean Principle

As we saw at the beginning of the chapter, infinitesimals require a somewhat different arithmetic than real numbers. Real numbers obey the *Archimedean principle*. Given any real number s, however small, and any arbitrarily large real number L, you can always add the small number to itself enough times to make it larger than the large number: $s + s + \ldots + s > L$, for some sufficiently large number of additions. Equivalently, there is a real number r, such that $r \cdot s > L$.

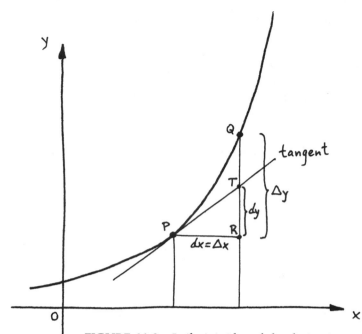

FIGURE 11.2   Leibniz's idea of the derivative at a point P on a curve was, as for Newton, the slope of the tangent to the curve at that point. But his reasoning was different. He reasoned that, in an infinitesimally small region around P, PT and PQ would be virtually identical. Thus, if the change in x (i.e., Δx) were infinitesimally small, the change in y (i.e., Δy) would be infinitely close to the rise of the tangent (dy). Thus, $\Delta y/\Delta x$ would be infinitesimally close to dy/dx—the slope of the tangent.

But this is not true of infinitesimals. Infinitesimals are so small that there is no real number large enough so that $r \cdot \partial > L$. You can't add up enough infinitesimals to get any real number, no matter how small that real number is and no matter how many infinitesimals you use.

Another way of thinking of this has to do with the concept of commensurability—having a common measure. Take two real numbers r_1 and r_2. There is always a way of performing arithmetic operations on r_1 with other real numbers to yield the number r_2. For example, given π and $\sqrt{2}$, you can multiply π by the real number $\sqrt{2}/\pi$ to get $\sqrt{2}$. But, given a real number r and an infinitesimal ∂, there are no arithmetic operations on r using the real numbers alone that will allow you to get ∂. The infinitesimals and the reals are therefore incommensurable.

The fact that the infinitesimals are not Archimedean and that they are incommensurable with the real numbers has made many mathematicians wary of them—so wary that they will not grant their existence. After all, all of classical mathematics can be done in the Newtonian way, without infinitesimals. If they are not needed, why even bother to consider them? Citing Occam's Razor, the a priori philosophical principle that you should not postulate the existence of an entity if you can get by without it, many mathematicians shun the infinitesimal numbers.

Why Monads Have No Least Upper Bounds

Infinitesimals interact with real numbers: They can be added to, subtracted from, and multiplied and divided by real numbers. Given any real number r, you can add to it and subtract from it any number of infinitesimals without reaching any other real number greater or smaller than r. This means that around each real number r there is a huge cluster of infinitesimals. That cluster was referred to by Leibniz as a *monad*.

Can a monad around a real number reach a monad around a neighboring real number? The answer is no. Here is the reason. Suppose the monads around two real numbers, r_1 and r_2, such that $r_1 < r_2$, were to overlap. This means that there would be two infinitesimal numbers ∂_1 and ∂_2 such that $r_1 + \partial_1 > r_2 - \partial_2$ even though $r_1 < r_2$. This would mean that by adding $\partial_1 + \partial_2$ to r_1 one could go beyond r_2. But this, in turn, would mean that the Archimedean principle worked for infinitesimals, which it cannot. Therefore, the monads around two real numbers can never overlap. No monad can ever reach any other monad.

Let us now ask: Can the monad around a real number r—call it $M(r)$—have a least upper bound? Suppose it did. Let the least upper bound of $M(r)$ be L. If L were real, L could have a monad of infinitesimals around it. In that monad would be an infinitesimal copy of the negative half of the real line, with each number on it less than L but still greater than all the members of $M(r)$. This means that each number in the negative half of the monad around L is an upper bound of $M(r)$ and less than L. Thus, L cannot be a least upper bound of $M(r)$. Now suppose that the least upper bound, L, of $M(r)$ were not a real number but a member of a monad around some real number r. The same argument would apply but in a smaller scale. L would have a tiny monad around it, and each number in the negative half of the monad around L would be less than L and greater than every member of $M(r)$. Thus each would be an upper bound of $M(r)$ but less than L, and L could not be a least upper bound.

Therefore, when the infinitesimals are combined with the real numbers, the least upper bound property cannot hold in general. However, because the least upper bound property holds of sets of real numbers, it will also hold of sets of numbers on infinitesimal copies of the real line.

The other nine axioms of the real numbers can apply to infinitesimals without contradiction.

An Implicit Use of the BMI

Though there is no *arithmetic* way to get the infinitesimals from the reals, there is a metaphorical mechanism that will do just that: the Basic Metaphor of Infinity. Newton explicitly used the notion of the limit, which implicitly uses the BMI, to characterize calculus. A different use of the BMI is implicit in Leibniz's development of calculus using infinitesimals.

We can get the infinitesimals from the reals using the BMI as follows: Form the following sequence of sets starting with the real numbers, as characterized by the first nine axioms for the real numbers in Chapter 9. (Recall that the Archimedean principle is *not* one of these axioms.) In the rest of the chapter, we will speak of a set of numbers, "satisfying" certain axioms. This is a shorthand meaning: the set of those numbers, such that they, together with their additive and multiplicative inverses and the identity elements 1 and 0, form a larger set that satisfies the specified axioms.

> The set of all numbers bigger than zero and less than 1/1, satisfying the
>   first nine axioms for the real numbers.
> The set of all numbers bigger than zero and less than 1/2, satisfying the
>   first nine axioms for the real numbers.
> The set of all numbers bigger than zero and less than 1/3, satisfying the
>   first nine axioms for the real numbers.

And so on.

Let this be the unending process in the BMI. Using the BMI, we get:

> The set of all numbers greater than zero and smaller than *all* real numbers,
>   satisfying the first nine axioms for the real numbers.

This is the set of infinitesimals.

Here is the special case of the BMI that gives rise to the infinitesimals:

INFINITESIMALS	
Target Domain ITERATIVE PROCESSES THAT GO ON AND ON	*Special Case* INFINITESIMALS
The beginning state (0) $\Rightarrow$	Consider sets of numbers that are greater than zero and less than $1/n$, satisfying the first nine axioms for the real numbers.

State (1) resulting from the initial stage of the process	⇒	$S(1)$: the set of all numbers greater than zero and less than $1/1$, satisfying the first nine axioms for the real numbers
The process: From a prior intermediate state (n-1), produce the next state (n).	⇒	From $S(n$-1), form $S(n)$.
The intermediate result after that iteration of the process (the relation between n and n–1)	⇒	$S(n)$: the set of all numbers greater than zero and less than $1/n$, satisfying the first nine axioms for real numbers.
"The final resultant state" (actual infinity "∞")	⇒	**$S(\infty)$: the set of all numbers greater than zero and less than *all* real numbers, satisfying the first nine axioms for the real numbers**
Entailment *E*: The final resultant state ("∞") is unique and follows every nonfinal state.	⇒	**Entailment *E*: $S(\infty)$ is unique.**

These infinitesimal numbers are metaphorical creations of the BMI. In the iterations, the numbers get smaller and smaller but are always real numbers at every finite stage. It is only at the final stage, characterized by the BMI, that we get a set of numbers greater than zero and smaller than *all* real numbers.

Because infinitesimals satisfy the first nine axioms for the real numbers, arithmetic operations apply to them. Thus you can multiply real numbers by infinitesimals—for example, $3 \cdot \partial = \partial \cdot 3 = 3\partial$. There is a multiplicative inverse:

$$1/\partial, \text{ such that } \partial \cdot 1/\partial = 1.$$

You can also take ratios of infinitesimals:

$$3\partial/4\partial = (3{\cdot}\partial)/(4{\cdot}\partial) = 3/4 \cdot \partial/\partial = 3/4 \cdot 1 = 3/4.$$

Thus, ratios of infinitesimals can be real numbers, as required by Leibnizian calculus.

Since the concept of the limit and the concept of the infinitesimals are special cases of the BMI, it is no surprise that Newton and Leibniz got equivalent results. They *both* implicitly used the Basic Metaphor of Infinity to extend the real numbers and the operations on them, but they used it in different ways. Newton used the BMI to define a new *operation*—taking the limit—but not new metaphorical entities. Leibniz used the BMI to define new *entities*—infinitesimals—but not a new metaphorical operation.

As for the principle of Occam's Razor, one might as easily apply it to kinds of operations as to kinds of entities. Thus, one might argue, against the Newtonian method, that it required a new kind of operation—limits that could be avoided if you had infinitesimals. Occam's Razor cuts both ways. A pro-infinitesimal anti-limit argument is just as much an Occam's Razor argument as a pro-limit anti-infinitesimal argument. Occam's Razor (if one chooses to accept it) in itself thus does not decide which use of the Basic Metaphor of Infinity, Newton's or Leibniz's, should be accepted for characterizing calculus. For historical reasons, Newton's use of the BMI was ultimately accepted, but both versions were used for a long time.

Robinson and the Hyperreals

In the late nineteenth century, the German mathematician Karl Weierstrass developed a pure nongeometric arithmetization for Newtonian calculus (see Chapter 14). This work effectively wiped out the use of infinitesimal arithmetic in doing calculus for over half a century. The revival of the use of infinitesimals began in the 1940s, with research by Leon Henkin and Anatolii Malcev. There is a theorem in mathematical logic called the Compactness theorem, which says:

> In a logical system L that can be satisfied by (i.e., mapped consistently onto) a set-theoretical model M, if every finite subset of a collection of sentences of L is satisfied in M, then the entire collection is satisfied in some model $M^{\star}$.

Here is the reason that this theorem is true.

- Every finite subset of sentences that is satisfied in M can be consistently mapped onto M.
- If a finite subset of sentences can consistently be mapped onto a model M, then that subset of sentences is consistent; that is, it does not contain contradictions.
- If every finite subset of sentences is satisfied in M, then every finite subset is consistent.
- If every finite subset is consistent, then the entire collection is consistent and therefore has some model $M^{\star}$, which may not be the same as M.

From a conceptual perspective, we can see the Compactness theorem as another special case of the BMI. Here are the BMI parameters that allow us to see the Compactness theorem as a form of the BMI:

The Logical Compactness Frame

The sentences of a formal logic are denumerable; that is, they can be placed in a one-to-one correspondence with the natural numbers. If every finite subset of a set of sentences S has a model, then the subset $S(n)$ consisting of the first n sentences of S has a model $M(n)$.

The BMI for Logical Compactness

Target Domain ITERATIVE PROCESSES THAT GO ON AND ON		*Special Case* LOGICAL COMPACTNESS
The beginning state (0)	$\Rightarrow$	The logical compactness frame, with models for all finite sets of sentences
State (1) resulting from the initial stage of the process	$\Rightarrow$	There is an ordered pair $(S(1), M(1))$, consisting of the first sentence and a model that satisfies it.
The process: From a prior intermediate state $(n - 1)$, produce the next state (n).	$\Rightarrow$	Given $(S(n - 1), M(n - 1))$, there is a model $M(n)$ satisfying $S(n)$.
The intermediate result after that iteration of the process (the relation between n and $n - 1$)	$\Rightarrow$	There is an ordered pair $(S(n), M(n))$, consisting of the set $S(n)$ of the first n sentences and the model $M(n)$ that satisfies it.
"The final resultant state" (actual infinity "∞")	$\Rightarrow$	**There is an ordered pair $(S(\infty), M(\infty))$, consisting of the set $S(\infty)$ of the *entire* infinite sequence of sentences and a model $M(\infty)$ that satisfies $S(\infty)$.**

Since the subject matter of this special case is the existence of some model or other, not a particular model, the uniqueness entailment applicable to other special cases of the BMI does not arise here.

This version of the BMI works as follows: Suppose we know that every finite set of sentences of S has a model. Since the sentences are countable, they can be ordered. Form the sequence of the first i subsets of sentences and the model paired with them. At each stage, there is one more sentence and a simple extension of the previous model to satisfy that sentence, given the others. The process of forming larger and larger subsets of sentences and larger and larger extensions of the previous models to satisfy those sentences goes on indefinitely. The BMI adds an endpoint at which all the sentences are considered together and an ultimate model is constructed that satisfies not just all finite subsets of sentences but the infinite set of all the sentences.

Of course, this is a cognitive characterization, not a mathematical characterization of the Compactness theorem. The mathematical proof uses the notion of a least upper bound/greatest lower bound (Bell & Slomson, 1969), which, from a cognitive viewpoint, is another version of the BMI (see Chapter 9). The mathematical version of the Compactness theorem can now be used to generate a model in which there is at least one infinitesimal. The result is due to Henkin.

Henkin considered the following set of sentences for some constant, C:

C is a number bigger than zero and less than $1/2$.
C is a number bigger than zero and less than $1/3$.
C is a number bigger than zero and less than $1/4$.
And so on.

For any finite subset of such sentences, there is some number C such that the finite subset of sentences is satisfied in any model of the axioms for the real numbers. According to the Compactness theorem, there is a model M^* in which all such sentences are satisfied. In that model, C must be greater than zero and smaller than any real number. As such, C is an infinitesimal. Hence, model theory shows that there are set-theoretical models of the real numbers in which at least one infinitesimal exists.

Since this is a result in classical mathematics, the conclusion came as something of a shock to those who thought that classical mathematics could get along without the existence of infinitesimals. The Compactness theorem requires that infinitesimals exist, even within the bounds of classical mathematics. In other words, Occam's Razor or no Occam's Razor, if there are least upper bounds, there are infinitesimals.

Of course, once you have a set-theoretical structure with any infinitesimals in it, the axioms of Infinity (another special case of the BMI) and Union (Chapter 7), taken together, allow for the formation of infinite unions of sets, so that there can then be set-theoretical models containing hugely more infinitesimals.

Once infinitesimals were shown to "exist" relative to the assumptions of classical mathematics and formal model theory, an entire full-blown formal theory of infinitesimal mathematics and calculus using them was constructed, mostly due to the work of Abraham Robinson. A branch of mathematics called *nonstandard analysis* has come out of Robinson's (1966, 1979) work.

It is not our intention here to review the field of nonstandard analysis. What we want to do instead is give the reader a conceptual feel of what infinitesimal numbers are, how they work, and what deep philosophical consequences follow from their "existence." We will use the Basic Metaphor of Infinity to show what the structure of infinitesimal arithmetic is like and how infinitesimals operate. Then we will show the consequences.

The remainder of this chapter will take the following form:

- First, we will use the BMI to generate "the first infinitesimal," which we will call ∂.
- Second, we will show that ∂ together with the reals forms a number system we will call the *granular numbers.*
- Third, we will form a system of numerals for the granular numbers, using "∂" as a numeral for the first infinitesimal number.
- Fourth, we will show that an understanding of this number system can be grounded in an understanding of the concept of a *speck.*
- Fifth, we will show a truly remarkable result. In the granular numbers, the sum of an infinite series *is not equal to its limit!* Moreover, the difference between the limit and the infinite sum can be precisely calculated in the granular numbers.
- Sixth, we will indicate how all of calculus can be done with the granular numbers. Indeed, there is a one-to-one translation of classical proofs in calculus without infinitesimals to a corresponding proof using the granular numbers.
- Seventh, we will then use the BMI again to show that starting from the "first infinitesimal," we can generate a countable infinity of "layers" of infinitesimals. We then gather these together, using the BMI to form an infinite union to produce an infinite set of layers of infinitesimals. Whereas there is a numeral system for the number system generated by the first infinitesimal, and numeral systems at any finite layer of infinitesimals, there is no system of numerals that can describe the full set of infinitesimals. This gives us the unsurprising result: *There are number systems with no possible system of numerals.*

The Granular Numbers

We use the term "granularity" as a metaphorical extension of its use in photography. In a photograph, background details that look like specks may be sufficiently important that you want to zoom in on them and get a more "fine-grained" print that allows you to see those details with greater resolution. The degree to which you can zoom in or out is the degree of "granularity" of the print. Since our understanding of infinitesimals is grounded in our understanding of specks, it seems natural to use the term "granular numbers" for those infinitesimals that we want to zoom in on and examine in detail. What links our *experiential grounding* for infinitesimals to an *arithmetic* of infinitesimals is the metaphor that Multiplication by an Infinitesimal Is Zooming In. For example, multiplying the numbers in the real line by an infinitesimal allows us to zoom in on an infinitesimal copy of the real line.

We will approach the infinitesimals conceptually, rather than through the model-theoretic methods used by Robinson, and we will build them up gradually so as to get a better look at their structure. We have already discussed the entire set of infinitesimals via the Basic Metaphor of Infinity, using the unending process of constructing "the set of all numbers bigger then zero and less than $1/n$, satisfying the first nine axioms for the real numbers." We then used the BMI to add a metaphorical end to the process. This time we proceed in a different way, in order to give the reader both a sense of the structure of the infinitesimals and an idea of how slightly different special cases of the BMI can yield radically different results. Where our previous use of the BMI iteratively formed sets of numbers and finally formed the set of infinitesimal numbers, we will use a version of the BMI to iteratively form individual numbers, not sets. The result will be *a single infinitesimal number*, which we call ∂.

In this use of the BMI, we will use the following unending process of number formation, where n is a natural number:

- $1/1$ is a number greater than zero, satisfying the first nine axioms for real numbers, and of the form $1/n$ where n is an integer.
- $1/2$ is a number greater than zero, satisfying the first nine axioms for real numbers, and of the form $1/n$ where n is an integer.
- $1/3$ is a number greater than zero, satisfying the first nine axioms for real numbers, and of the form $1/n$ where n is an integer.
- $1/4$ is a number greater than zero, satisfying the first nine axioms for real numbers, and of the form $1/n$ where n is an integer.

As the integers n get larger, the numbers $1/n$ get smaller endlessly. The Basic Metaphor of Infinity conceptualizes this unending process as having an endpoint with a unique result: A number, ∂, of the form $1/n$, which is greater than zero, satisfies the first nine axioms for real numbers—and, via the BMI, is less than all real numbers. We will pronounce "∂" as "delta" and refer to it as "the first infinitesimal number produced by the BMI." We will include the symbol ∂ in the list of numerals 0, 1, 2, 3, and so on.

Since ∂ satisfies the first nine axioms for the real numbers, it combines freely with them. And since ∂ is also a numeral, we can write down all the combinations that define "granular arithmetic." Moreover, ∂ is of the form $1/k$, where k is an integer. We will call that integer H for "huge number," a number bigger than all real numbers. Note that H = $1/\partial$. Also note that the symbol H is not in italics, since H is a fixed number, not a variable.

Incidentally, the real numbers are contained among the granulars; expressions with no ∂'s are just ordinary real numbers. Note also that since the granulars satisfy the first nine axioms of the real numbers, they are linearly ordered.

For the sake of completeness, here is the special case of the BMI for forming ∂.

THE FIRST INFINITESIMAL

Target Domain ITERATIVE PROCESSES THAT GO ON AND ON		*Special Case* THE GRANULAR NUMBERS
The beginning state (0)	$\Rightarrow$	Individual numbers of the form $1/k$, each greater than zero and meeting the first nine axioms of the real numbers (where k is a positive integer)
State (1) resulting from the initial stage of the process	$\Rightarrow$	$1/1$: It is of the form $1/k$, is greater than zero, and meets the first nine axioms of the real numbers.
The process: From a prior intermediate state $(n - 1)$, produce the next state (n).	$\Rightarrow$	Given $1/(n - 1)$, form $1/n$. It is of the form $1/k$, which is greater than zero and meets the first nine axioms of the real numbers.
The intermediate result after that iteration of the process (the relation between n and $n - 1$)	$\Rightarrow$	$1/n$, with $1/n < 1/(n - 1)$
"The final resultant state" (actual infinity "∞")	$\Rightarrow$	**The infinitesimal number $1/H = \partial$, where ∂ is greater than zero and meets the first nine axioms of the real numbers. H is an integer greater than all the real numbers.**
Entailment *E*: The final resultant state ("∞") is unique and follows every nonfinal state.	$\Rightarrow$	**Entailment *E*: $\partial = 1/H$ is unique and is smaller than all $1/k$, where k is a natural number.**

Granular Arithmetic

By using normal arithmetic operations on ∂ and the reals, we arrive at an arithmetic of granular numbers. Since ∂ satisfies the first nine axioms for the real numbers, it has the normal properties that real numbers have.

$$0 + \partial = \partial + 0 = \partial \qquad\qquad \partial + (-\partial) = 0$$
$$1 \cdot \partial = \partial \cdot 1 = \partial \qquad\qquad \partial/\partial = 1$$
$$0 \cdot \partial = \partial \cdot 0 = 0 \qquad\qquad \partial^1 = \partial$$
$$\partial^0 = 1 \qquad\qquad\qquad\qquad \partial^{-1} = 1/\partial$$
$$\partial^n/\partial^m = \partial^{n-m}$$

Here, ∂ is acting like any nonzero real number.

The Special Property

What makes ∂ an infinitesimal is the special property it acquired by being created via the BMI: *Any multiple of ∂ and a real number is greater than zero and smaller than any positive real number.*

$$0 < \partial \cdot r < s, \text{ for any positive real numbers } r \text{ and } s.$$

Since $\partial/\partial = 1$, the ratio of two real multiples of ∂ is a real number. For example, $2\partial/5\partial = 2/5$, since $\partial/\partial = 1$.

Since ∂ can be multiplied by itself, we can have $\partial \cdot \partial = \partial^2$. Since multiplying anything by ∂ results in a quantity that is infinitely smaller, ∂^2 is infinitely smaller than ∂. As one continues to multiply by ∂, one gets numbers that are progressively infinitely smaller: ∂^{100}, $\partial^{1,000,000}$ and so on, without limit. As a consequence, all real multiples of ∂^2 are less than all real multiples of ∂.

The granulars also form infinite polynomials based on ∂. Thus, there are granulars like: $3\partial + 7\partial^2 + 64\partial^3 + \ldots$, where the sequence continues indefinitely.

The Monads of Real Numbers

∂ and its multiples can be added to or subtracted from any real number. Thus, we have granular numbers like $47 + 7\partial - 57\partial^2$ or $53 - 5\partial + 87\partial^5$. Thus any multiple of a real number times ∂ or of ∂ times itself occurs added to and subtracted from the real numbers. This has two remarkable consequences.

Suppose we extend the metaphor that Numbers Are Points on a Line to the granular numbers. We can do this, since the granular numbers are linearly ordered (because they satisfy the first nine axioms for the real numbers). That means that every granular number can be put in correspondence with a point on a line. The result is the *granular-number line*, which contains within it the points corresponding to all the real numbers. The first remarkable consequence is that on the granular-number line, the real numbers are rather sparse. Most of the numbers are granulars.

The fact that multiples of ∂ are less than any positive real number has a second remarkable consequence: the formation of monads, as described earlier. For convenience, we repeat the principal points of that discussion here.

Pick a real number and add to it and subtract from it any sum of multiples of ∂^n, where n is a natural number. By doing this you will not reach any other real number. Around each real number on the granular-number line, there is an infinity of granular numbers extending in both directions. This is the monad around that real number.

Let us look at the structure of each such monad around a real number. We can get a sense of that structure by performing the following construction:

- Take the real-number line and multiply every number in it by ∂. Call this the ∂^1 number line.
- Take the ∂^1 number line and multiply every number in it by ∂. The result consists of each real number multiplied by ∂^2. Call this the ∂^2 number line.
- And so on for ∂^n.
- Now pick a real number r and add every number in the ∂^1 number line to it. This will give you an infinitely small copy of the real-number line around r. Let us call this the ∂^1 monad around r.
- Now, to every number in the ∂^1 monad around r, add every number in the ∂^2 number line. This will give you an infinitely smaller copy of the real-number line around r and around each multiple of ∂ times a real number; that is, in the monad around r there will be a real number of real-number lines. Now keep doing this indefinitely for ∂^n. You will get more and more.
- Around every real number r there will be a monad of infinitely small copies of the real number line, with infinitely smaller and smaller copies of the real-number line embedded in it.
- The monads around any two distinct real numbers do not overlap.

In other words, on the granular-number line, each real number is separated from every other real number by a vast structure of granulars stretching indefinitely in both directions but never reaching a member of another monad. By adding infinitesimals to a real number, you can at best get within an infinitesimal distance of another real number. That's how small infinitesimals are!

The Huge Numbers

Recall that ∂, as we constructed it, has the form $1/n$, where n is an integer. Recall that we are calling this integer H, for "huge number." H is the reciprocal of ∂. Thus,

$$\partial = 1/H, \ \partial \cdot H = 1, \ H = 1/\partial, \ \partial = H^{-1}, \ \text{and} \ H = \partial^{-1}.$$

Since ∂ is less than all real numbers, H is greater than all real numbers. Just as ∂ is infinitesimal, H is infinite. Since H satisfies the first nine axioms for real numbers, it can enter into all arithmetic operations. Thus, there are numbers H + 1, H − 1, H + 2, H − 2, and so on. There are also 2H, 3H, and so on, as well as H^2, H^3, . . . , H^H, and even

$$H^{H^{H^{\cdots}}}.$$

Thus there is no bound to the hugeness of the huge numbers. What we get are higher and higher levels of infinity. Similarly, we can have

$$\partial^{H^{H^{\cdots}}},$$

which defines infinitely smaller and smaller levels of infinitesimals.

As an integer, H has successors. There is H + 1, H + 2, and so on indefinitely. It also has predecessors: H − 1, H − 2, and so on indefinitely. However, all the granular numbers of the form H − n, where n is a real integer, are still huge. You can't get back down to the real numbers by subtracting a real number from H. Similarly, by performing the successor operation—the successive addition of 1 to a finite integer—you can never count all the way up to H or even H − r, for any real r. That will give you some idea of just how "huge" H is. H and the natural numbers are not connectable via the successor operation. And yet, via the operation of the Basic Metaphor of Infinity, H is a metaphorical integer.

So far we have defined ∂^n in terms of self-multiplication. But the exponential function, which is not self-multiplication (see Case Study 2), naturally extends to the granulars, so that expressions like ∂^π, e^∂, and e^H are well defined. As a result, we can raise ∂ and H to the power of any real or granular number. This gives us a *continuum* of ∂ levels. This is a granular continuum, not just a real continuum; that is, there is a ∂ level—a level of granularity—not just for every real number on the real-number line but also for every granular number on the granular-number line!

Why Infinite Sums Are Not Equal to Their Limits in Granular Arithmetic

Within the real numbers alone—without granulars or infinitesimals—infinite sums are *defined* in terms of limits. For example, consider the infinite sum of $9/10^n$—that is, 0.9999. . . . As we saw in Chapter 9, this infinite sum is usually defined—via the Infinite Sums Are Limits metaphor—as *being* its limit:

$$\lim_{n \to \infty} \sum_{i=1}^{n} 9/10^i = 1.$$

Since the limit is conceptualized via the special case of the BMI for limits, this is a metaphorical sum. What the expression above says is that 0.999999. . . = 1.

Now, for any finite series of the form

$$\sum_{i=1}^{n} 9/10^i$$

where there is a last term n, there is a difference between the last term and the limit. That difference is

$$1 - \sum_{i=1}^{n} 9/10^i = 1/10^n.$$

For example, if $n = 5$, then the sum is 0.99999 and the difference between the limit 1 and the sum is $1 - 0.99999$, which $= 0.00001$, or $1/10^5$. If the sum over the real positive integers is taken to infinity, the difference at stage n is of the form $1/10^n$. Taking the limit as n approaches ∞, $1/10^n$ converges to 0. At the limit—the only place where the sum is infinite—there is no difference between the infinite sum and the limit. The infinite sum *equals* the limit.

The granular numbers give us a very different perspective on what an infinite sum is. In the granular numbers, H is an integer larger than all real numbers. Relative to the real numbers, therefore, H is infinite. Indeed, since H is beyond all the real numbers, a sum up to H includes additional terms beyond a mere real-number sum approaching ∞.

To see this, let us look closely at H. As an integer, H occurs in a sequence of integers: ... , H − 3, H − 2, H − 1, H, H + 1, H + 2, ... H is so huge that subtracting any finite real number from H still yields a number incommensurably larger than any real number. Now consider the unending sequence of natural numbers: 1, 2, 3, ... No matter how big this sequence gets, every member of this sequence must not only be smaller than H but *infinitely* smaller than H. We can see this rather easily. Suppose we simultaneously start counting up from 1 and down from H. We get the following two sequences, each of which is open-ended:

$$1, 2, 3, \ldots, 10^{100}, \ldots \qquad \ldots H - 10^{100}, \ldots, H - 3, H - 2, H - 1, H$$

No matter how far we go in both directions, no contact can be made. Thus, H is infinitely beyond the infinity of natural numbers.

Since H is a specific granular number, since it is infinite, and since it is an integer, we can create an infinite sum whose *last term* is the specific granular integer H. This is an unusual feature of the granular numbers. Series can have an infinite number of terms, yet also have a *last* term. We could, in principle, have an infinite series with more than H terms—say, H + 1 or H^3—but for our purposes H will suffice to make the series not just infinite but longer than any infinite series whose terms correspond to natural numbers alone. In other words, a sum up to H is not just an infinite sum but it has infinitely more terms than a sum over all the natural numbers!

Now consider the series

$$\sum_{i=1}^{H} 9/10^i.$$

This is the series $9/10 + 9/100 + 9/1000 + \ldots + 9/10^H$. The last term of this series is $9/10^H$. Therefore, the difference between the limit 1 and the sum to H terms is $1/10^H$. This is a specific granular number. It is an infinitesimal. With the granular numbers, then, this infinite series is not only less than its limit but *measurably* less than its limit. It is *exactly* $1/10^H$ less than its limit.

Now consider another series with a limit of 1—namely,

$$\sum_{i=1}^{n} 1/2^i.$$

This is the sum $1/2 + 1/4 + 1/8 + \ldots$, whose limit is 1. At any term n, the difference between the limit and the nth term is $1/2^n$. For example, the sum for $n = 5$ is $1/2 + 1/4 + 1/8 + 1/16 + 1/32$, which equals $1 - 1/32$. Now take the sum of this series up to H.

$$\sum_{i=1}^{H} 1/2^i.$$

The precise difference between the limit 1 and the sum to H terms is $1/2^H$.

Note that $1/10^H$ is smaller than $1/2^H$. Thus the infinite sum (up to H terms) of $9/10^n$ is larger than the infinite sum of $1/2^n$, even though they both have the same limit. The precise difference between the two numbers is the granular number $1/2^H - 1/10^H$.

What limits do is to ignore precise infinitesimal differences—differences that can be calculated with precision in the granular numbers.

The Real Part

Since the real numbers are embedded in the granulars, we can define an operator R that maps each granular number onto its "closest" real number. R maps all the infinitesimals—the granulars whose absolute value is less than any positive real number—onto 0. R maps all the positive huge numbers onto ∞ and all the negative huge numbers onto $-\infty$. We will call R the *real-approximation operator*. From the perspective of the granular numbers, each real number is an extremely gross approximation to an infinity of granular numbers.

In addition, R is defined so that for granulars a and b, $R(a + b) = R(a) + R(b)$ and $R(a \cdot b) = R(a) \cdot R(b)$. In other words, the real approximation to the sum is the sum of the real approximations, and the real approximation to the product is the product of the real approximations. Here is an example of the effect of R. Consider the granular number $37 - 14\partial + 19\partial^3$.

$$
\begin{aligned}
R(37 - 14\partial + 19\partial^3) &= R(37) - (R(14) \cdot R(\partial)) + (R(19) \cdot R(\partial^3)) \\
&= 37 - (14 \cdot 0) + (19 \cdot 0) \\
&= 37
\end{aligned}
$$

Or take $H - 19$.

$$
\begin{aligned}
R(H - 19) &= R(H) - R(19) \\
&= \infty - 19 \\
&= \text{undefined}
\end{aligned}
$$

Suppose we apply R to the sums of infinite series up to H. Consider the two sums we just discussed:

- $1 - 1/10^H$ and
- $1 - 1/2^H$

Their real parts are

$$R(1 - 1/10^H) \text{ and } R(1 - 1/2^H),$$

which equal

$$R(1) - R(1/10^H) \text{ and } R(1) - R(1/2^H),$$

which equal

$$1 - 0 \text{ and } 1 - 0,$$

which equal

$$1 \text{ and } 1.$$

The two sums differ from 1 by an infinitesimal. The real-approximation operator ignores that infinitesimal and makes both sums equal to their limits. In other words, the real-approximation operator has the same effect as the notion of a limit. That effect is to ignore precise differences that are visible in the granulars. It also has the same effect as the notion of a limit for huge granulars. $R(H - 19)$ has the same effect as $\lim_{n \to \infty} (n - 19)$. The former is undefined, as is the latter, which technically has no limit since $n - 19$ grows indefinitely as n approaches ∞.

Incidentally, Robinson's treatment of infinitesimals does not include the granulars and differs in certain significant ways from the discussion given here. In a later section, we will compare the two. For a full treatment of calculus from Robinson's perspective, see H. J. Keisler (1976a, 1976b).

Calculus Becomes Arithmetic

Within the granular numbers, defined merely by the first infinitesimal ∂, one can do all of calculus without the notion of a limit. Calculus simply becomes arithmetic.

The derivative can be defined as follows:

$$f'(x) = R\left(\frac{f(x + \partial) - f(x)}{\partial}\right)$$

Compare this to the definition for limits.

$$f'(x) = \lim_{\Delta x \to 0} \frac{f(x + \Delta x) - f(x)}{\Delta x}$$

Instead of the variable Δx in the limit definition, we have a fixed constant ∂. Instead of a limit operator, $\lim_{\Delta x \to 0}$, we have a real-approximation operator R that has the effect of ignoring all infinitesimals. The results are always the same. Indeed, for any proof with limits, there is a corresponding proof with ∂'s.

Compare the treatment of $f(x) = x^2$ with limits and with ∂'s. With limits we have

$$\lim_{\Delta x \to 0} \frac{(x + \Delta x)^2 - x^2}{\Delta x} \qquad = \lim_{\Delta x \to 0} \frac{x^2 + 2x\Delta x + \Delta x^2 - x^2}{\Delta x}$$

$$= \lim_{\Delta x \to 0} \frac{2x\Delta x + \Delta x^2}{\Delta x}$$

$$= \lim_{\Delta x \to 0} 2x + \Delta x$$

At this point, we have to apply the special case of the BMI for limits and treat Δx, which never literally gets to zero, as if it did. The (metaphorical) limit of $2x + \Delta x$ as Δx approaches zero is $2x$.

The derivation is similar, but a bit simpler, with infinitesimals.

$$R\left(\frac{(x + \partial)^2 - x^2}{\partial}\right) \qquad = R\left(\frac{x^2 + 2x\partial + \partial^2 - x^2}{\partial}\right)$$

$$= R\left(\frac{2x\partial + \partial^2}{\partial}\right)$$

$$= R(2x + \partial)$$
$$= R(2x) + R(\partial)$$
$$= 2x + 0$$
$$= 2x$$

As should be clear, one can construct an algorithm to convert any derivation from limit notation to granular notation and vice versa. In the granulars, there are no limits, just arithmetic and the real-approximation operator.

Integrals

The definition of the integral in the granular numbers is correspondingly straightforward:

$$\int_a^b f(x)dx = R\left(\sum_{n=1}^{H(b-a)} f(x_n) \cdot \partial\right)$$

The integral divides up the interval from a to b into subintervals of length ∂. $f(x_n)$ is the height of the curve at the nth division into subintervals. The product $f(x_n) \cdot \partial$ is the area of the rectangle of width ∂ and height $f(x_n)$. The sum is the sum of all the rectangles in the interval from a to b (see Figure 11.3).

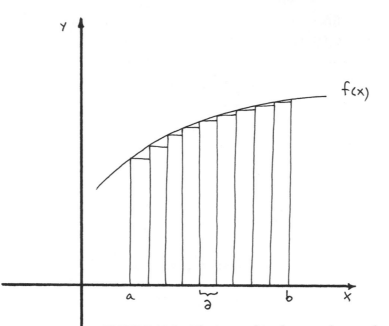

FIGURE 11.3   The integral in the granular numbers: The interval $[a, b]$ is divided into subintervals of length ∂. At the nth division into subintervals the value of the function is $f(x_n)$, and the area of each rectangle is $f(x_n) \cdot \partial$. The area under the curve is given by the sum of the areas of all the rectangles in the interval $[a, b]$.

Since $\partial H = 1$, there are H subintervals for every interval of length 1. The number of such subintervals in the interval of length $(b - a)$ is $H \cdot (b - a)$.

This sum will differ from the area under the curve by an infinitesimal. The operator R eliminates that infinitesimal difference. The result is the same as in standard calculus.

For example, consider the integral of $f(x) = 2x$ from $x = a$ to $x = b$. For the sake of simplicity, suppose that a equals zero and b is positive (see Figure 11.4). It is obvious from the figure what the right answer should be. Since the length b and the height $2b$ form a rectangle of area $2b^2$, the area of the triangle under the curve is half that: b^2. The definition of the integral in terms of an infinite sum of rectangles of infinitesimal width will give exactly the same result. Here's how:

Let us add up the area of the columns of rectangles under the curve in Figure 11.4. Each column has a width of ∂. There are $H \cdot b$ columns. The first column is composed of only one rectangle of area $\partial \cdot 2\partial = 2\partial^2$. The second column is composed of two rectangles of area $2\partial^2$. The total area of the second column is $2 \cdot 2\partial^2$. Similarly, the third column is composed of three rectangles and has the area of $3 \cdot 2\partial^2$. In general, the nth column has the area of $n \cdot 2\partial^2$. And the last

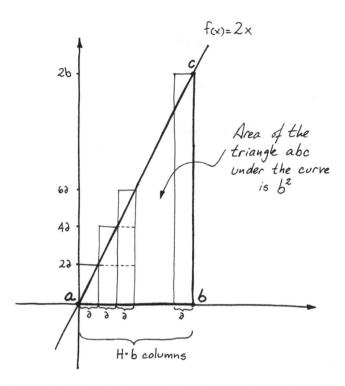

$f(x) = 2x$

$2b$

6∂

4∂

2∂

a b

∂ ∂ ∂ ∂

Area of the triangle abc under the curve is b^2

$H \cdot b$ columns

FIGURE 11.4   The integral of $f(x) = 2x$ from a to b in the granular numbers (where $a = 0$ and b is positive). The interval $[a, b]$ is divided into $H \cdot b$ columns, each with the width ∂. In general, the nth column has the area of $n \cdot 2\partial^2$, and the last column has the area of $Hb \cdot 2\partial^2$. The sum of all columns is $2\partial^2 \cdot (1 + 2 + \ldots + Hb)$, which after a little algebra yields $b^2 + \partial b$.

column has the area of $Hb \cdot 2\partial^2$. The sum of all the columns therefore is $2\partial^2 \cdot (1 + 2 + \ldots + Hb)$. That is,

$$\sum_{n=1}^{Hb} 2\partial^2 \cdot n$$

It is well known that the sum of the first n integers is $\frac{n(n+1)}{2}$. Thus the sum of the areas of the columns is

$$2\partial^2 \cdot \frac{Hb(Hb + 1)}{2}$$

which equals

$$2\partial^2 \cdot \frac{H^2b^2 + Hb}{2}$$

which equals

$$\partial^2 (H^2b^2 + Hb),$$

which equals

$$\partial^2\, H^2\, b^2 + \partial^2 H b.$$

Since $\partial H = 1$, we have

$$b^2 + \partial b.$$

Applying the real part operator R, we get

$$R(b^2 + \partial b),$$

which equals

$$R(b^2) + R(\partial b),$$

which equals

$$b^2 + 0 = b^2.$$

This is essentially the same as doing the definite integral via limits. The corresponding limit expression for this integral would be

$$\int_a^b f(x)dx = \lim_{\Delta x \to 0} \sum_{i=1}^{\infty} f(x_i) \cdot \Delta x$$

With granular numbers, there is only one sum and it is computed by simple arithmetic. With limits, there is an infinity of finite sums. The reason is that limits are defined in terms of sequences of sums, with the number of sums approaching infinity.

Remember that in the granular numbers, ∂ and H are fixed numbers whose product is 1. The number of divisions into columns under the curve is a fixed number, $H(b - a)$. There is just one figure with a sum of column areas to be computed, whereas in the case of limits there is an infinity of such figures and sums, since the number of divisions into columns—namely, n—approaches infinity.

In the granular numbers, the infinitesimal widths and the infinite number of divisions are fully arithmetized. In the case of the limit, there is no such arithmetization. With limits, we do not even know how many elements of width Δx are being summed in each sum, since Δx is a variable, nor do we know the exact width of any Δx.

As Δx gets small, the number of elements being summed gets indefinitely large, but we do not know exactly how large, because there is no arithmetization of infinity. Within the granular numbers, there is precise arithmetization of everything.

The Hyperreals

The system of granular numbers is vast. It dwarfs the real numbers. Half the granulars are infinitesimal and half are huge. Do the infinitesimal granulars and the huge granulars exhaust all the infinitesimal and huge numbers? The answer is no. There are more infinitesimal and huge numbers outside the granulars.

Recall that we formed the granulars in such a way that they constitute the smallest extension of the real numbers to infinitesimals. The way we did this was to apply the Basic Metaphor of Infinity to the process of forming numbers, $1/n$, such that "$1/n$ is a number greater than zero and less than $1/(n-1)$, satisfying the first nine axioms for real numbers." The BMI completes this process by yielding a unique infinitesimal number, which we called ∂. The rest of the granulars arose by arithmetic operations on ∂ and the reals.

Recall also that we had previously characterized the set of all infinitesimals by applying the BMI to a different process—a process of forming *sets*, not individual numbers. The sets were of the form: "the set of all numbers bigger than zero and less than $1/n$, satisfying the first nine axioms for the real numbers." When the BMI completes that process, it yields the set of *all* infinitesimals. The multiplicative inverses of those infinitesimals are included, because the axioms for the real numbers must be satisfied and those axioms require inverses. Those inverses are *all* the huge numbers. This set of *all* infinitesimals and huge numbers is called the *hyperreals*.

It is easy to see that the granulars do not by any means exhaust the hyperreals. The reason is that once we form the first infinitesimal via the BMI, we can use the BMI again to characterize more infinitesimals that are not in the granulars.

Here's how we do it. Recall the following: $\partial = 1/H$. It is infinitesimal, because it is divided by an infinite integer. $\partial^H = 1/H^H$. Now, ∂^H is infinitely smaller than even ∂ is, because it equals 1 divided by H^H, which is infinitely larger than H is. Within the granular numbers, we can go on forming successive powers of powers, such as

$$\partial^{H^H} \text{ and } \partial^{H^{H^H}}$$

We will write these as $\partial^{H \,(2\,\text{levels})}$ and $\partial^{H \,(3\,\text{levels})}$. In general, we can form an arbitrary number of levels of powers, $\partial^{H \,(n\,\text{levels})}$.

We can now form an unending process for creating infinitesimal numbers greater than zero:

Form $\partial^{H \,(1\,\text{level})}$; raise $\partial^{H \,(1\,\text{level})}$ to the power H, yielding $\partial^{H \,(2\,\text{levels})}$; . . . ; raise $\partial^{H \,(n-1\,\text{level})}$ to the power H, yielding $\partial^{H \,(n\,\text{levels})}$.

We now let this be the unending process in the BMI. The BMI conceptualizes this as a completed process with an endpoint and a new result: $\partial^{H\ (H\ levels)}$, an infinitesimal smaller than any granular infinitesimal, yet greater than zero. We will call it the second infinitesimal and write it as $\partial^{H\ (H\ levels)} = {}^2\partial$, or "delta super-two".

Using ${}^2\partial$, we can now form another layer of granular numbers, with infinitesimals smaller than any of the first granulars and huge numbers larger than any in the first layer of granulars. Once we have the layer of granulars defined by ${}^2\partial$, we can use the BMI in the same way again to form ${}^3\partial$, and so on indefinitely. Every layer defined by a ${}^n\partial$ is a subset of the hyperreals, which is the set of *all* infinitesimals and huge numbers.

Each time we get another ${}^n\partial$—the nth infinitesimal number—we can create a numeral for it—say, "${}^n\partial$". We started with the numerals "0," "1," "2," "3," "4," "5," "6," "7," "8," "9" for the real numbers and added "∂" to give a numeral system that could symbolically represent every granular number. We then added "${}^2\partial$," "${}^3\partial$," "${}^4\partial$," ..., "${}^n\partial$" to give a finite numeral system in which every number in the first n layers of granulars could be symbolically represented. But there is no end to the ${}^n\partial$'s, and so there is no end to the numerals that would have to be created to represent these numbers in a system of numerals. To represent all the layers of granulars in numerals would require more than a finite number of symbols. Thus, it is impossible to design a numeral system to symbolically represent all infinitesimals and huge numbers.

Recall that we can do calculus just fine with only one layer of granulars, so only one added numeral—"∂"—suffices. But if for theoretical reasons we wanted to have a numeral system for all the infinitesimals, we would not be able to achieve that goal.

Beyond All Granulars

The formation of more and more ${}^n\partial$'s via the BMI is an unending process itself. We can now apply the BMI to that unending process. That is, we can apply the BMI recursively. Here is the unending process:

From ${}^{n-1}\partial$, apply the BMI to yield ${}^n\partial$.

Applying the BMI to this unending process yields a metaphorical completion of the process. The result is a new infinitesimal number smaller than any of the ${}^n\partial$'s—a superinfinitesimal: ${}^H\partial$. This, too, is of the form $1/k$ (where k is an integer). It is smaller than any number in any layer of granulars, satisfies the first nine axioms for the real numbers, combines freely with all reals and granulars in every layer, and produces a new set of numbers that are linearly ordered with respect to all the other granulars.

We can now begin the process all over again, producing another infinity of layers, applying the BMI recursively to that, and on and on endlessly. Each time we use the BMI, we create another layer of infinitesimal and huge numbers that is a subset of the hyperreals.

This gives us a sense of just how big the hyperreals are and what kind of sub-structures exists within the hyperreals. Because *all* the hyperreals can be linearly ordered (since they satisfy the first nine axioms for the real numbers), they can all be mapped onto the line via the metaphor Hyperreal Numbers Are Points on a Line. Notice that the reals are such a small subset of the hyperreals that they are barely noticeable. Virtually the whole line is taken up by nonreals.

Calculus in the Hyperreals

The field known as nonstandard analysis does calculus using the hyperreals, not the granulars. Because there can be no numeral system for the hyperreals, there can be no finite symbol system for the arithmetic of the hyperreals. Where we used arithmetic and arithmetic notation to do calculus, those scholars using the hyperreals must take another approach. Their approach is to use variables over the infinitesimals and huge numbers (which they call *infinite numbers*).

In H. J. Keisler's calculus text based on Robinson's treatment (Keisler, 1976a, 1976b), Keisler uses ε and δ as symbols for *variables* over the infinitesimals in the hyperreals, and H and K as *variables* over the huge, or "infinite," numbers. (As variables, they are symbolized in italics.) Since nonstandard analysis does not recognize the existence of the granular numbers, it cannot name individual infinitesimals and huge numbers at the granular level. Moreover, it cannot have statements in granular arithmetic like $H \cdot \partial = 1$, which implies that $H^2 \cdot \partial = H$ and $H \cdot \partial^2 = \partial$. For Keisler, the product $K \cdot \varepsilon$ of a variable K over an infinite number and a variable ε over the infinitesimals is "indeterminate." The reason this has to be true for Keisler, as for Robinson, is clear. The K is a variable and so, in granular terms, could be varying over any huge number at all (e.g., H or H^2), whereas the ε could be varying over any infinitesimal at all (e.g., granular ∂ or ∂^2). Thus, Keisler's "$K \cdot \varepsilon$" could be varying over the granular numbers H $\cdot \partial$, $H^2 \cdot \partial$, or H $\cdot \partial^2$. The first is 1, the second is H, and the third is ∂ in the granulars. Keisler, who cannot "see" the granulars, can only conclude that the result in the hyperreals is indeterminate—not clearly finite, infinite, or infinitesimal.

We can now begin to see the conceptual disadvantages of using the hyperreals without the granulars. Imprecision results. The numbers cannot be named. The only way to even discuss the product of an infinite and an infinitesimal number in hyperreals is with variables, but the result can only be called indeterminate.

There are further problems with the way Robinson happened to develop his conception of the hyperreals. He made a three-way category distinction between "infinitesimal," "finite," and "infinite" numbers. He also made a distinction between the "standard part," analogous to what we called the real-approximation operator, and the "nonstandard part" of the hyperreal numbers. The "infinitesimals" have an absolute value smaller than any positive real number. This makes zero an "infinitesimal" for Robinson and Keisler. In their development of calculus, they must distinguish between zero and nonzero infinitesimals so that there is never division by zero. In the granular numbers, we keep zero separate from the infinitesimal granulars, for reasons that will become clear in the next chapter.

Here's how Robinson and Keisler characterize differentiation using variables ranging over nonzero infinitesimals. The derivative is defined as follows, using "st" for the standard-part operator:

$$f'(x) = \text{st}\left(\frac{f(x + \Delta x) - f(x)}{\Delta x}\right)$$

Here, the Δx's are variables over nonzero infinitesimals. From here on, the differential calculus is just like it is for the granular numbers. And as with the granular numbers, there is a correlate for every proof in standard differential calculus. Keisler defines the definite integral—the area under a curve between two points—using variables over infinitesimals and infinite numbers. Given the interval on the x-axis $[b, a]$ with length $b - a$, that length is divided into H subintervals, each of length δ, but where H is some unspecified "hyperinteger" and δ is some unspecified infinitesimal. In Keisler's text, the result is a set of H rectangles, each of width δ and height $f(x)$. Then the definite integral is defined as the standard part of the sum of the areas of that rectangle. In short, the strategy is again to use variables over hyperreals rather than specific hyperreal numbers. The results, again, are exactly the same as in granular calculus and standard calculus.

A Comparison

Calculus can be done equally well either the standard way with limits, or using granulars with a real-approximation operator, or using hyperreals with a standard-part operator. From our cognitive perspective, the granulars have certain advantages over the other two. Compared with real numbers and limits, the granulars allow us additionally to see that infinite sums are not equal to their limits, unless the real-approximation operator applies. This allows us to better understand the *nature* of infinite sums. The granulars also allow us to precisely calculate infinite sums with the same limit and compare their size. Part of this job can be done with the hyperreals. In the hyperreals, you could show that there is a difference between the limit and the sum to H terms, where H is not a fixed number but a variable over hyperreal integers. But you cannot precisely calculate the exact infinite sum as a specific number, nor can you calculate the difference between the infinite sums as a specific number.

Compared with the hyperreals, the granulars allow us to use a numeral system and to do calculus in terms of the arithmetic of particular granular numbers. Using the granulars, every expression in calculus is an arithmetic expression with particular numbers, rather than an algebraic expression with variables. Moreover, the granulars have a conceptual advantage: The precise structure of the number system you are using is well defined and revealed through the notation.

The Mathematical Importance of Ignoring Differences

The study of infinitesimals teaches us something extremely deep and important about mathematics—namely, that *ignoring certain differences is absolutely vital to mathematics!* This idea goes against the view of mathematics as the supreme exact science, that science where precision is absolute and differences, no matter how small, should never be ignored.

Whether we are using real numbers and limits, the granular numbers with the real-approximation operator, or the hyperreals with the standard-part operator, *calculus is defined by ignoring infinitely small differences*. This has not always been an uncontroversial view. In the eighteenth century, the Irish philosopher and bishop George Berkeley, for example, rejected calculus as a form of mathematics on the grounds that mathematics required *exact* precision and calculus ignored infinitesimals.

Berkeley's argument is revealing. As Davis and Hersh (1981, p. 244) observe, discussing Berkeley,

> After all, dt is either equal to zero, or not equal to zero. If dt is not zero, then $32 + 16dt$ is not the same as 32. If dt is zero, then the increment in distance ds is also zero, and the fraction ds/dt is not $32 + 16dt$ but a meaningless expression $0/0$.

Berkeley's own comment (in *The Analyst*, 1734) was as follows:

> For when it is said, let the increments vanish, i.e., let the increments be nothing, or let there be no increments, the former supposition that the increments were something, or that there were increments, is destroyed, and yet a consequence of that supposition, i.e., an expression got by virtue thereof, is retained. Which is a false way of reasoning.

Speaking of Newton's "fluxions," or derivatives, Berkeley wrote,

> What are these fluxions? The velocities of evanescent increments. And what are these same evanescent increments? They are neither finite quantities, nor quantities infinitely small, nor yet nothing. May we not call them the ghosts of departed quantities?

Bishop Berkeley was being a literalist, as usual. As we have seen, there is always a metaphor involved in differentiation—which we have called the BMI—whether in the form of limits or of infinitesimal numbers. Literally—that is, without accepting any use of the BMI—Berkeley was right. Differentiation *does* require ignoring certain differences—infinitely small differences. Using limits to define infinite sums also involves ignoring infinitely small differences. Thus, accepting 0.9999. . . = 1 also requires ignoring infinitely small differences. If you think the very definition of mathematics requires absolute precision—never ignoring differences, no matter how small—then you, too, with Berkeley, should reject calculus as a form of mathematics.

But calculus is central to mathematics, as is the study of infinite series, as are dozens of uses of the BMI. Let us think for a moment about what would be lost to mathematics if infinitesimal differences were *not* ignored, at least in certain well-defined situations.

We can see this most clearly with the granular numbers. Suppose the derivative—the rate of change of a function—were still defined by the metaphor that Instantaneous Change Is Average Change Over an Infinitely Small Interval. But now suppose that instead of taking the real approximation of the resulting expression, we leave in the ∂'s. Let us call this the cumulative derivative "$f!$", in contrast to the normal derivative f'.

NORMAL DERIVATIVE f'	CUMULATIVE DERIVATIVE $f!$
$f'(x) = R(\frac{f(x + \partial) - f(x)}{\partial})$	$f!(x) = \frac{f(x + \partial) - f(x)}{\partial}$
$f'(x^2) = 2x$	$f!(x^2) = 2x + \partial$
$f'(x^3) = 3x^2$	$f!(x^3) = 3x^2 + 3x\partial + \partial^2$
$f'(\sin(x)) = \cos(x)$	$f!(\sin(x)) = \frac{\sin(x + \partial) - \sin(x)}{\partial}$
$f'(e^x) = e^x$	$f!(e^x) = \frac{e^{(x + \partial)} - e^x}{\partial}$

First, with cumulative derivatives of polynomials, the ∂ expressions accumulate. Second, there is no function that is its own cumulative derivative, since the ∂ terms keep accumulating. For example, e^x would no longer be a function that is its own derivative. Indeed, there would be no such function! Third, consider periodic functions like sine and cosine. Whereas the derivative of $\sin(x) = \cos(x)$, the *cumulative* derivative of $\sin(x) \neq \cos(x)$, but equals a much longer expression. Moreover, while the second derivative of $\sin(x) = -\sin(x)$, the second *cumulative* derivative of $\sin(x) \neq -\sin(x)$, but equals another very long expression. And the fourth *cumulative* derivative of $\sin(x)$ is not $\sin(x)$ itself but an extremely long expression. Thus, the values of sine and cosine do not repeat after four derivatives but become increasingly complex.

If there were only cumulative derivatives, a great many of the beauties of classical mathematics would cease to exist, as would the usefulness of classical calculus in actual computation. In short, ignoring infinitesimal differences of the right kind in the right place is part of what makes mathematics what it is.

The Closure Engine

Calculus is ubiquitous in advanced mathematics curricula. But calculus with infinitesimals is barely taught anywhere, despite the fact that calculus was first formulated with infinitesimals and infinitesimals were used for doing calculus for two hundred years. Imaginary and complex numbers are a commonplace in high school mathematics classes. Infinitesimal numbers are not. Hardly anyone has ever heard of them. Why?

The main impetus for extending our number system over centuries has come from the demands of closure (see Chapter 4). Zero and negative numbers were needed for closure under addition, rational numbers for closure under multiplication, the real numbers for closure under self-multiplication, and the complex numbers for closure with the square roots of negative numbers. In each case, closure was needed to solve basic polynomial equations: $x + 1 = 1$; $x + 5 = 3$; $7x = 2$; $x^2 = 2$: $x^2 + 1 = 0$.

The infinitesimal numbers are not needed to solve basic polynomial equations. They are not necessitated by the drive for closure. Indeed, they are not needed for any practical mathematical purpose such as solving equations and doing calculations. Moreover, they have a certain stigma. Given the construction of natural numbers by the successor operation—the successive addition of 1 starting with 1—you cannot reach the infinite integers, which are inverses of certain infinitesimals. This is counterintuitive for most people. How can there be an "integer" that you can't count up to? If you don't admit the existence of the infinitesimals, you don't have to face this problem.

What about the granulars? We invented the granulars by applying the BMI as shown. In our search through the literature, we could not find any mention of a number system generated by a "first infinitesimal" and fully symbolizable by the addition of one numeral, in which all of calculus could be done as arithmetic. Why not? Certainly the remarkable mathematicians who have worked on the hyperreals could have developed such a number system easily. Why didn't they?

We believe that the reason has to do with the formal Foundations of mathematics movement and its values (see Chapter 16). Technically, the hyperreals arise within the confines of formal foundations because of an important technical property of model theory: namely, the existence of nonstandard models—models that happen to satisfy a set of axioms even though you didn't intend them to. Henkin found in the late 1940s that there was a model containing *all* the hyperreals, which unexpectedly satisfied the first nine axioms for the real numbers.

But suppose you don't want a model containing *all* the hyperreals. Suppose you want a model containing the granulars and you want to use the techniques of model theory to get it. Your model would have to contain a particular entity—the "first infinitesimal," which we have called ∂. The way that mathematicians generate such particular entities is via additional axioms. But what was attractive about the hyperreals was that you could get *all* infinitesimals without adding any extra axioms. Why add extra axioms to get some infinitesimals when you can get them all for free?

Nonstandard models are, of course, things that classical mathematicians tend not to want to talk about, and calculus has been one of the most classical of the classical subject matters of mathematics.

These are some of the reasons that infinitesimals are generally not taught. We will discuss others in the next chapter.

Two Types of Infinite Numbers

Cantor's transfinite cardinals and ordinals are infinite numbers. So are the huge numbers in the granulars and the infinite numbers in the hyperreals. Do they have anything to do with each other? They appear to be two different types of infinite numbers. What does it mean for there to be two different types of infinite numbers?

The answer is quite interesting: The different types of infinite numbers differ in their conceptual structures. The transfinite numbers have a conceptual structure that is imparted by

- the BMI applied to the formation of sets of natural numbers to form the set of all natural numbers,
- Cantor's metaphor, and
- the results that power sets have "more" members (in Cantor's sense) than the sets they are formed from.

The huge granulars and the infinite hyperreals have structure imparted to them by first using the BMI to form infinitesimal numbers and then taking the inverses of the infinitesimals—which are guaranteed to exist, since the infinitesimals must satisfy the first nine axioms for the real numbers.

Conceptually, these are two utterly different structures, leading to two utterly different notions of "infinite numbers." It is not surprising that Cantor did not believe in infinitesimals. After all, the infinitesimals provided a different notion of infinite number than his transfinite numbers—one that characterized infinity and degrees of infinity in a completely different fashion.

How can there be two different conceptions of "infinite number," both valid in mathematics? By the use of different conceptual metaphors, of course—in this case, different versions of the BMI.

Some Contributions of Mathematical Idea Analysis to the Discussion of Infinitesimals

As we observed, Abraham Robinson characterized infinitesimals in formal mathematical terms. What is different in the mathematical idea analysis of infinitesimals?

First, it makes the whole idea much less mysterious. Mathematical idea analysis shows that the concept is grounded in our everyday experience with specks. And via the use of the BMI, it shows how the infinitely small is linked to the infinitely large. Moreover, it shows explicitly how huge numbers differ from transfinite numbers. It provides a characterization of the granular numbers with an explicit notation that turns calculus into simple arithmetic. Finally, the analysis shows how ordinary human ideas give rise naturally to such an apparently abstruse notion.

Part IV

Banning Space and Motion: The Discretization Program That Shaped Modern Mathematics

12

Points and the Continuum

YOU MIGHT THINK THAT POINTS, lines, and space are simple concepts. They aren't. There are no deeper concepts in mathematics. The complexity of these concepts are reflected in three results that we will discuss in the course of this chapter.

1. "The real-line" is not a line.
2. "Space-filling curves" do not fill space.
3. "The Continuum hypothesis" is not about the continuum.

These results are not in any way paradoxical or contradictory. The expressions in quotes all refer to concepts that are characterized using conceptual metaphors from within mathematics. Those concepts are at odds with our ordinary concepts of curves, lines, and space as expressed by the language not in quotes. There is nothing surprising about this; any technical discipline develops metaphors that are not in the everyday conceptual system. In order to teach mathematics, one must teach the difference between everyday concepts and technical concepts, making clear the metaphorical nature of the technical concepts.

This chapter is about the role of the Basic Metaphor of Infinity in the modern conceptualization of space, lines, and points in mathematics. The BMI, as we shall see, is the crucial link between discrete mathematics (set theory, logic, arithmetic, algebra, and so on) and "continuous" mathematics (geometry, topology, analysis, and so on). This relationship is profound and anything but obvious. Indeed, it is the source of a great deal of confusion not only among students and the lay mathematical public, but in the philosophy and epistemology of mathematics itself. We will try, from the perspective of cognitive science, to sort out those conceptual confusions as we look into how the BMI has permitted a reconceptualization of the continuous in terms of the discrete.

Two Conceptions of Space

Space has been conceptualized in two very different ways in the history of mathematics. Prior to the mid-nineteenth century, space was conceptualized as most people normally think of it—namely, as naturally continuous. Here is how we all think about space in everyday life.

NATURALLY CONTINUOUS SPACE

Space is absolutely continuous. Space does not consist of objects. Rather, it is the background setting that objects are located *in*. Space exists independently of, and prior to, any objects located in space. Planes, too, are absolutely continuous. They, too, are not made up of objects but have locations on them where objects can be situated. Similarly, a line or curve is absolutely continuous, like the path traced by a moving point. Lines and planes also exist independently of, and prior to, any objects located on them.

Points are locations in space, on lines, or on planes. They are not objects that can exist independently of the line, plane, or space where they are located. Dimensionality is a property of a space, a plane, or a line.

Why Naturally Continuous Space Became a "Problem" for Mathematics

Descartes's invention of analytic geometry changed mathematics forever. His central metaphor, Numbers Are Points on a Line (see Case Study 1), allowed one to conceptualize arithmetic and algebra in geometric terms and to visualize functions and algebraic equations in spatial terms. The conceptual blend of the source and target domains of this metaphor lets us move back and forth conceptually between numbers and points on a line, and between n-tuples of numbers and points in n-dimensional spaces. Just as it lets us visualize functions as curves, it lets us conceptualize higher-dimensional spaces in terms of n-tuples of numbers. Ultimately, it led to a precise calculus of change—especially of movement and acceleration—which began with a geometric visualization of change in spatial terms (tangents of curves) and resulted in calculus, the arithmetization of the idea of change.

Two mental constructs were crucial in this development: the association of discrete numbers with discrete points, and the association of discrete symbols with discrete numbers. Arithmetization and symbolization allowed for precise calculation and the constructions of algorithms for differentiation and integration. The study of change became arithmetizable and therefore formalizable. Calculus became as precise and calculable as arithmetic.

What made this possible was Descartes's Numbers-As-Points metaphor, which matched discrete numbers (having discrete symbols) to discrete points on lines and in space. Thus began what we will call the program of *discretization*. Analytic geometry and calculus began the process of reconceptualizing naturally continuous space and naturally continuous change in terms of the discrete: points, numbers, symbols, and algorithmic rules for calculation.

As we shall see, much of modern mathematics has been defined as the progressive discretization of mathematics.

THE DISCRETIZATION PROGRAM SINCE
THE LATE NINETEENTH CENTURY

1. The arithmetization program, which sought to fully arithmetize calculus to eliminate any notion of naturally continuous space.
2. The formalization program, which sought to reconceptualize mathematics as the manipulation of discrete symbols.
3. Symbolic logic, which sought to discretize reason itself, using discrete symbols and precisely formulated algorithms that employed only discrete symbols.
4. Logicism, which sought to discretize all of mathematics through the claim that mathematics could be "reduced" to symbolic logic and the theory of sets, where sets and members of sets are both discrete.
5. Point-set topology, which sought to reconceptualize all understanding of naturally continuous space in terms of discrete points, sets of points, and discrete, symbolizable operations on sets of points.

What propelled the discretization program was the success of analytic geometry and calculus and the idea that anything not formalizable was "vague," "intuitive" (as opposed to "rigorous"), and imprecise. In late-nineteenth-century Europe, mathematics had gained an important stature: the discipline that defined the highest form of reason, with precise, rigorous, and indisputable methods of proof. The discretization program was seen as crucial to preserving that stature for mathematics.

Even so-called intuitionists and constructivists were part of the discretization program. Indeed, they were even more radical discretizers, since they insisted on "constructive" methods alone: algorithmic proofs that were "direct" (disallowing reductio ad absurdum proofs, which use a negative of a negative to "prove" a positive) and finite algorithms (rather than those using what we have called the BMI). This was a further discretizing, in the direction of an even more constrained formalization and an even narrower concept of "rigor."

The discretization program not only made sense in social and historical terms, but it made important mathematical sense as well. It could be seen as asking: What are the limits of discretization? How far can one go in discretizing naturally continuous space? How many naturally continuous concepts (like change or likelihood) can be reasonably discretized and brought into the realm of discrete algorithmic processes? After Gödel, this became the question "What is computable by discrete algorithmic processes?"—a vitally important question in the age of digital computers.

Conceptualizing the Naturally Continuous
in Terms of the Discrete

From the perspective of cognitive science, the discretization program is fraught with difficulty from the outset. The continuous and the discrete are conceptual opposites. To conceptualize the continuous—naturally continuous space, motion, and change—in terms of its opposite, the discrete, is at the very least a formidable metaphorical enterprise. It is at worst conceptually impossible, and at best extremely challenging. What we, as cognitive scientists, have been particularly impressed by is how far the mathematical community has come with this program. As we shall see, the discretization program has not been completed and may not be completable. But the progress it has made is stunning. Through the remarkably creative use of conceptual metaphors, a huge amount of new discretized mathematics has been created. Our task in this chapter and the next two will be to give some detailed initial understanding of what the cognitive processes implicit in the program have been, what the triumphs are, and what has been left out or left undone, possibly because it is undoable.

Carrying Out the Discretization Program

There is a central metaphor at the heart of the discretization program:

A SPACE IS A SET OF POINTS

Source Domain A SET WITH ELEMENTS		Target Domain NATURALLY CONTINUOUS SPACE WITH POINT-LOCATIONS
A set	→	An n-dimensional space— for example, a line, a plane, a 3-dimensional space
Elements are members of the set.	→	Points are locations in the space.
Members exist independently of the sets they are members of.	→	Point-locations are inherent to the space they are located in.
Two set members are distinct if they are different entities.	→	Two point-locations are distinct if they are different locations.
Relations among members of the set	→	Properties of space

This is a radical departure from the commonplace concepts of lines, planes, and space. Here is a description of it.

THE SET-OF-POINTS CONCEPTION OF SPACE

A space is just a set of elements with certain relations holding among the elements. There is nothing inherently spatial about a "space." What are called "points" are just elements of the set of any sort. They are discrete entities, distinct from one another.

Like any members of sets, the points exist independently of any sets they are in. Spaces, planes, and lines—*being* sets—do not exist independently of the points that constitute them.

A line is a set of points with certain relations holding among the points. A plane is a set of points with other relations holding among the points. A geometrical figure, like a circle or a triangle, is a subset of the points in a space, with certain relations among the points. There is thus nothing inherently spatial about a circle or triangle. For example, a circle is just a subset of the elements of the space with certain relations to one another.

What we ordinarily understand as the properties of spaces and figures are characterized, according to this metaphor, as relations among elements in the set, or as functions that assign numbers to elements or *n*-tuples of elements. For example, the dimension of a space is a number assigned by a "dimension function" to a set of elements called "points." A distance between "points" is a number assigned by a function to pairs of set elements. The definitions of "dimension" and "length" are given by formal statements characterizing these functions.

What is the curvature of a line or curve at a point? If the curve is a set of points, then the degree of "curvature" at each point is a number assigned by a function to each point in the set. For example, each point in a circle has the same curvature assigned to it—namely, the number equal to $1/r$, where r is the radius. A "straight" line is a set of points meeting the axioms for a line, where the curvature function assigns the number 0 to each point.

Since a "point in space" is metaphorically conceptualized as just an element of a set, any kind of mathematical entity that can be a member of a set can be seen as a "point," provided that an abstract "distance function" can be defined over the set. For example, there are "function spaces" in which a set of functions is seen as a set of "points" constituting a "space" with a distance "metric"— a function assigning a nonnegative real number to each pair of "points" (i.e., functions).

Given the metaphor A Space Is a Set of Points, "points" are not necessarily spatial in nature but can be any kind of mathematical entities at all. Spaces, lines, and planes are therefore not inherently spatial in nature. They are sets of elements that meet certain axioms and to which certain functions apply, like those assigning numbers indicating metaphorical "dimensionality" and "curvature," which are also formal mathematical notions and not conceptually spatial notions.

To see what this means, consider what a circle is, given these metaphors. A "circle" is a set of "points" in a "plane" that are all at the same distance from a single point *C* called the "center. " That is, for every point *P* in the set of points constituting a "circle," there is a "distance" function that maps the ordered pairs of points (C, P) onto a single real number, called the "radius." The "points" constituting the "circle," the "center," and the "plane" need not be spatial at all; they can be any entities whatever, provided they bear the appropriate relations.

Comparing the Two Conceptions of Space

The Space-As-Set-of-Points metaphor, which is ubiquitous in contemporary mathematics (but which did not exist two centuries ago) provides a conception of points, lines, planes, and space that is quite different from our ordinary one and anything but obvious to unsophisticated students of mathematics. Indeed, the two conceptions are inconsistent. In one, spaces, lines, and planes *exist independently of* points, while in the other they *are constituted by* points. In one, properties are inherent; in the other they are assigned by relations and functions. In one, the entities are inherently spatial in nature; in the other, they are not.

The first—naturally continuous space—is our normal conceptualization, one we cannot avoid. It arises because we have a body and a brain and we function in the everyday world. It is unconscious and automatic. The second has been consciously constructed to suit certain purposes. It is a reconceptualization of the first, via conceptual metaphor.

Even professional mathematicians think using the naturally continuous concept of space when they are functioning in their everyday lives and communicating with nonmathematicians. It takes special training to think in terms of the Set-of-Points metaphor. Moreover, one must learn which kinds of mathematical problems require which metaphors.

The set-of-points conception is the one taken for granted throughout contemporary mathematics. When you are thinking about mathematics, you are dealing simultaneously with two utterly different conceptualizations of our most basic geometrical concepts. It is crucial to keep them straight. Failing to do so may lead to apparent paradoxes. Becoming a professional mathematician requires learning how to operate in this dual fashion.

Indeed, one of the most paradoxical-sounding concepts in twentieth-century mathematics is a result of just such a confusion: the concept of so-called *space-filling curves*. We shall see shortly that such "curves" do not "fill space" at all. But in order to see exactly why, we will have to look at what a point is and how points are related to lines, planes, and numbers.

How Is a Point Conceptualized?

To understand how modern mathematics differs from classical mathematics, we must understand how points, lines, and spaces are conceptualized in the discretization program, where spaces are conceptualized as *being* sets of points. Let us begin with the question of how points are commonly conceptualized, and later turn to the question of what conceptualization is needed to understand discretized mathematics.

Euclid defined a surface as "that which has length and breadth only," a line as "breadthless length," and a point as "that which has no part." Euclid used the ordinary concept of a *lack*: A surface lacks thickness, a line lacks breadth and thickness, and a point (which is made up of no parts) lacks all of these.

In modern mathematics, the lack of a feature is conceptualized metaphorically as the presence of that feature with value zero. Thus, a lack of length is conceptualized as a *particular length*—zero. From the perspective of the modern discretized mathematics, a "point" is an abstract object. Three functions—length-, width-, and thickness-functions—might each assign the number 0 to the "point." This is one way to render Euclid's idea in terms of the Properties Are Functions metaphor.

How are we to understand a point as an object with length, width, and depth equal to zero? Or, to make it simpler, a point in a plane as an object with length and width equal to zero? A common way to conceptualize points is to start with a blob, or a disc, and then make it smaller and smaller and smaller until it is as small as it can get. That's how most people learn to think about what a point is. In other words, a point is infinitely small. And to conceptualize *infinite* small-ness, one needs to make use of the Basic Metaphor of Infinity (see Chapter 8).

Start with a disc of some unit radius 1. Keep shrinking the size of the disc to half its previous diameter. The result is an indefinitely large set of discs:

Shrink the disc from diameter 1 to diameter $1/2^1$.
Shrink the disc from diameter $1/2^1$ to diameter $1/2^2$.
$$\vdots$$
Shrink the disc from diameter $1/2^{n-1}$ to diameter $1/2^n$.

This is an unending process. It produces an infinite sequence of discs, each smaller than the previous one. Let it be the unending process in the BMI. The BMI supplies a metaphorical completion to the process and a metaphorical final result: a "disc" of diameter "0." Such a "disc" is a point in the plane. Of course, instead of nested discs in the plane, we could have used nested intervals on the line (see Chapter 9).

This is a common way to conceptualize a point, and every such conceptualization uses some such version of the BMI to create an infinitely small entity. But this application of the BMI produces a conceptual problem. Here is the frame semantics for a disc.

THE FRAME FOR A DISC

Roles: Center, Circumference, Interior, Diameter, where
    Center ≠ Circumference ≠ Interior
Parts: Interior, Center, Circumference
Constraints: (a) Center is in Interior. (b) Distance from Cen-
    ter to the Circumference is the same for all points of the
    Circumference.

Conceptually, a disc has the following parts: a center, a circumference, and an interior, with the center in the interior and distinct from the circumference—

and both center and circumference distinct from the interior. For this to be true, the diameter has to be larger than zero. If the diameter is zero, the center, interior, and circumference are no longer distinct. They collapse into the same entity, violating the frame semantics for what a disc is. Literally, a disc cannot be a disc and have diameter zero.

But one of the interesting things about human conceptual systems is that we can form metaphorical blends (as discussed in Chapter 2). That is how we typically use the BMI applied to discs. We form a blend of the Disc frame and a Line-Segment frame (see Figure 12.1). At each stage of the BMI, we have a pair of frames (Disc, Line-Segment), where the length in the Line-Segment frame is set identical to the diameter in the Disc frame.

This linking of the two frames forms a conceptual blend. In the BMI, the length in the Line-Segment frame gets shorter and shorter, resulting in the diameter of the disc getting smaller and smaller. At the final resultant state of the BMI, the length in the Line-Segment frame is zero. At this final stage, there are still a pair of frames (Disc, Line-Segment), with the diameter in the Disc frame still set equal to the length in the Line-Segment frame. That is, we are conceptualizing a "disc" with zero diameter, even though, if we consciously put together all that we know, the constraints of the Disc frame are violated. However, as human beings, we do not always put together all that we know; we can focus our attention separately on the Disc frame (a point is a disc) and the Line-Segment frame (the length is zero). What is logically a contradiction may not be recognized by a functioning human being unless attention is focused on both at once.

Notice that the following table has two-headed arrows. This is to differentiate the tables for blends from the tables we have used to characterize conceptual metaphors which have unidirectional arrows.

The Disc/Line-Segment Blend

Element 1 The Disc Frame		*Element 2* The Line-Segment Frame
A disc, with roles: center, circumference, interior, and diameter, where center ≠ circumference ≠ interior	↔	A line segment, with roles: endpoint A, endpoint B, center, interior, and length, where endpoint A ≠ endpoint B ≠ center ≠ interior
Diameter	↔	Length
Center	↔	Center
Opposite points on circumference	↔	Endpoints A and B

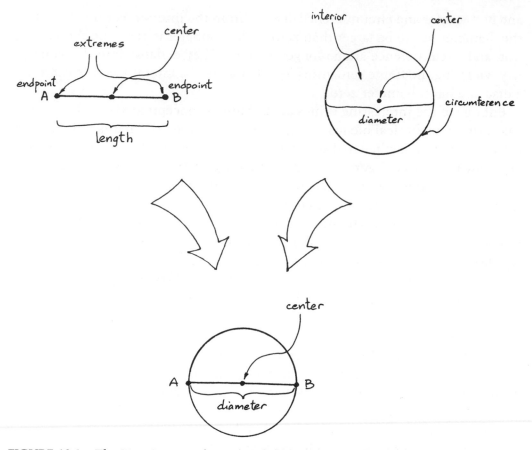

FIGURE 12.1   The Line-Segment frame (top left) and the Disc frame (top right) form a conceptual blend (bottom) in which the line segment is made identical to the diameter of the disc, while the center of the line segment is made identical to the center of the disc. In the blend there are new inferences that weren't available in the frames taken separately.

A point, as we conceptualize it, is a metaphorical entity. It is conceptualized via the Basic Metaphor of Infinity and the Disc/Line-Segment blend. If you try to think of it as a literal entity—one you could encounter in the world—contradictions will arise.

The Infinitesimal Point

Points in textbooks are very often drawn as little discs (see Figure 12.2). This is not a coincidence. It is common to think of a point as a disc made as small as possible.

This is a special case of the Basic Metaphor of Infinity, which applies to produce something infinitely small. But as we have seen, there are two kinds of

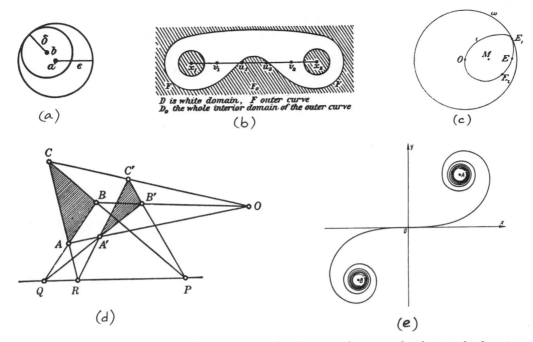

FIGURE 12.2   Points as discs. Here are examples from mathematics books in which points are represented as small discs. Drawings (a) and (b) are taken from a text on the topology of plane sets of points (Newman, 1992); (c), from Hilbert's (1971) famous *Foundations of Geometry*; (d), from Courant and Robbins's (1941) classic *What Is Mathematics?*; and (e), from an Eastern European handbook of mathematics (Bronshtein & Semendyayev, 1985).

things that are "infinitely small": things of size zero and things of an infinitesimal size. To understand the concept of a point better, it is useful to contrast a point of zero diameter with a point that has an infinitesimal diameter. This can be produced by the BMI as follows.

Let us start again with a disc of diameter 1. Again we shrink the disc to half its diameter. But we do one thing different in the process: We keep the diameter greater than zero at all stages in this special case of the BMI. The unending process, then, is:

Shrink the disc from diameter 1 to diameter $1/2$, which is > 0.
Shrink the disc from diameter $1/2^{n-1}$ to diameter $1/2^n$, which is > 0.

The BMI completes the process, producing a disc not with diameter zero but with an infinitesimal diameter (the granular $1/2^H$ as discussed in Chapter 11), which is smaller than any real number! We now have a disc that has an infinitely small diameter but is still a normal disc, with its center distinct from its circumference and interior. If we call this a "point," it will have radically different properties from Euclid's "point."

Child psychologists Jean Piaget and Bärbel Inhelder, in experiments as far back as the 1940s, established the following for children around the age of seven or younger. When you tell a child to imagine a disc (or a dot or circle) and make it smaller and smaller and smaller until it's as small as it can get, the child will conceptualize something that is as small as it can be but is still a bona fide disc, with a center separate from the circumference (Piaget & Inhelder, 1948). Suppose a child is told: Start with one of various figures—a circle, triangle, or square—and make it as small as it can get. What the child gets is a point. Now make the point larger and larger. What does it look like? The answer is the same figure the child started with—the circle, triangle, or square. The figures do not lose their integrity and collapse to the mathematical point. Moreover, when children were asked to shrink an object with volume—say, a ball—to a point, they said that points had volume (see Núñez, 1993c).

We can now see why many children are confused about what a point is. The procedure they are told to use to understand a point leads to a contradiction. They avoid the contradiction and come up with something like the infinitesimal point. This is not the Euclidean point at all.

Does this matter? Is the infinitesimal point all that different from Euclid's point? As long as both are infinitely small, what difference can it make?

All the difference in the world.

If points conceptualized as infinitesimal discs touch, then they must touch *at a point*. What is that point? Well, you could say that that point is an infinitesimal disc at the next smallest level of infinitesimals—what we called $^2\partial$ in the previous chapter. But then the same problem arises again at that level of infinitesimals. It's turtles all the way down. There is no final characterization of a point, if it is an infinitesimal point.

It is very hard not to think of a point in one of these two ways—either as a zero-diameter disc or as an infinitesimal disc. There are two immediate problems. Either there is a contradiction between the very concept of a disc and the length of its diameter being zero, or there is no way to characterize a point without infinite regress.

Yet people do think in these ways, as we shall see immediately.

Do the Points on a Line Touch?

We have received two kinds of answers to this question. The most common answer, sometimes from professionals who have studied college mathematics, is something like, "Yes. Of course they touch. If they didn't touch, the line wouldn't be continuous. There would be gaps between the points."

The other answer, usually given by mathematicians, is, "Of course not. If two points on a line touched, there would be no distance between them and so they would be the same point."

Those in the first group seem to be thinking of points as being like either zero-diameter discs or infinitesimal discs—infinitesimal discs visualized as little dots or beads on a string. If the line is to be continuous—if it is not to have gaps—then the points have to touch, don't they? The image is of a line made up of discs of zero or infinitesimal diameter, with the circumference of one disc bumping up against the circumference of the next. Such an image of points is inconsistent with a conception of points of zero diameter, which are indeed the same point if they "touch."

But the image of points of zero diameter raises a problem: If two such points touch, they are the same point, but if they don't, there must be a gap between them. And if there is a gap, isn't the line discontinuous—not just in a place or two, but everywhere?!

What Is "Touching?"

Suppose a friend touches your shoulder with his hand. Then there is some location on your shoulder where the distance between your shoulder and his hand is zero. Does this mean that your shoulder and his hand share a common part? No, of course not. Your shoulder and his hand are distinct. There is just no distance between them. Using the metaphor for arithmetizing the concept of "no distance," the distance between shoulder and hand is zero.

Is this what is meant in discretized mathematics by two geometric entities "touching"—that the distance between them is zero? The answer is no.

Imagine two circles touching "at one point." If one adopts the Spaces Are Sets of Points metaphor, then each circle is a set of points. Let us call it a circle-set. When two circles "touch," they are sharing a point in common. That is, there is a single object—a point P—that is a distinct, existent entity (independent of any sets it is a member of) and a member of both circle-sets. That point plays a role in constituting both circle-sets.

In other words, when two circles "touch," it is not the case that the two circles are distinct sets of points, with zero distance between two points—Point A in the first circle and Point B in the second. Points A and B are the *same point*—the same object in both circle-sets.

This brings us to another way in which discretized space differs from our ordinary conception of space. We normally conceptualize geometric figures as if they were objects *in* space; for example, it is common to think of a circle in the plane as if it were something like a circle painted on the floor. The floor exists independently of the circle; you can erase the circle and the floor is still there—with not one point of the floor changed. The circle is painted on top of the floor and is distinct from the floor.

But this is not the case in discretized geometry. There, geometric figures are sets of points *that are part of the space itself!* The circle "in" the plane consists of a subset of the same points that make up the plane. Discretized space is not something to be "filled"; it is constituted by point-objects. Figures are not "in" space; they are part of space. For this reason, the only way in discretized mathematics to model our ordinary idea of figures "touching" is for a point on one figure and a point on the other figure to *be the same point*. The same is true for points on a line. The only way two points can "touch" is if they are the same point.

This is not like our ordinary notion of touching at all. It would be like saying that two lovers' lips can touch only if they shared common skin. The image is a bit creepy. But that's the metaphor that mathematicians are forced to use in order to carry out the program of discretizing space.

When we think or speak of figures as "touching," we are using our ordinary everyday, naturally continuous notion of space, where geometric figures are metaphorically thought of as objects "in" space and points are conceptualized in their normal way as locations. This is what we are doing, for example, when we think of a tangent line as "touching a curve *at* one point." In thinking this way, we are using the commonplace everyday conceptual metaphor that Geometric Figures Are Objects in Space; we are using our normal metaphor system, which was not developed for discretized mathematics. To think in terms of discretized space is to think using metaphors developed for technical reasons in the discretization program.

This is why mathematicians tend to wince when asked if two points on a line can touch and then say, no. They wince because, in the discretized mathematics in which they are trained, "points" are not the kinds of things that can either touch or not touch. Since they take the Set-of-Points metaphor as the correct way of conceptualizing space, they see the question as an improper question to ask. If points are just elements of sets that can be taken as anything at all—if they are not inherently spatial and not seen technically as objects in space—then the idea of their touching or not touching makes no sense.

As noted, it is common for people who first approach the study of mathematics to conceptualize points as infinitesimally small discs, either zero-diameter discs or infinitesimal discs. Indeed, this is how most students are taught what a point is, and how diagrams in textbooks represent points—namely, as small disclike dots. Yet this conception of a point is radically at odds with the discretization program. From the perspective of that program, points are very different entities—entities that cannot literally touch.

For this reason, many mathematics educators consider the understanding of points as infinitely small discs—either zero or infinitesimal—as a "misconception." Yet this "misconception" is a perfectly natural concept—indeed, an unavoidable one. It is the only natural way that most people have of understanding what a point is. If you are going to teach what a point is in discretized mathematics, you will have to cope with this aspect of normal human cognition. Your students will inevitably understand a point in this way initially. It is important that you and they both understand that such a conception of a point is natural, but that it does not fit the metaphors defining discretized mathematics— metaphors that defy our ordinary intuitions about space.

How Can We Conceptualize a Point in Discretized Mathematics?

It is not the job of mathematics to characterize how people think, or how people conceptualize (or should conceptualize) mathematical ideas. That is the task of the cognitive science of mathematics as a subject matter. The job of mathematicians taking part in the discretization program is to provide precise symbolic expressions "defining" all the notions used in discretizing space: limit points, accumulation points, neighborhoods, closures, metric spaces, open sets, and so on. Our job as cognitive scientists is to comprehend how these "definitions" in symbols are conceptualized, to give an account in terms of human cognition of the ideas that the symbols are meant to express. The question we turn to now is how a point, given the metaphor Spaces Are Sets of Points, is conceptualized within discretized mathematics.

The key to understanding how space is conceptualized in the discretization program is the BMI. The reason is that the strategy of the discretization program was to replace naturally continuous space with infinite sets of points. The theoretical mechanisms that were used all make implicit use of the BMI.

Let us look at the details.

The Centrality of the BMI in the Discretization of Space

The program for discretizing naturally continuous space had a strategy:

1. Pick out the necessary properties of naturally continuous space that can be modeled in a discretized fashion and model them.
2. Model enough of those necessary properties to do classical mathematics as it was developed using naturally continuous space.
3. Call the discretized models "spaces," and create new discretized mathematics replacing naturally continuous space with such "spaces."

For this strategy to be successful, there have to be some small number of such necessary properties that can be successfully modeled in a discretized fashion. The remarkable thing about the discretization program is that such a small number of discretizable properties of naturally continuous space were found and precisely formulated.

This does *not* mean that such properties *completely* model the everyday concept of naturally continuous space. As far as we can tell, it would be impossible to model a naturally continuous concept completely in terms of its opposite. But complete modeling is not necessary for mathematical purposes. All that is necessary is sufficient modeling to achieve a purpose: in this case, being able to discretize classical mathematics with a new notion of a "space" and then create new discretized mathematics using that notion.

Let us now look at the small number of necessary properties of naturally continuous space that have successfully been discretized and are sufficient for the purposes of classical mathematics.

Consider the following properties of naturally continuous space as we normally conceptualize it:

1. *The metric property:* Our everyday concept of "distance" in naturally continuous space has the following necessary properties:
   - The distance between any two distinct locations is greater than zero.
   - If the distance between two locations is zero, "they" are the same location.
   - The distance from location A to location B is the same as the distance from B to A.
   - The distance from A to B is less than or equal to the distance from A to another location C plus the distance from C to B.
2. *The neighborhood property:* "Nearness" implicitly uses the concept of distance. Though nearness is relative to context, in any context, one can pick a distance that is "near." Call that distance *epsilon*. Everything that is closer than epsilon to a point is "near" that point. A "neighborhood" of a point P is the collection of all points near P, in that sense of "near."

3. *The limit point property:* Near every point, there are an infinity of other points, no matter how close you take "near" to be.
4. *The accumulation point property:* Given any point, you can find at least one other point near it.
5. *The open set property:* In any naturally continuous space, there are bounded regions (conceptualized via the Container schema, as discussed in Chapter 2). The interior of such a bounded region (the container minus the boundary) is an open set containing all the spatial locations in that bounded region.

Within discretized mathematics, each of these properties is made precise, starting with the Spaces Are Sets of Points metaphor. These properties have been picked out by mathematicians for the purpose of constructing a discretized mathematics; as such, they are not necessarily the properties that most of us would ordinarily think of.

Given the metric property, the other properties can all be conceptualized via the BMI. Our strategy in demonstrating this will be to construct a special case of the BMI that can be used to characterize all these concepts at once, showing their relationships and just how they implicitly make use of the BMI. We will do this by defining the *infinite nesting property for sets* in terms of a special case of the BMI, and then showing that these concepts can all be characterized using the infinite nesting property.

The basic idea of the infinite nesting property is this.

- Consider a set S with a distance metric—that is, a relation $d(x,y)$ holding among all the members of the set. The distance metric has the following properties: $d(x,y) = d(y,x)$; $d(x,y) = 0$ if and only if $x = y$; and $d(x,y) + d(y,z) \geq d(x,z)$.
- Consider an element C, which may or may not be in S but to which the distance metric applies.
- Characterize an "epsilon disc" around C as a set of points in S within a distance epsilon of C.
- Using the BMI, construct an infinity of nested epsilon discs around C containing members of S.

If this is possible for S and C, we will say that S has the infinite nesting property with respect to C. The main objectives are:

- First, to show how this property can be seen as a special case of the BMI.
- Then, to show how *limit points* and *accumulation points* can be straightforwardly defined as special cases of this special case of the BMI.

We begin by characterizing the concept of an epsilon disc in terms of a conceptual frame, with the semantic roles "distance function," epsilon, and C, the center of the disc.

THE EPSILON-DISC FRAME

A set S with a distance function, or "metric," d.
An element C, either in S or not, to which the distance metric applies.
Epsilon: a positive number.
An epsilon disc around C: a subset of members, x, of S such that $d(C, x) <$ epsilon.

We then characterize what it means for a set S to have the infinite nesting property with respect to an element C, using a special case of the BMI.

NESTED DISCS

Target Domain ITERATIVE PROCESSES THAT GO ON AND ON		*Special Case* A SET OF ELEMENTS WITH THE INFINITE NESTING PROPERTY
The beginning state (0)	⇒	The ε_0-disc: the epsilon-disc frame, with epsilon $= \varepsilon_0$ D_1: the set with the ε_0-disc as a member
State (1) resulting from the initial stage of the process	⇒	Choose ε_1 arbitrarily smaller than ε_0. Result: the set $D_2 = D_1 \cup \{$the ε_1-disc$\}$
The process: From a prior intermediate state (n–1), produce the next state (n).	⇒	Given an ε_{n-1}, choose ε_n arbitrarily smaller than ε_{n-1}. Form $D_{n+1} = D_n \cup \{$the ε_n-disc$\}$.
The intermediate result after that iteration of the process (the relation between n and n–1)	⇒	The set D_{n+1}, containing all ε_i-discs, for $0 \le \varepsilon_i \le n$
"The final resultant state" (actual infinity "∞")	⇒	**The set D_∞ containing all ε_i-discs, for $0 \le \varepsilon_i \le n$, where n is finite.** **Property 1: Every ε_i-disc in D_∞ contains an infinite number of members of S.** **Property 2: Every ε_i-disc in D_∞ contains both C and a member x of S such that $x \ne C$.**

Let us now look at a set S that has the infinite nesting property with respect to element C, which may or may not be in S.

- Being an ε_i-disc in D_∞ characterizes the notion of being "a neighborhood of C in S." That is, it characterizes the *neighborhood* property.
- Property 1 characterizes the *limit point* property.
- Property 2, when C is in S, characterizes the *accumulation point* property.
- An *open set* is a subset, O, of S such that every member of O is the center C of an ε_i-disc in D_∞.

A neighborhood of an element C is usually characterized simply as an epsilon disc around C. However, the prototypical uses of neighborhoods are in $\mathfrak{R}^n$, the n-dimensional Cartesian space with points as n-tuples of real numbers. As we saw in Chapter 9, the real numbers are defined relative to some use of the BMI—either least upper (greatest lower) bounds, infinite intersections, infinite decimals, infinite polynomials, or some other use. These uses of the BMI to characterize real numbers all impose the infinite nesting property on $\mathfrak{R}^n$, since the infinite nesting property can be seen as a generalization over all those cases: They all involve infinite nesting of some sort. Suppose one takes into account these uses of the BMI in characterizing the reals, as one should. The conceptualization of a neighborhood of C in $\mathfrak{R}^n$ will then implicitly involve a use of the infinite nesting property. Any such neighborhood of C in $\mathfrak{R}^n$ will therefore be an epsilon disc—that is, an ε_i-disc in D_∞, as defined above. Such an epsilon disc will contain an infinity of other epsilon discs around C in $\mathfrak{R}^n$, which is just the property that neighborhoods of a point C have in $\mathfrak{R}^n$. That is why the BMI is implicit in the concept of a neighborhood in $\mathfrak{R}^n$.

The infinite nesting property is a conceptual property intended to show how the BMI is used in the cognitive characterization of all the other properties we have given. It is intended not to be part of discretized classical mathematics as formulated in contemporary mathematics but, rather, to characterize a cognitively plausible account of how such properties can be characterized using the mechanisms of human conceptual systems—for example, frames, conceptual metaphors, conceptual blends, image schemas, and so on.

We will delay until the next chapter an account of how the properties just discussed are used to characterize central notions like continuity in discretized mathematics. For the sake of the remainder of this chapter, there are a number of morals to bear in mind.

The Morals So Far

Here's what we can conclude at this point in our discussion.

- Don't confuse the ordinary, naturally continuous concept of space with the discretized concept characterized by the Spaces Are Sets of Points metaphor.
- Don't confuse the discretized notion of a point with either a spatial location, an infinitesimal disc, or a zero-diameter "disc."
- The discretized notion of a geometric figure (e.g., a circle) contradicts our ordinary notion of a geometric figure in naturally continuous space. The discretized concept of a geometric figure—the one used in contemporary formal mathematics—is therefore *not* a generalization over the concept of a figure used in our ordinary conceptualization. It is a very different kind of conceptual entity, with very different properties.
- Both our ordinary concept of a point and the concept of a point in discretized mathematics make implicit use of the Basic Metaphor of Infinity.
- Conceptualizing a point metaphorically as a disc of zero diameter or an interval of zero length contradicts our everyday conceptual system.
- In thinking about contemporary discretized mathematics, be aware that your ordinary concepts will surface regularly and that they contradict those of discretized mathematics in important ways.
- Trying not to think about your ordinary concepts is like trying not to think of an elephant. You just can't do it when words like "point," "line," and "space" are used. Just be aware that your everyday concepts are there unconsciously and can interfere with your understanding of the metaphorical mathematical concepts.

What Is a Number Line?

Actually, there are two number lines. There is the one you learn in grammar school, where the line is just the ordinary, everyday, naturally continuous concept of a line. This is the line you understand when you examine graphs in the newspaper for temperature readings or stock fluctuations, say. This number line is conceptualized via the metaphor Numbers Are Points on a Line.

NUMBERS ARE POINTS ON A LINE
(FOR NATURALLY CONTINUOUS SPACE)

Source Domain POINTS ON A LINE		Target Domain A COLLECTION OF NUMBERS
A Point P on a line	$\rightarrow$	A Number P'
A point O	$\rightarrow$	Zero
A point I to the right of O	$\rightarrow$	One
Point P is to the right of point Q	$\rightarrow$	Number P' is greater than number Q'
Point Q is to the left of point P	$\rightarrow$	Number Q' is less than number P'
Point P is in the same location as point Q	$\rightarrow$	Number P' equals number Q'
Points to the left of O	$\rightarrow$	Negative numbers
The distance between O and P	$\rightarrow$	The absolute value of number P'

This metaphor constitutes our nontechnical understanding of numbers as points on a line. The number line we learn in elementary school is a conceptual blend—the Number-Line blend—of the source and target domains of this metaphor, in which the entities are *simultaneously numbers and points* (see Figure 12.3).

But there is a second number line as well, one requiring a technical understanding of mathematics, where space is fully discretized—that is, conceptualized using the Spaces Are Sets of Points metaphor. Those with such a technical understanding have a conceptual blend of the source and target domains of that metaphor, which we will call the Space-Set blend.

THE SPACE-SET BLEND

Target Domain NATURALLY CONTINUOUS SPACE WITH POINT-LOCATIONS SPECIAL CASE: THE LINE		Source Domain A SET WITH ELEMENTS
The line	$\leftrightarrow$	A set
Points are locations on the line.	$\leftrightarrow$	Elements are members of the set.
Point-locations are inherent to the line they are located on.	$\leftrightarrow$	Members exist independently of the sets they are members of.

| Two point-locations are distinct if they are different locations. | ↔ | Two set members are distinct if they are different entities. |
| Properties of the line | ↔ | Relations among members of the set |

In the Space-Set blend, space is conceptualized as a set of elements in which necessary properties of natural continuous space are conceptualized as formal relations on elements of the set.

Those who conceptualize space in this way have a more technical version of the metaphor that Numbers Are Points on a Line. In their metaphor, the source domain is the Space-Set blend, where space is discretized. That more technical

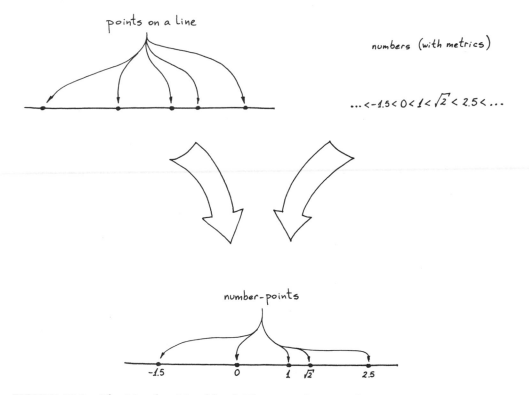

FIGURE 12.3   The Number-Line blend. The metaphor Numbers Are Points on a Line sets up a correspondence between points on a line and numbers. When the source and target domains of this metaphor are both active, the result is a conceptual blend in which the corresponding elements of the metaphor become single entities—number-points. The figure illustrates the source domain of points on a line at the top left and the target domain of numbers (ordered and with a metric) at the top right. The resulting blend is shown at the bottom.

metaphor can be stated as follows, taking the real numbers as special cases of numbers and the line as a special case of a space.

NUMBERS ARE POINTS ON A LINE (FULLY DISCRETIZED VERSION)

Source Domain			Target Domain
THE SPACE-SET BLEND			
NATURALLY CONTINUOUS SPACE: THE LINE		SETS	NUMBERS
The line	↔	A set	→ A set of numbers
Point-locations	↔	Elements of the set	→ Numbers
Points are locations on the line.	↔	Elements are members of the set.	→ Individual numbers are members of the set of numbers.
Point-locations are inherent to the line they are located on.	↔	Members exist independently of the sets they are in.	→ Numbers exist independently of the sets they are in.
Two point-locations are distinct if they are different locations.	↔	Two set members are distinct if they are different entities.	→ Two numbers are distinct if there is a nonzero difference between them.
Properties of the line	↔	Relations among members of the set	→ Relations among numbers
A point O	↔	An element "0"	→ Zero
A point I to the right of O	↔	An element "1"	→ One
Point P is to the right of point Q.	↔	A relation "P > Q"	→ Number P is greater than number Q.
Points to the left of O	↔	The subset of elements x, with 0 > x	→ Negative numbers
The distance between O and P	↔	A function d that maps (0,P) onto an element x, with x > 0	→ The absolute value of number P

This is a complex metaphor for conceptualizing numbers in terms of discretized space. The discretized number line is a conceptual blend of the source and target domains of this metaphor. Since the source domain is a metaphorical blend itself, the complete blend is a blend of *three domains*. One can see why this is not so easy to teach or to learn.

It is crucial for what is to follow that the reader understand the vast difference between these two metaphorical conceptions of the number line. In the discretized version based on the Space-Set blend, the source domain is fully discretized. The "points" in "space" are discrete elements in a set. In the naturally continuous version, the source domain contains point-locations in a naturally continuous space—a continuous medium. Change in the discretized version can be represented only as a sequence of discrete set elements, while change in the first is a naturally continuous, holistic path with discrete point-locations on it.

Why the "Real Line" Is Not a Line

When mathematicians speak technically of the "real line," they do not mean our ordinary, naturally continuous line with real numbers sprinkled along it; that is the nontechnical "real line." The technical "real line" is fully discretized, via the Space-Set blend. The "real line" is a set of discrete elements (of an unspecified nature) constrained by various relations among the elements. There is no naturally continuous line in the technical "real line." From the perspective of our ordinary concept of a line as naturally continuous, the "real line" of discretized mathematics is not a line.

A Crucial Property of Naturally Continuous Number Lines

There is a crucial difference between the two conceptions of the number line. In the Discretized Number-Line blend, there is a one-to-one correlation between the point-elements of the set in the source domain and the numbers in the target domain. In this mapping, everything in the source domain is mapped: All the elements of the set are mapped onto all the numbers.

This is not true in the naturally continuous Number-Line blend. There, only certain point-locations on the naturally continuous line are mapped onto numbers. But the whole naturally continuous line—the *medium* in which the points are located—is not mapped. Since the point-locations are zero-dimensional and have no magnitude, no point-location "takes up" any amount of that medium. Point-locations don't take up space. (Recall that in the Space-Set blend, the points *constitute* the space. That is not true here.) This means that no matter what number systems point-locations are mapped onto, the medium of space that is left unmapped will have an unlimited supply of point-locations "left." The point-locations of a naturally continuous number line are never exhausted.

This is not true of the discretized number line, which is a conceptual blend of three conceptual domains—space, sets, and numbers. The set domain plays the central role in the blend. Because sets constitute the source domain of the Spaces Are Sets of Points metaphor, the target domain of space is structured by the ontology of sets. In the Numbers Are Points on a Line metaphor, the "points" are set elements. The kind of number system that the set maps onto depends on the formal relations (e.g., the axioms for the real numbers in Chapter 9) that constrain what the members of the set can be and how they are related to one another.

The real-number system is "complete." Once the reals are characterized, the members of the set are completely determined. This means that the "points" constituting the line are fully determined by the axioms that define the real numbers. For this reason, it is true of the Discretized Real-Number-Line blend that *the real numbers exhaust it*. Of course they do. Since there can be a point in this blend only if a real number maps onto it, the real numbers determine what the points in this blend can be.

But in the naturally continuous real-number line, *the real numbers do not exhaust the points on the line*. The point-locations that map onto the real numbers are not all the possible point-locations on the line. The reason is that this line is a naturally continuous background medium. There is no limit at all to the number of zero-length point-locations that can be assigned to such a background medium. For example, in the case of the naturally continuous hyperreal line, the real numbers are relatively sparse among the hyperreals on that line. On the hyperreal line, a huge number of points surround each real number. The naturally continuous line has no problem whatsoever accommodating all the hyperreal numbers in excess of the real numbers.

Thus, the question of whether or not the real numbers "exhaust the line" is not formulated precisely enough; it depends, unsurprisingly, on what you take the "line" to be.

The Real, Granular, and Hyperreal-Number-Line Blends

Here are two statements one might be tempted to make:

- It is only with the nontechnical, nondiscretized, naturally continuous number line that the real numbers do not exhaust the points on the line. Since the line is naturally continuous, there are always more points of zero length.

- On the line as understood technically by professional mathematicians—the discretized number line, which is a set of points—the real numbers always *do* exhaust the points on the line.

Tempting as such a pair of claims might be, they are not true.

The reason is quite simple. The hyperreal and granular numbers (see Chapter 11) form linearly ordered sets (since they satisfy the first nine axioms for the real numbers). Consider the hyperreals once more. The sets that model the hyperreals are linearly ordered. For this reason, there can be a discretized hyperreal-number line. Just let a set modeling the hyperreals be the set in the metaphor that Spaces Are Sets. Then there will be a discretized hyperreal-number line. On this number line, the real numbers are incredibly sparse. Therefore, the real numbers do not exhaust even a discretized number line—the technical number line used by mathematicians.

Why "Space-Filling Curves" Do Not Fill Space

Just what is a "space-filling curve"?

Imagine a function with the following properties.

- Its domain is the closed interval [0, 1] of the discretized real-number line (i.e., $\Re^1$).
- Its range is the unit square of $\Re^2$, that portion of the Cartesian plane with x- and y-coordinates both within [0, 1].
- The mapping is surjective (onto, but not one-to-one) and is "continuous."

Looking ahead to Chapter 13, we will take "continuity" for discretized functions to mean that the function *preserves closeness*. This can technically be achieved by a condition like the following: The inverse image of the function f^{-1} maps open sets in the range onto open sets in the domain. Since neighborhoods are open sets, what this conditions says is that neighborhoods around each $f(x)$ come from neighborhoods around x.

In other words, the "continuous" discretized line [0, 1] is to be mapped via a "continuous" function onto a "continuous" curve in the square. (Technically, a "curve" can be a straight line.)

Now the question arises: Can such a mapping from the unit line to a curve in the unit square "fill" the entire square? Giuseppe Peano constructed the first such

FIGURE 12.4   Hilbert's "space-filling" curve, as described in the text. Shown here are some of the first few members of the sequence of curves, which the interval [0, 1] on the real line is mapped onto. As the sequence gets longer, the resulting curves are said to "fill" more and more of the real-valued points in the unit square. Since the function is real-valued, the points with hyperreal values remain "unfilled." Hilbert's curve, characterized via the BMI, "fills" only the real-valued points in the unit square. The moral here is that when spaces are understood as sets of points, the term "space-filling" means "mapping onto every member of the relevant set of points." "Space" does not mean naturally continuous space. Therefore, Hilbert's "space-filling" curve doesn't *fill space*.

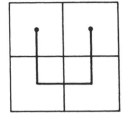

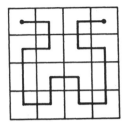

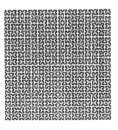

curve in 1890, with others following: David Hilbert in 1891, E. H. Moore in 1900, W. Sierpinsky in 1912, and George Pólya in 1913 (for details, see Sagan, 1994). To give you an idea of what such a "curve" is like, Hilbert's "space-filling" curve is shown in Figure 12.4. Hilbert's curve, like Peano's and all the others, is constructed via an infinite sequence of ordinary curves, as follows. Consider the simple curve in the top box—the first step in the construction of the function. The mapping at this step has the following parts:

1. When x varies from 0 to 1/3 in the interval [0, 1/3], $f(x)$ varies continuously along the left vertical line going from top to bottom.
2. When x varies from 1/3 to 2/3 in the interval (1/3, 2/3], $f(x)$ varies continuously along the horizontal line from left to right.
3. When x varies between 2/3 and 1 in the interval (2/3, 1], $f(x)$ varies continuously along the right vertical line from bottom to top.

At stage 2, the square is divided into 16 smaller squares. The line [0, 1] is divided into 15 subintervals: [0, 1/15], (1/15, 2/15], . . . , (14/15, 1].

1. When x varies from 0 to 1/15 in the interval [0, 1/15], $f(x)$ varies continuously along the upper left horizontal line going from left to right.
2. When x varies from 1/15 to 2/15 in the interval (1/15, 2/15], $f(x)$ varies continuously along the connecting vertical line from top to bottom.

And so on.

At each stage n, there are 4^n subsquares and $4^n - 1$ divisions of the unit line [0, 1] mapped continuously onto $4^n - 1$ connected line segments going from the center of one subsquare to the center of the adjacent subsquare. At each stage, the mapping is "continuous."

As we go from stage to stage, the curve gets longer and longer, with the interval [0, 1] being mapped onto more and more points of the square "filling up" progressively more of the square.

The limit of this sequence of functions is defined in the following way.

- The process defines a sequence of functions $f_1, f_2, \ldots$ from [0, 1] onto the square.
- Notice that a given point p in the domain [0, 1] will be mapped onto different points $f_1(p), f_2(p), \ldots$ at successive stages of the process.
- For each such point p, there will therefore be a sequence of points $f_1(p)$, $f_2(p), \ldots$ in the square.
- Each such sequence will converge to a limit. Let us call that limit $f_\infty(p)$. (Of course, this can be conceptualized via the BMI.)
- Now consider the set of all $f_\infty(x)$, for all x in [0, 1]. It is a function from [0, 1] to the unit square. (This is the set of final resultant states of the BMI.)
- It is "continuous" since every open set in the unit square is mapped by the inverse function $f_\infty^{-1}(x)$ onto an open set in [0, 1].
- Finally, $f_\infty(x)$ "fills" the unit square. That is, for every point in the unit square, there is at least one point in [0, 1] that maps onto it.

This is what Hilbert proved.

Note that as the subsquares get smaller, the boundaries of the subsquares get closer and closer to the center of each subsquare. Consequently, as n gets larger, the points in each subsquare not on the curve get closer and closer to points on the curve. As n approaches infinity, what happens? Look at the distance within each subsquare between points in the subsquare on the curve and points in the subsquare *not* on the curve. As n approaches infinity, that distance approaches zero. At limit stage, where we have the function f_∞, that distance *is* zero. The subsquares have zero area, so that no subsquare can have a point in the plane that is not on the curve. In other words, every point in the square *is* a point on the curve, because the subsquares have become the points on the curve!

At first, this seems bizarre. A curve is one-dimensional. It is infinitely thin. A plane is two-dimensional. How can a one-dimensional curve fill a two-dimensional space?

First, notice that the curve is a metaphorical "curve." Each limit operation for each point in [0, 1] is an instance of the Basic Metaphor of Infinity. The range of the function f_∞ is an infinite set, each member of which is conceptualized via the BMI. In that sense, the function is *infinitely metaphorical*—every one of its infinity of values arises via the Basic Metaphor of Infinity.

Second, notice that the domain and range of the function are metaphorical, since they are not a naturally continuous line segment and a naturally continuous square but, rather, are discretized—produced via the Spaces Are Sets of Points metaphor.

Third, notice that the "continuity" of each f_i and of f_∞ is also metaphorical. It is not the natural continuity of a path of motion. It is the metaphorical "continuity" for discrete "space" characterized in terms of preservation of closeness, as noted. (We will give a detailed analysis of continuity in chapters 13 and 14.)

Fourth, notice that from the perspective of natural continuity, the space of the unit square is not "filled." A naturally continuous square is a holistic background medium. The range of the function is only the set of points in the unit square with real-valued coordinates.

However, the naturally continuous unit square must be able to accommodate not just points with real-valued coordinates but also points with hyperreal-valued coordinates. Thus, the real-valued point-locations do not exhaust the naturally continuous holistic background space of the unit square and therefore the "space-filling curve" does not "fill" the naturally continuous unit square.

Fifth, suppose that we rule out the naturally continuous plane and think only in terms of the discretized plane. Suppose further that we assume that the function maps the real numbers within the hyperreal interval [0, 1] onto the real-valued coordinates within the hyperreal unit square.

In this case, the function f_∞ is exactly the same as it was before. It maps the same real numbers in the interval [0, 1] onto the same points with real-valued coordinates in the unit square. The difference is that now most points in the hyperreal unit square have coordinates that are *not* real-valued but hyperreal-valued. Since the function f_∞ (as we have defined it) does not apply to hyperreal values in the interval [0, 1], it does not "fill" any of the points in the unit square with hyperreal coordinates. Indeed, since the real-valued points in the hyperreal unit square are incredibly sparse, most of the square is not "filled."

In summary, the so-called space-filling curve f_∞ does not "fill space" at all. It only maps discrete real-valued points in the interval [0, 1] onto discrete points with real-valued coordinates in the unit square. That's all.

Moreover, it does not "fill space" under either assumption of what "space" is—naturally continuous space or discretized space.

What Is "The Continuum"?

Before the nineteenth century, the *continuum* referred to the naturally continuous line. The discretization program changed the meaning of the term. In discretized mathematics, "the continuum" became a discretized "line" characterized via the Spaces Are Sets of Points metaphor; that is, it became a set of abstract elements. Given the discretized number line, "the continuum" was seen as a set of numbers.

The question as to whether the real numbers "exhaust" the continuum is the same question as whether the real numbers exhaust the points on a line. It is not a well-put question. It depends on what you mean by "the continuum." The term can mean all of the following:

1) A *naturally continuous line*.
2) A *discretized line*—a "line" conceptualized via the Spaces Are Sets of Points metaphor, consisting of a set of elements that characterize *all* the points on the line.
3) A *Discretized Number-Line blend*, where the "points" are elements of sets, which in turn correspond one-to-one to numbers. The numbers then determine what the metaphorical "points" are. Suppose the number system is "complete," as the real numbers are; that is, they are closed under arithmetic operations and contain the limits of all the infinite sequences. Since the numbers determine what the points are, a complete set of numbers exhausts *all* the points defined by those numbers. And since the real numbers are complete and define what a "point" is in a Discretized Number-Line blend, they, of course, exhaust all the points in that blend. That does not mean that they exhaust all the points on a naturally continuous line.

The claim dating from the late nineteenth century that the real numbers "exhaust the continuum" is usually a claim about case 3. However, many people assume that the claim is about case 1 and somehow true of case 1. This is, of course, a mistake arising from the confusion about what "the continuum" means.

Unfortunately, the mathematical community has done little to disabuse their students and the public of this misunderstanding. As a result, many people have been led to believe (incorrectly) that it has been proved mathematically that the naturally continuous line—taken *to be* the continuum—can somehow be exhausted by the real-number points, which are discrete entities without any length. This seems a wondrous, even mystical idea. Indeed, it is. But it isn't true. And the continued used of the term "continuum" doesn't help clarify the situation, since it suggests the line that most people understand in their ordinary lives—the naturally continuous line.

This misunderstanding can be traced back at least to the German mathematician Richard Dedekind (1831–1916), one of the founders of the discretization program. When he conceptualized a real number as a *cut* on the line, the line he had in mind was naturally continuous (case 1), not just a set of elements. (For cognitive evidence of this claim, see the next chapter.) On the other hand, it was Dedekind who defined real numbers in terms of sets. The image of the "cut" and the definition in terms of sets do not give a consistent picture. It leads to a confusion between case 1 and case 2.

As far as we can tell, that misunderstanding has continued to the present day, even in some segments of the mathematical community (see Weyl, 1918/1994; Kramer, 1970; Longo, 1998).

Is the "Continuum Hypothesis" About the Continuum?

George Cantor, as we saw in Chapter 10, used the term "*C*" for "continuum" to name the cardinality of the real numbers. Most likely, he did this because he took it for granted that the real numbers exhausted the points on the continuum. This meant to him that the real numbers could be put in one-to-one correspondence with the points on the continuum and, by Cantor's metaphor, implied that there were the same number of real numbers as there were points on the continuum—strongly suggesting that even Cantor believed (incorrectly) that the real numbers exhausted the points on the naturally continuous line.

Suppose that you take "the continuum" to mean the naturally continuous line. And suppose you then ask the question, "How many points can fit on the continuum?" The answer, as we have seen, is "As many as you like." If by "the continuum" you mean the one-dimensional, naturally continuous, holistic background space, and if by "point" you mean a location in that background medium with zero length, then you don't have to stop with the real numbers;

you can go into the hyperreals. The number of such potential point-locations is unlimited.

Cantor's Continuum hypothesis was about the cardinality of the set of real numbers—the transfinite *number* of real numbers. He had proved that there were "more" reals than rationals or natural numbers (in the Cantor-size sense). As we saw, he had named the cardinality of the natural and rational numbers "$\aleph_0$" and the cardinality of the reals "C". He called the next largest transfinite number "$\aleph_1$". The Continuum hypothesis said that there was no set of cardinality larger than that of the natural numbers ($\aleph_0$) and smaller than that of the reals (C). That, if true, would mean that the cardinality of the reals was $\aleph_1$. In other words, the Continuum hypothesis can be stated as: $C = \aleph_1$.

Is the Continuum hypothesis about the continuum? It is certainly about the real numbers. It is not about the naturally continuous line. For those people who believe that the term "the continuum" refers to the naturally continuous line, the Continuum hypothesis is not about the continuum.

There is nothing strange or paradoxical or wondrous about this. It is simply a matter of being clear about what you mean.

Closure and the Continuum

The passion for closure pushed the mathematical community to find *all* the linearly ordered numbers that could be solutions to polynomial equations—that is, the system of all the linear numbers closed under the operations of addition and multiplication, including all the limits of sequences of such numbers.

The principle of closure then had another effect. It tended to keep the mathematical community from looking further at ordered number systems that were not required for closure—systems such as the hyperreals. Though infinitesimals were used successfully in mathematics for hundreds of years, the drive for closure of a linear number system led mathematicians to stop with the reals. Closure took precedence over investigating the infinitesimals.

Achieving closure for the real numbers meant "completing" the integers and the rationals. Since numbers were seen as points on the line, that meant "completing" the line—the entire "continuum," finding a "real" number for every possible point. The cardinality of the real numbers—the numerical "size" as measured by Cantor's metaphor—became the "number of points on the line." And the Continuum hypothesis, which is about numbers, was thought of as being about points constituting the naturally continuous line.

What was hidden was that "completeness"—closure with limit points—under the discretization program wound up *defining* what a "point" on a "line" was to be.

The Difference That Mathematical Idea Analysis Makes

All this becomes clear only through a detailed analysis of the mathematical ideas involved. Without such an analysis, false interpretations of crucial ideas are perpetuated and mystery reigns.

- The naturally continuous line is not clearly distinguished from the line as a set of discrete "points," which are not really points at all but arbitrary set members.
- The real numbers are not clearly distinguished from the points on the line.
- The so-called real line is misnamed. It isn't a line.
- So-called space-filling curves are also misdescribed and misnamed. They do not "fill space."
- What is called the continuum is not distinguished adequately from a naturally continuous line.
- And perhaps most important, the discretization program goes undescribed and taken as a mathematical truth rather than as a philosophical enterprise with methodological and theoretical implications. The myth that the discretization program makes classical mathematical ideas more "rigorous" is perpetuated. That program abandons classical mathematical ideas, replacing them with new, more easily symbolizable but quite different ideas.

What suffers isn't the mathematics itself but the understanding of mathematics and its nature.

Mathematical idea analysis removes the mystery of "space-filling" curves. It clarifies what is meant by the "real line" and the "continuum." And most important, it places in clear relief the mathematical ideas used in the philosophical program of discretizing mathematics—a program that introduces new ideas, rather than formalizing old ones, as it claims to do.

Students of mathematics—and all of us who love mathematics—deserve to have these mathematical ideas made clear.

13

Continuity for Numbers: The Triumph of Dedekind's Metaphors

THE BASIC METAPHOR OF INFINITY, as we have just seen, is at the heart of the discretization program for reconceptualizing the continuous in terms of the discrete. From the cognitive perspective, it is the BMI that permits the appropriate conceptualization in purely discrete terms of infinite sets, infinite sequences, limits, limit points, points of accumulation, neighborhoods, open sets, and (as we shall see in this chapter) *continuity*.

To appreciate the remarkable achievement of this enterprise and the fundamental role of metaphor in it, let us turn to a historic moment in modern mathematics. The turning point came in the early 1870s, with the classic work of Richard Dedekind and Karl Weierstrass. They successfully launched the movement toward discretization, a reaction against the use of geometric methods via analytic geometry. Descartes had seen the real numbers as points on a naturally continuous line. Newton, linking physical space with naturally continuous mathematical space, had created calculus in a way that depended on geometric methods.

Dedekind showed, through a dramatic use of conceptual metaphor, that the real numbers did not have to be seen as points on a naturally continuous line. And through an implicit use of the Basic Metaphor of Infinity, he showed how to construct the real numbers using sets (infinite sets, of course) of discrete elements. Weierstrass, making another implicit use of the BMI, showed how to make calculus discrete, eliminating all naturally continuous notions of space and movement.

An important dimension of the discretization program was the concept of "rigor." This meant the use of discrete symbols and precisely defined, systematic algorithmic methods, allowing calculations that were clearly right or wrong. They could be written down step by step and checked for correctness. The prototypical cases were the methods of calculation in arithmetic. Those methods of calculation provided for certainty and precision, which were taken as the hallmarks of mathematics in nineteenth-century Europe.

Geometry involved visualization and spatial intuition. But mathematics in nineteenth-century Europe increasingly became a field that valued certainty, order, and precise methods over imagination and visualization. Indeed, visualization and spatial intuition were eventually disparaged. By the end of the century, they were regarded negatively—that is, not for what they could contribute but for how they threatened certainty, objectivity, precision, and the sense of order those mathematical virtues brought. Geometric methods came to be seen as vague, imprecise, not independently verifiable, and therefore unreliable. What could be worse than unreliability in a field that valued certainty, reliable methods, and objective truth?

It was thus imperative in nineteenth-century European mathematics to arithmetize calculus and eliminate geometry from the idea of number in arithmetic. Calculus had to be rescued from geometric methods and "put on a secure foundation"—a discrete and fully symbolizable and calculable foundation. If Weierstrass could arithmetize calculus—reconceptualize calculus as arithmetic—he could eliminate from calculus all visualization and spatial intuition, and thereby bring it into the realm of "rigorous" methods.

This meant conceptualizing naturally continuous space in terms of discrete entities and holistic motion in terms of *stasis* and *discreteness*. From a cognitive perspective, the task required conceptual metaphor—a way to conceptualize the continuous in terms of the discrete. Combined with other fundamental metaphors, the Basic Metaphor of Infinity permitted the understanding of naturally continuous space in discrete terms.

The central issue was continuity. The place and time were Germany in 1872.

Dedekind Cuts and Continuity

Like many other mathematicians of his era, Dedekind was profoundly dissatisfied with the way calculus had been developed—namely, through geometric notions like secants and tangents. Calculus was, after all, part of the subject matter of arithmetic functions and so ought properly to be understood in terms of arithmetic alone and not geometry. But continuous functions were understood at the time in geometric terms, as naturally continuous curves. How, he asked, could continuity be understood in arithmetic terms? The key, he believed, was the real numbers.

Here is Dedekind writing in his 1872 classic, *Continuity and Irrational Numbers* (translated into English in 1901).

> The statement is so frequently made that the differential calculus deals with continuous magnitude, and yet an explanation of this continuity is nowhere given; even the most rigorous expositions of the differential calculus do not base their proofs upon continuity but, with more or less consciousness of the fact, they either appeal to geometric notions or those suggested by geometry, or depend upon theorems which are never established in a purely arithmetic manner. (p. 2)

Dedekind further assumed that all arithmetic should come out of the natural numbers. The arithmetic of the rational numbers had been characterized in terms of the arithmetic of the natural numbers.

> I regard the whole of arithmetic as a necessary, or at least natural consequence of the simplest arithmetic act, that of counting. . . . The chain of these numbers . . . presents an inexhaustible wealth of remarkable laws obtained by the introduction of the four fundamental operations of arithmetic. (p. 4)

The idea here is that the full development of arithmetic should come from the same laws governing the natural numbers.

> Just as negative and fractional rational numbers are formed by a new creation, and as the laws of operating with these numbers must and can be reduced to the laws of operating with positive integers, so we must endeavor completely to define irrational numbers by means of the rational numbers alone. (p. 10)

In other words, irrational numbers must be understood in terms of rationals, which in turn must be understood in terms of the natural numbers.

> . . . [N]egative and rational numbers have been created by the human mind; and in the system of rational numbers there has been created an instrument of even greater perfection. . . . [T]he system *R* [the rational numbers] forms a well-arranged domain of one dimension extending to infinity on two opposite sides. What is meant by this is suffi-

ciently indicated by my use of expressions from geometric ideas; but just for this reason it will be necessary to bring out clearly the corresponding purely arithmetic properties in order to avoid even the appearance as if arithmetic were in need of ideas foreign to it. . . . (p. 5)

Even though numbers were at that time understood in geometric terms, the geometry had to be eliminated in order to characterize the *essence* of arithmetic. Dedekind then describes in great detail (pp. 6–8) what we have called the Number-Line blend (see Chapter 12), and goes on to say,

This analogy between rational numbers and the point of a straight line, as is well known, becomes a real correspondence when we select upon the straight line a definite origin or zero-point and a definite unit length for the measurement of segments. (pp. 7–8)

Dedekind's notion of "real correspondence" is fateful. He means a one-to-one correspondence—not merely that there is a point for every number but also that *there must be one and only one number for every point!* What Dedekind has described is essentially what we have called the Discretized Number-Line blend, which is a conceptual blend that uses two metaphors: Spaces Are Sets and Numbers Are Points on a Line. The blend, as we saw in the previous chapter, contains three domains precisely linked by these metaphorical mappings: Space, Sets, and Numbers.

Here we see Dedekind in the process of constructing this metaphorical blend. In the process, he goes back and forth, attending first to one domain and then another: first to space and then to numbers and sets. Back and forth. If naturally continuous space is to be "eliminated," the crucial elements of space must be picked out and modeled by corresponding entities in the domains of numbers and sets.

Dedekind's activity in constructing his version of the Discretized Number-Line blend is taken for granted in his reasoning from here on. He observes that when the rational numbers are associated with points on the line, there are points not associated with numbers, which creates a problem.

Of the greatest importance, however, is the fact that in the straight line *l* there are infinitely many points that correspond to no rational number. If the point *p* corresponds to the rational number *a*, then, as is well-known, the length *op* is commensurable with the invariable unit of measure used in the construction, i.e., there exists a third length, a so-called common measure, of which these two lengths are integral multiples. But the ancient Greeks already knew and had demonstrated that there are lengths incommensurable with the given unit of length. If we lay off such a length from the point *o* upon the line, we obtain an endpoint which corresponds to no rational number. Since it further can be easily shown that there are infinitely many lengths that are incommensurable with the unit of length, we may affirm: The straight line *l* is infinitely richer in point-individuals than the domain *R* of rational numbers in number-individuals.

If now, as is our desire, we try to follow up arithmetically all phenomena in the straight line, the domain of rational numbers is insufficient and it becomes absolutely necessary that the instrument *R* constructed by the creation of the rational numbers be essentially improved by the creation of new numbers such that the domain of all numbers shall gain the same completeness, or as we may say at once, the same *continuity*, as the straight line. (p. 9)

Here we have the heart of Dedekind's link between the real numbers and continuity. The line is absolutely continuous. When arithmetic functions are characterized in the Cartesian plane, continuous functions are represented, metaphorically, by continuous lines (see Case Study 1). A function maps numbers (which are discrete) onto other numbers. If a function in the Cartesian plane is a continuous curve, then the collection of *numbers* corresponding to points on a line must be *continuous*, too. This is a metaphorical inference in the version of the Discretized Number-Line blend that Dedekind was constructing. The naturally continuous line is part of the space domain in the blend. The line in the blend is (naturally) continuous. Points on the line are mapped one-to-one to numbers via the metaphors that Numbers Are Points on a Line and A Line Is a Set of Points. The inference in Dedekind's developing version of the blend is clear: If the line, seen as made up of points, is continuous, then the set of numbers must be continuous, too.

The above comparison of the domain *R* of rational numbers with a straight line has led to the recognition of the existence of gaps, of a certain incompleteness or discontinuity of the former, while we ascribe to the straight line completeness, absence of gaps, or continuity. (p. 10)

This is a crucial inference! The rational numbers, by themselves, have no *gaps*. In the set of rational numbers, rational numbers are all there is. Rational numbers are ratios of integers. Among the ratios of integers, there are no gaps; that is, every ratio of integers is in the set of rational numbers (see Figure 13.1).

Where does the very idea of a "gap" in the rational numbers come from? The answer is metaphor and conceptual blending. When the Number-Line blend is formed on the basis of the metaphor that Numbers Are Points on a Line, a

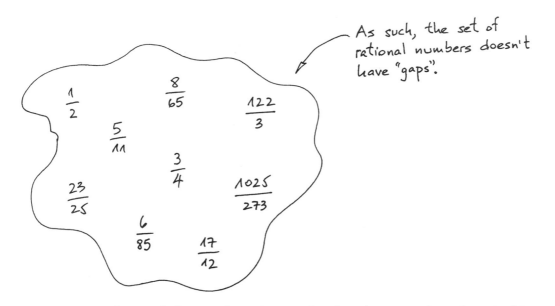

As such, the set of rational numbers doesn't have "gaps".

FIGURE 13.1   The set of all rational numbers—and nothing but rational numbers. Within *this* set, by itself, there are no "gaps" between the rational numbers, since there are no other entities being considered. For there to be a "gap," other entities must be under consideration. So-called gaps between the rational numbers appear under two additional conditions: (1) in the Number-Line blend, where all points on a line are assumed to correspond to a number; and (2) with the added idea of closure, where arithmetic operations (e.g., taking the square root) must yield a number as output (when such a number can be well-defined).

metaphorical correspondence is set up between numbers and points. Since this conceptual blend of space and arithmetic is used for measurement and built into measuring instruments (like rulers), it is taken for granted as objectively true: There *is* a true correspondence between points and numbers, as seen in the act of measuring and in the instruments for doing so.

Dedekind accepts this metaphorical conceptual blend as a truth—as many mathematicians have since the Greeks—because it is instantiated in physical measuring instruments. Those instruments have been (for centuries) constructed in order to be consistent with that conceptual blend. The instrument and the conceptual blend are, of course, different things. The instrument—as a measuring instrument—makes no sense in the absence of the conceptual blend of lines and numbers. And of course, there are things in the blend that are not in the instrument. For example, real numbers are not on any physical measuring tape or thermometer. Measuring tapes and thermometers are physical instruments in the world, conceived to be consistent with, and interpreted through, conceptual metaphorical blends. But that does not mean that all the entities in the conceptual blend (like the real numbers, in this case) are in the instrument or in the world. To believe so is to impute reality to something conceptual. That is just what Dedekind was doing when he concluded that there was a "gap" in the rational numbers. *It is only in the Number-Line blend that a "gap" exists*—a "gap" in points on a naturally continuous line that are paired with numbers. The "gap" is in the space domain, where there are points on a naturally continuous line that are not paired with rational numbers. This "gap" makes sense only in the metaphorical conceptual blend that Dedekind was constructing and we have inherited.

Measurement and Magnitude

As we have just seen, the Number-Line blend is not an arbitrary blend. It arises from an activity that has a physical as well as a conceptual part—the activity of measuring real objects in the world. In measuring, numbers are associated with pointlike locations in the physical world. Those physical points are not like metaphorical geometric points of length zero. They are physical positions of small but positive physical dimensions. And in measuring along a line, two distinct physical positions correspond to two different distances from a chosen origin point. Those distances are measured by numbers, and so each distinct "point" corresponds to a different number. This intuitive understanding is characterizable as a folk theory.

THE FOLK THEORY OF MEASUREMENT AND MAGNITUDE

- A line segment from one point to another has a magnitude.
- When a line segment is assigned as a unit of measure and set equal to the number 1, then any magnitude can be measured by a positive number.
- The Archimedean principle: Given numbers A and B (where A is less than B) corresponding to the magnitudes of two line segments, there is some natural number n such that A times n is greater than B.

Notice that the Archimedean principle is built into our everyday folk theory of measurement and magnitude. It entails that, given a measuring stick, even a very small one, you can measure a magnitude of any finite size by marking off the length of the measuring stick some finite number of times. This folk theory, including the Archimedean principle, is implicit in Dedekind's account of the ancient Greek ideas about number lines, as well as in his version of the Number-Line blend.

All this is taken for granted when Dedekind says, "we ascribe to the straight line completeness, absence of gaps, or continuity." Here he is creating yet another metaphor—one that has come down to us today: *Continuity Is Gaplessness*—or, more precisely, Continuity for a Number Line Is Numerical Completeness, the absence of gaps between numbers. Not gaps between points per se, but gaps between *numbers!*

Here is the great break with the everyday concept of natural continuity, the concept that had been used in mathematics for two millennia! Dedekind's new metaphor, *Continuity Is Numerical Completeness*, represented a change of enormous proportions. The "natural" continuity imposed by motion is gone. The metaphor that A Line Is a Set of Points is implicitly required. Continuity no longer comes from motion but from the completeness of a number system. Since each number is discrete and each number is associated one-to-one with the points on the line, there is no longer any naturally continuous line independent of the set of points. *The continuity of the line—and indeed of all space—is now to come from the completeness of the real-number system, independent of any geometry or purely spatial considerations at all*. Dedekind asks,

> In what then does this continuity consist? Everything must depend on the answer to this question, and only through it shall we obtain a scientific basis for the investigation of all continuous domains. By vague remarks on the broken connection in the smallest parts, obviously nothing is gained; the problem is to indicate a precise characteristic of continuity that can serve as a basis for valid deductions. . . . It consists in the following. . . . [E]very point p of the straight line produces a separation of the same into two portions such that every point on one portion lies to the left of every part of the other portion. I find the essence of continuity in the converse, i.e., in the following principle:
>
> "If all points of the straight line fall into two classes such that every point of the first class lies to the left of every point of the second class, then there exists one and only one point which produces this division of all points into two classes, this severing of the straight line into two portions." (p. 11)

This is Dedekind's celebrated characterization of the real numbers in terms of what has come to be called *the Dedekind cut*. This is a profound mathematical idea! It seems obvious at first, but as we shall see, it isn't. To show the true richness of Dedekind's idea, we need to apply the tools of mathematical idea analysis.

· · · · ·

Dedekind's Idea of the Cut is built out of three conceptual parts: the Cut frame, the Geometric Cut metaphor, and the Arithmetic Cut metaphor. Dedekind begins with the Number-Line blend for the rational numbers and defines a new and original structure on it, which we will call *Dedekind's Cut frame*.

DEDEKIND'S CUT FRAME

The Number-Line blend for the rational numbers, with a point *C* (the "cut") on the line, dividing all the rationals into two disjoint sets, *A* and *B*, such that every member of *A* is to the left of, and hence less than, every member of *B*.

Given this precise notion of a "cut" defined relative to the Rational-Number-Line blend, Dedekind goes on to propose a new geometric metaphor for irrational numbers. By this means, he extends the Rational-Number-Line blend to metaphorically create the Real-Number-Line blend.

DEDEKIND'S GEOMETRIC CUT METAPHOR

Source Domain THE RATIONAL-NUMBER-LINE BLEND, WITH THE CUT FRAME	*Target Domain* THE REAL-NUMBER-LINE BLEND, WITH THE CUT FRAME
Case 1: *A* has a largest rational *R*, or *B* has a smallest rational *R*. $\rightarrow$ *C* (the "cut")	*R*
Case 2: *A* has no largest rational and *B* has no smallest rational. $\rightarrow$ *C* (the "cut")	*I*, an irrational number

The idea here is this: Either the "cut" will fall at a rational point or it will not. If it falls at a rational point, that rational number will either be the smallest number in *B* or the largest number in *A*. In that case, we conceptualize the rational number *R* as being the cut *C*. This is not really anything new. The innovation of the metaphor comes in case 2, where *A* has no largest rational and *B*, no smallest rational. The "cut" in the rational number line then falls "between" the rationals. The great innovation is to metaphorically "define" an irrational number as *being* the "cut" between the two sets of rationals. Notice that there are no irrational numbers in the source domain of the metaphor but only a "cut"—a point on the line in the blend. The irrationals are conceptualized metaphorically as *being* such cuts.

The brilliance of Dedekind's Geometric Cut metaphor is that the cut characterizes a *unique* irrational number. The uniqueness is not obvious. It follows from three assumptions:

- Assumption 1: Dedekind's Number-Line blend, in which there is a one-to-one correspondence between points and numbers. This guarantees that there will be a number for every point.
- Assumption 2: the Archimedean principle, which holds in his version of the Number-Line blend. This guarantees that the number will not be infinitesimal.

These two assumptions guarantee that if there are two irrational points characterized by (A, B), then there has to be a finite distance between them.

- Assumption 3: the denseness property for the rationals—namely, that there is a rational number on every finite interval in the Rational-Number-Line blend.

Assumption 3 guarantees that if there were two distinct irrational points defined by (A, B) and separated by a finite distance, then there would be a rational point between them, and hence not in either A or B. But this contradicts the assumption that A and B jointly contain *all* the rational points. Given this contradiction, it follows that the irrational point I must be unique.

Notice that it is only by making assumptions 1 and 2 that this conclusion follows. Assumptions 1 and 2 are not necessary assumptions for doing mathematics, as we have seen in previous chapters. Dedekind thus got his uniqueness result by making assumptions that others do not, and need not, make.

But Dedekind achieved only half his job with the Geometric Cut metaphor. What he did was to extend the Rational-Number-Line blend to the Real-Number-Line blend, creating the reals from the rationals using only one frame and one metaphor. But his ultimate goal was to define the real numbers in terms of the rationals, using no geometry at all. Having constructed the Real-Number-Line blend, which has geometry in it, he now had to construct a metaphor to eliminate the geometry—that is, to conceptualize his Real-Number-Line blend only in terms of numbers and sets. Below is the conceptual metaphor with which he did so. The basic idea here is to conceptualize the cut C in the Number-Line blend in terms of the ordered pair of sets of rational numbers (A, B).

DEDEKIND'S ARITHMETIC CUT METAPHOR

Source Domain RATIONAL NUMBERS AND SETS		*Target Domain* THE REAL-NUMBER-LINE BLEND, WITH THE GEOMETRIC CUT METAPHOR DEFINING IRRATIONALS
An ordered pair (A, B) of sets of rational numbers, where $A \cup B$ contains all the rationals and every member of A is less than every member of B	$\rightarrow$	The point C (the "cut") on the line dividing all the rationals into two sets A and B, such that every member of A is to the left of, and hence less than, every member of B
Case 1: A has a largest rational R, or B has a smallest rational R. $(A, B) = R$	$\rightarrow$	Case 1: A has a largest rational R, or B has a smallest rational R. C (the "cut") $= R$
Case 2: A has no largest rational, and B has no smallest rational. $(A, B) = I$, not a rational.	$\rightarrow$	Case 2: A has no largest rational, and B has no smallest rational. $C = I$, an irrational number.

Here the geometric cuts are "reduced" via the metaphor to sets and rational numbers alone. In each case, the set (A, B) is mapped onto the geometric cut C. The cut C is therefore conceptualized in terms of *sets and numbers alone*. Irrational numbers, which were shown to be uniquely characterized via the Geometric Cut metaphor, are now also uniquely characterized via the Arithmetic Cut metaphor.

In short, the Dedekind Arithmetic Cut metaphor states: *A real number is an ordered pair of sets (A, B) of rational numbers, with all the rationals in A being less than all the rationals in B, and all the rational numbers are in $A \cup B$.*

Here is the triumph of metaphor at a crucial moment in the birth of the discretization program. We can see in its full beauty the depth and subtlety of Dedekind's metaphorical idea of the cut.

The Geometric Basis of Dedekind's Completeness for the Real Numbers

Dedekind's two cut metaphors, taken together, have an extraordinary entailment: The real numbers characterized in this way constitute *all* the real numbers! This entailment does not come from arithmetic alone; it relies on the

Geometric Cut metaphor. This is an important point and is worth going into in some detail.

Dedekind assumed that for each point on a line, there was a real number that could measure the distance of that point from an origin point (given the choice of a unit). This defines what we will call the *Measurement Criterion for Completeness* of the real numbers. If you can characterize numbers for all such points, then you have characterized *all* the real numbers. The Measurement Criterion for Completeness is a geometric criterion.

Dedekind sought an arithmetic characterization of the real numbers—a characterization that would meet this geometric criterion. If he could find it, he could be sure he had characterized *all* the real numbers. To meet this geometric criterion, Dedekind needed *both* the Geometric and Arithmetic Cut metaphors.

It is important to realize that the Measurement Criterion for Completeness of the real numbers uses assumptions 1 and 2 above: Dedekind's Number-Line blend and the Archimedean principle. It is the Number-Line blend that gets you from (a) assigning numbers to all the points measured to (b) characterizing all the real numbers. It is the Archimedean principle that guarantees that two points will not be separated by an infinitesimal.

Thus, Dedekind's characterization of the completeness of the real numbers depended on geometry after all, since he took a geometric criterion to define what "completeness" was to be. Notice that he did not use the algebraic criterion explicitly in arguing for completeness, though he knew from the work of Carl Friedrich Gauss and others that the real numbers constituted a complete ordered field. The algebraically complete real numbers are those that are both (1) roots of polynomial equations and (2) linearly ordered. Apparently, algebraic completeness was not enough for Dedekind. He did not argue for the completeness of the real numbers on algebraic grounds (*that* had been proved by Gauss) but only on geometric grounds.

It is somewhat ironic that the man given credit for arithmetizing the completeness of the real numbers did so using a *geometric* criterion and a *geometrically* based metaphor.

Continuity as Numerical Completeness

Dedekind's Arithmetic Cut metaphor is stated in such a way as to guarantee that there will be no "gaps" in the real numbers. Every real number meeting the Measurement Criterion for Completeness will be characterized in terms of the rational numbers via the Arithmetic Cut metaphor.

"Continuity" for the real-number line can then be conceptualized purely in terms of numbers, which are discrete entities, via the following metaphor: Continuity for the Number Line Is Arithmetic Gaplessness. The result is a metaphorical "continuum" of discrete entities: the "real-number continuum"—that is, the "gapless" sequence of all real numbers defined metaphorically using only numbers and sets as "cuts"—with the geometry eliminated by the Arithmetic Cut metaphor.

Continuity for *space* can now be characterized in terms of *metaphorical "continuity"* for numbers. Dedekind adds,

> The assumption of this property of the line is nothing else than an axiom by which we attribute to the line its continuity, by which we find continuity in the line. If space has at all a real existence, it is *not* necessary for it to be continuous; many of its properties would remain the same even were it discontinuous. And if we knew for certain that space was discontinuous, there would be nothing to prevent us from filling up its gaps, in thought, and thus making it continuous; this filling would consist in the creation of new point-individuals and would have to be effected in accordance with the above principle. (p. 12)

It is remarkable that he says, "*If space has at all a real existence. . . .* " If he can eliminate space from mathematics, maybe it doesn't exist. He goes on, "*If space has at all a real existence, it is* not *necessary for it to be continuous; many of its properties would remain the same even were it discontinuous.*" Here it is clear that by "continuous" space he means our ordinary concept of naturally continuous space. If he can create a mathematics that does as well as the old mathematics, but is discrete rather than naturally continuous, maybe real space (if it exists!) is also discrete. Maybe the world fits the mathematics he is creating . . . .

The bottom line for Dedekind is this: Continuity for space in general is now to be conceptualized in terms of continuity for numbers. Space must therefore be made up of discrete points, which are not mere locations in a naturally continuous medium. The points *constitute* the space, the plane, the line, and all curves and surfaces and volumes. *What makes lines and planes and all space continuous is their correspondence to numbers in a complete, "continuous" number system.*

And what do Dedekind's metaphors—his continuity metaphor and his two cut metaphors—imply about infinitesimal numbers? The metaphorical entailment is immediate: *Infinitesimal numbers cannot exist!* If the real numbers use up all points on the line, and if the real numbers form a complete system—with the principle of closure and the measurement criterion requiring no further extensions—then there is no room in the mathematical universe, either the universe of arithmetic or that of geometry, for infinitesimals. As we have seen, this comes from the Archimedean principle, which is built into the folk theory of measurement and magnitude, which in turn is implicit in Dedekind's concept of the number line. The Archimedean principle was absolutely necessary both to guarantee the uniqueness of the real numbers as Dedekind constructed them and to guarantee the completeness of the real numbers according to the measurement criterion.

Infinitesimal numbers had been used for two centuries by hundreds of mathematicians to do calculus. But the arithmetization of calculus that Dedekind envisioned could not make use of infinitesimal numbers, because it used a version of the Number-Line blend with an implicit Archimedean principle. If calculus was to be arithmetized in terms of numbers alone and not geometry, only the real numbers would do. Moreover, Dedekind's Continuity metaphor did not even leave room for infinitesimals in space. All space was to be structured by the real numbers alone.

What we see in these pages from Dedekind is one of the most important moments in the history of modern mathematics. Not only is arithmetic freed from the bonds of the Cartesian plane, but the very notion of space itself becomes reconceptualized—and mathematically redefined!—in terms of numbers and sets. In addition, the real-number line becomes reduced to the arithmetic of rational numbers, which had already been reduced to the arithmetic of natural numbers plus the use of classes. This sets the stage for the Foundations movement in modern mathematics, which employs the Numbers Are Sets metaphor to reduce geometry, the real line, the rational numbers, and even the natural numbers to sets. All mathematics is discretized—even reason itself, in the form of symbolic logic.

All that was needed was the reduction of calculus to arithmetic. Such a program of reduction had begun with the work of Augustin-Louis Cauchy in 1821. Cauchy had given preliminary accounts of the real numbers, continuity, and derivatives in purely arithmetic terms using what has come to be called the epsilon-delta characterization of a limit. It remained for Karl Weierstrass to complete the job.

14

Calculus Without Space or Motion: Weierstrass's Metaphorical Masterpiece

BETWEEN 1821 AND THE 1870S, the need for an arithmetization of calculus had reached crisis proportions. The reason had to do with the discovery of a number of functions called monsters. What made these functions "monsters" was a common understanding of arithmetic functions and calculus in terms of three interlocking ideas forming a conceptual paradigm, which we will call the *Geometric Paradigm*.

We will discuss the "monsters" below. In order to appreciate them and their effect on the mathematical community, it is important first to have a clear idea of the geometric paradigm, and then to understand the arithmetization of calculus that came to replace it. Here is an outline of the geometric paradigm.

THE GEOMETRIC PARADIGM

- The Cartesian plane, with Descartes's metaphor that A Mathematical Function Is a Curve in the Cartesian Plan
- Newton's geometric characterization of calculus in terms of a sequence of secants with a tangent as a limit
- The understanding of a curve in terms of natural continuity and motion

Central to the geometric paradigm was the notion of a curve. An excellent characterization was given by James Pierpont, professor of Mathematics at Yale, in addressing the American Mathematical Society. Pierpont correctly and insightfully listed what a cognitive scientist would now refer to as eight "prototypical" properties of what he called a *curve*—that is, a line in three-dimensional space that is either straight or curved (Pierpont, 1899, p. 397):

PIERPONT'S PROTOTYPICAL PROPERTIES OF A CURVE

1. It can be generated by the motion of a point.
2. It is continuous.
3. It has a tangent.
4. It has a length.
5. When closed, it forms the complete boundary of a region.
6. This region has an area.
7. A curve is not a surface.
8. It is formed by the intersection of two surfaces.

Take note of the year: 1899—a quarter of a century after the successful launching of the discretization program by Dedekind and Weierstrass. Pierpont, as we shall see shortly, was both attracted and repulsed by the program—attracted by the prospect of increased rigor, repulsed by the loss of the tools of visualization and spatial intuition. Pierpont's reaction was like that of many of his colleagues. The discretization program was still controversial, but it had won.

The geometric paradigm had formed the intuitive basis of what mathematicians of that era took for granted as being *the* characterization of algebraic and trigonometric functions. It characterized the conceptual framework of mathematicians of the day; the assumption was that all functions should fit this paradigm.

Discretization and "Triumph" over the Monsters

Part of the reason why discretization won out over the geometric paradigm had to do with what were called *monster functions*. What made a monster function a "monster" was that it violated the geometric paradigm, which characterized the idea of what a function should be. The monsters violated the very conceptual framework that mathematicians had learned to think in. That framework was very largely geometric, while the functions themselves were functions from numbers to numbers. In the functions themselves, there was no inherent geometry.

The geometric paradigm allowed mathematicians of the day to use their geometric intuitions in the study of functions from numbers to numbers. The monsters were cases where the geometric paradigm failed. The failure was falsely, we believe, attributed to the failure of "intuition"—geometric intuition. The "failure" was one of expecting too much of a particular paradigm—one that works perfectly well for the cases it was designed to work for, but which cannot be generalized beyond those. What made the geometric paradigm appear to be a failure was a search for *essence*—for a single unitary characterization of all numerical functions.

1872 Again, Elsewhere in Germany: Weierstrass and the Arithmetization of Calculus

The person usually credited with devising an acceptable replacement for the geometric paradigm was Karl Weierstrass, whose ideas from lectures are cited by H. E. Heine, in his *Elemente* in 1872. This was the same year that Dedekind published his *Continuity and Irrational Numbers*. Weierstrass's theory was an updated version of Cauchy's earlier account (which had certain flaws).

Weierstrass, like Dedekind and Cauchy, sought to eliminate all geometry from the study of numbers and functions mapping numbers onto numbers, including derivatives and integrals in calculus. This project required many changes in the geometric paradigm, essentially the same changes proposed by Dedekind for pretty much the same reasons:

- Natural continuity had to be eliminated from the concepts of space, planes, lines, curves, and geometric figures. Geometry had to be reconceptualized in terms of sets of discrete points, which were in turn to be conceptualized purely in terms of numbers: points on a line as individual numbers, points in a plane as pairs of numbers, points in *n*-dimensional space as *n*-tuples of numbers.
- The idea of a function as a curve defined in terms of the motion of a point had to be completely replaced. There could be no motion, no direction, no "approaching" a point. All these ideas had to be reconceptualized in purely static terms using only real numbers.
- The geometric idea of "approaching a limit" had to be replaced by static constraints on numbers alone, with no geometry and no motion. This is necessary for characterizing calculus purely in terms of arithmetic.
- Continuity for space was to be reconceptualized as "continuity" for numbers.
- Continuous functions also had to be reconceptualized purely in terms of numbers.

- Calculus had to be reformulated without either geometric secants and tangents or infinitesimals. Only the real numbers could be used.

How Weierstrass Did It

How do you accomplish such a complete paradigm shift? How do you utterly change from one mode of thought to another, from one system of concepts to another? The answer is, via conceptual metaphor.

Implicit in Weierstrass's theory are some of the conceptual metaphors and blends we have already discussed: Spaces Are Sets of Points, Numbers Are Points on a Line, The Number-Line blend (including the folk theory of measurement and magnitude), Continuity for a Number Line Is Arithmetic Gaplessness. Like Dedekind, Weierstrass metaphorically conceptualized the real numbers as aggregates of rationals, though in a way that was technically different than Dedekind's cuts or Cantor's infinite intersections of nested intervals (see Boyer, 1968, p. 606; also see Chapter 9).

Weierstrass's achievement was to add to these conceptual metaphors and blends a new metaphor of his own, a way to metaphorically reconceptualize the continuity of naturally continuous space itself and continuous functions in it. That is, he had to come up with a remarkable metaphor, one conceptualizing the continuous in terms of the discrete. Via this metaphor, he had to be able to *redefine* what "continuity" could mean for sets of discrete elements.

Continuity

Weierstrass's redefinition of continuity for a function over sets of discrete numbers was perhaps his greatest metaphorical magic trick. Think of the job he had to do. He had to eliminate one of our most basic concepts—the natural continuity of a trajectory of motion—and replace it with a concept that involved no motion (just logical conditions), no continuous space (just discrete entities), no points (just numbers), with functions that are not curves in a plane (but, rather, sets of ordered pairs of numbers). And in the process he had to convince the mathematical community that this was the *real* concept of "continuity."

Weierstrass's strategy was this:

Look at continuous functions, conceptualized as motions through continuous space. Look at what is true of discrete point-locations along those continuous paths of motion. Try to reformulate in discrete, static terms what is true of the discrete point-locations on the continuous paths. Using the Number-Line blend, think of points converging to a spatial limit as numbers converging to a numerical limit.

Weierstrass saw that, with continuous functions, points that are close in the domain of the function tend to be close in the range of the function, though what counts as "close" might be different in the range than in the domain, and different at different values for the function. His idea was this:

- For a given point $f(a)$ in the range, pick a number that defines a standard of closeness for $f(a)$. Call it epsilon.
- Then, for the corresponding point a in the domain, pick another number that defines a standard of closeness for a. Call it delta.
- Suppose the function is naturally continuous over point-locations in naturally continuous space. Then points that are "close" to $f(a)$ (within distance epsilon) in the range will have been mapped from points that are "close" (within a corresponding distance delta) in the domain.

That is, for every standard of closeness in the range (epsilon) there will be corresponding standard of closeness in the domain (delta), and this will be true for every point-location $f(a)$ in the range and its corresponding point-location a in the domain. This is where the classical quantification statement "For every epsilon, there exists a delta" comes from and why it is important.

The Weierstrass program for discretizing continuity came with a minimal standard of adequacy: Begin with all of the prototypical classical continuous curves in the geometric paradigm. Replace these curves with sets consisting of the discrete points on each curve. To be minimally adequate, Weierstrass's new "definition" of "continuity" had to assign the judgment "continuous" to each set of points corresponding to a classical naturally continuous curve.

Weierstrass now used the Number-Line blend, in which there is a correspondence between point-locations and numbers, based on the metaphor Numbers Are Points on a Line. In the Number-Line blend, every choice of *point-locations* corresponds to a choice of *numbers*, and every *distance between point-locations* corresponds to an arithmetic *difference between numbers*. Using these metaphorical correspondences, Weierstrass observed that the idea of continuity itself, the defining property of naturally continuous space, could be stated in entirely discrete terms, using numbers in place of their corresponding point-locations in the Number-Line blend.

Weierstrass's observation that naturally continuous functions *preserved closeness* for point-locations, via the Number-Line blend, is characterized by the following conceptual metaphor:

WEIERSTRASS'S CONTINUITY METAPHOR

Source Domain NUMBERS		*Target Domain* NATURALLY CONTINUOUS SPACE
Discrete numbers	→	Discrete point-locations
Sets of numbers	→	Curves
Numbers in sets of numbers	→	Point-locations on continuous curves
Functions seen as mapping discrete numbers within sets to discrete numbers within sets	→	Functions seen as mapping points on continuous curves to points on continuous curves
Preservation of numerical closeness for functions over discrete numbers	→	Continuity for functions over continuous curves

Weierstrass's Implicit Use of the BMI: Epsilon-Discs and Infinite Nested Intervals

The discerning reader will note that these ideas can be conceptualized using the Epsilon-Disc frame and the special case of the BMI for infinite nested intervals, discussed in Chapter 12. Each number defines a standard of closeness, which is the radius of an epsilon disc. The full range of standards of closeness for the real numbers is infinite. This infinity can be conceptualized via the BMI using the special case of the infinite nesting property (see Chapter 12). The connection between the epsilon and delta standards of closeness can be conceptualized by a correspondence between the epsilon discs on the $f(x)$-axis and those on the x-axis.

Thus, using the Number-Line blend, the Epsilon frame, and two linked instances of the BMI for infinite nested intervals, we have the cognitive apparatus necessary to characterize Weierstrass's basic idea for redefining continuity in terms of arithmetic alone.

Limits for Functions

To arithmetize calculus, Weierstrass next had to arithmetize the concept of a limit for functions. We have already seen continuity defined in terms of linked standards of closeness across the x- and $f(x)$-axes. And we have seen how these can be conceptualized using the linked version of the BMI for infinite nested intervals. With this apparatus, the notion of a limit follows immediately.

THE LIMIT L OF A FUNCTION $f(x)$ AS x APPROACHES a

- An infinite set of nested discs with L as a center on the $f(x)$-axis.
- A corresponding infinite set of nested discs with a as a center on the x-axis.
- The radius of each disc on the $f(x)$-axis is $|L - \varepsilon_i|$.
- The radius of each corresponding disc on the x-axis is $|a - \delta_i|$.
- As each radius gets smaller on the $f(x)$ axis, the corresponding radius gets smaller on the x-axis.

We are now in a position to see how these ideas were written down by Weierstrass in common mathematical notation.

THE TRANSLATION INTO COMMON MATHEMATICAL NOTATION

- The infinite nested sequence of discs around L is covered by the quantifier expression "For every $\varepsilon > 0$."
- The correspondence across the two nested sequences of discs is given by the expression "There exists a $\delta > 0$" following the above quantifier expression.
- The radius of each disc on the $f(x)$-axis around L is "$|f(x) - L|$."
- The radius of a corresponding disc on the x-axis around a is "$|x - a|$".

Given these translations, the concepts just discussed can be understood as covered by the following formal definition.

WEIERSTRASS'S NOTATION FOR HIS CONCEPT OF LIMITS

Let a function f be defined on an open interval containing a, except possibly at a itself, and let L be a real number. The statement

$$\lim_{x \to a} f(x) = L$$

means that for every $\varepsilon > 0$, there exists a $\delta > 0$, such that if $0 < |x - a| < \delta$, then $|f(x) - L| < \varepsilon$.

Given this notation for Weierstrass's concept of the limit, we can use it to provide a statement expressing Weierstrass's notion of Continuity in common mathematical notation.

WEIERSTRASS'S NOTATION FOR HIS CONCEPT OF CONTINUITY

A function f is continuous at a number a if the following three conditions are satisfied:

1. f is defined on an open interval containing a,
2. $\lim\limits_{x \to a} f(x)$ exists, and
3. $\lim\limits_{x \to a} f(x) = f(a)$.

Here Weierstrass is exploiting the linkage between his concept of continuity and his concept of a limit. Note that the "function f [is] defined on an open set of real numbers containing a, except possibly at a itself." The assumption of an open set of real numbers presupposes something crucial. It builds in a hidden case of the BMI, as we saw in Chapter 12. The reason is that the open set of reals containing a has the nested interval property with respect to a. In other words, an open set containing a presupposes the concept of an infinite nested sequence of epsilon discs around a—conceptualized, of course, via the BMI.

The Result

Weierstrass's characterization of limits for functions and of continuity is a statement "redefining" what was meant by limits and continuity in the geometric paradigm. In the redefinition, there is no motion, no time, no "approach," no "limit." There are only numbers. "a" and "L" are numbers; "x," "δ" and "ε" are variables over numbers. There are no points or variables over points. Instead of motion through space, we have only sets of pairs of numbers. In the notation for the concept of a limit, there is no motion, no approaching, no endpoint; there is only a static logical constraint on numbers, a constraint of the form:

For every u, there exists a v such that, if $G(v)$, then $H(u)$.

Take this statement and make the following replacements: $u = \varepsilon$, $v = \delta$, $G(v) = 0 < |x - a| < v$, and $H(u) = |f(x) - L| < u$. The result will be Weierstrass's discretized account of continuity. Just numbers and logic. There is nothing here from geometry—no points, no lines, no plane, no secants or tangents. In place of the natural spatial continuity of the line, there is just the numerically gapless set of real numbers. There are no infinitesimals, just static real numbers. The function is not a curve in the Cartesian plane; it is just a set of ordered pairs of real numbers.

How does Weierstrass replace Newton's geometric idea of approaching a limit, if there is no motion, no space, no approach, and no limit? Weierstrass's work of genius was a clever use of conceptual metaphor.

Having reconceptualized "limit" and "continuity" in purely arithmetic terms without any geometry, Weierstrass had no difficulty characterizing a derivative in the same way. The Newtonian derivative was characterized using the metaphor that Instantaneous Speed Is Average Distance Traveled over an Infinitely Small Interval of Time. Newton had conceptualized this metaphorically in geometric terms: Taking the derivative (the instantaneous rate of change) of a function as a tangent, he conceptualized the tangent as a secant cutting across an infinitely small arc of the curve. That metaphor allowed him to calculate the tangent as the limit of an infinite series of values of the secant. Newton had arithmetized the tangent as:

$$f'(x) = \lim_{\Delta x \to 0} \frac{f(x + \Delta x) - f(x)}{\Delta x}$$

Given his new metaphor for approaching a limit, Weierstrass simply appropriated Newton's arithmetization of the idea of average change over an infinitely small time, jettisoning the idea and keeping the arithmetic.

Weierstrass's static, purely arithmetic idea of the derivative is as follows.

The derivative function $y = f'(x)$ is a set of pairs (x, y) meeting the condition: For every neighborhood of size ε around y, there is a neighborhood of size δ around x such that $\frac{f(x + \delta) - f(x)}{\delta}$ is in the neighborhood of size ε around y.

In other words, $y = f'(x)$ if y *preserves closeness* to $\frac{f(x + \delta) - f(x)}{\delta}$, for arbitrarily small values of δ. "Neighborhood" here is defined in arithmetic and set-theoretic terms as an open set of numbers. The ingenuity here is that y is defined not directly but in terms of what it is indefinitely close to.

Here again, there is no geometry, no motion, no curves, no secants or tangents, no approaching, and no limit. There are just numbers and logical constraints on numbers. $\frac{f(x + \delta) - f(x)}{\delta}$ is the arithmetic expression of Newton's idea of instantaneous change as average change over an infinitely small interval. But the idea and the geometrization are gone. Only the arithmetic is left. The ε-δ condition expresses *preservation of closeness*.

The Hidden Geometry in Weierstrass's Arithmetization

Weierstrass is given credit for having arithmetized calculus—and in a very important way, he did. Yet the geometry is still there. Calculus is about the concept of change. Change, in conceptual systems around the world as well as in classical calculus, is conceptualized in terms of motion. Motion in mathematics is conceptualized in terms of the ratio of distance to time. Time, in turn, is metaphorically conceptualized in terms of distance. Ratios are conceptualized in terms of the arithmetic operation of division. The Newton-Leibniz metaphor for instantaneous change is average change of distance over an interval (time conceptualized as distance) of an infinitely small size. This is arithmetized by the expression $\frac{f(x + \delta) - f(x)}{\delta}$. It is this collection of metaphors that links the concept of change to $\frac{f(x + \delta) - f(x)}{\delta}$.

In these metaphors, there is implicit geometry: the ratio of *distance* to *time*, where time is itself conceptualized metaphorically as *distance*. If mathematics is taken to include the ideas that arithmetic expresses—that is, if calculus is taken to be about something—namely, change—then Weierstrass did not eliminate geometry at all. From the conceptual perspective, he just hid it. From the perspective of mathematical idea analysis, no one *could* eliminate the geometry metaphorically implicit in the very concept of change in classical mathematics.

Weierstrass and the Monsters

It is now time to discuss the monsters, those functions that did not fit the geometric paradigm. It was the monsters that created the urgent need for a new nongeometric paradigm—a need that the work of Dedekind and Weierstrass filled. Weierstrass was said to have "tamed the monsters."

What makes a function a monster? Suppose you believed with Descartes, Newton, and Euler that a function could be characterized in terms of naturally continuous geometric curves in a classical Cartesian plane. Your understanding of a function would then be characterized by your understanding of curves. As Pierpont pointed out, the prototypical curve has the following properties:

1. It can be generated by the motion of a point.
2. It is continuous.
3. It has a tangent.
4. It has a length.
5. When closed, it forms the complete boundary of a region.
6. This region has an area.
7. A curve is not a surface.
8. It is formed by the intersection of two surfaces.

Monsters are functions that fail to have *all* these properties.

Imagine yourself assuming that a function *was* a curve in the Cartesian plane with all these properties. Then imagine being confronted by the monster functions we are about to describe. What we will do in each case is

- describe one or two monsters.
- show how each does not fit the geometric paradigm.
- describe what Weierstrass's paradigm says about each one.

It is often said that Weierstrass's account of "continuity" is (1) a formalization and (2) a generalization of "continuity" in the geometric paradigm—that is, natural continuity. We will also ask, in each case, whether this is true.

Here Come the Monsters

$$\text{Monster 1: } f(x) = \begin{cases} \sin{(1/x)} & \text{for } x \neq 0 \\ 0 & \text{for } x = 0 \end{cases}$$

$$\text{Monster 2: } f(x) = \begin{cases} x \sin{(1/x)} & \text{for } x \neq 0 \\ 0 & \text{for } x = 0 \end{cases}$$

These are represented in Figures 14.1 and 14.2, respectively.

As x gets close to zero, $1/x$ grows to infinity (when x is positive) and minus infinity (when x is negative). As x approaches zero and $1/x$ approaches positive or negative infinity, both monster functions oscillate with indefinitely increasing frequency. That is, there are more and more oscillations in the functions as x approaches zero from either the positive or negative side.

Monster 1 oscillates between −1 and 1 all the way up to (but not including) zero (see Figure 14.1). Monster 2 is more constrained. It oscillates between two straight lines each at a 45-degree angle from the x-axis and intersecting at the origin. But since x gets progressively smaller as it approaches zero, the function goes through progressively smaller and smaller oscillations (see Figure 14.2).

What happens when these monsters confront the geometric paradigm, in which all functions are supposed to be curves? Do monsters (1) and (2) have all the properties (1 through 8) of prototypical curves? Suppose we ask if the curves for these functions can be generated by the motion of a point. The answer is no. Such a point would have to be moving in a direction at every point, including zero. But as each function approaches zero, it oscillates—that is, changes direction—more and more often. The definition of the function has been gerrymandered a bit to artificially include a value at $x = 0$, because $1/x$ would not be

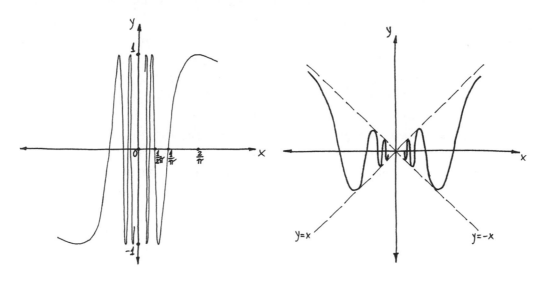

FIGURE 14.1   The graph of the function $f(x) = \sin(1/x)$.

FIGURE 14.2   The graph of the function $f(x) = x \sin(1/x)$.

defined for $x = 0$. The nongerrymandered part of the function does not "pass" through the origin at all. It just keeps getting closer and closer, changing direction more and more, but never reaching the origin. The added point makes it seem as if the function "passes" through the origin, but there is no motion. It is just a point stuck in there.

At the origin—the gerrymandered point—the oscillation, or change of direction, approaches infinity. What direction is the curve going in as it "reaches" the origin? There is no well-defined direction. Does it approach from above or below? There is no answer, because *at* the origin it isn't coming from any one direction. For this reason, the function cannot be a curve traced by the motion of a point.

Since no direction can be assigned to the function as it passes through zero, the function cannot be said to have any fixed tangent at zero. Additionally, consider an arc of each function in a region including zero—for example, an arc between $x = -0.1$ and $x = 0.1$. What is the length of such an arc? Because the function oscillates infinitely, the arc does not have a fixed length. In short, monsters (1) and (2) fail to have properties 1, 3, and 4 of curves.

What about property 2? Are the monster functions "continuous"? Continuity for a prototypical curve means "natural continuity." If natural continuity is characterized, as Euler assumed, by the motion of a point, then the answer is no. Since neither monster function can be so characterized, neither is naturally

continuous. Thus, they also fail to have property 2. Since they lack half the
properties of prototypical curves, and since properties 5 and 6 don't apply to
them, they lack four out of six of the relevant properties of curves. That is what
makes them monsters from the perspective of the geometric paradigm.

From the perspective of the Weierstrass paradigm, the purely arithmetic definitions of "continuity" (i.e., preservation of closeness) and differentiability
apply to such cases exactly as they would to any other function. In the geometric paradigm, neither case has a derivative at the origin, since neither has a tangent there. On the Weierstrass account, there is no derivative at zero, either (the
limit does not exist), but for an arithmetic reason: In neighborhoods close to
zero, the value of $\frac{\sin \frac{1}{x+\delta} - \sin \frac{1}{x}}{\delta}$ and $\frac{(x+\delta) \cdot \sin \frac{1}{x+\delta} - x \cdot \sin \frac{1}{x}}{\delta}$ vary so wildly that closeness to any value at zero is not preserved. The condition "$y = f'(x)$, if for every ε
> 0, there exists a $\delta > 0$, such that $|\frac{f(x+\delta) - f(x)}{\delta} - y| < \varepsilon$" is not met for any y at x
$= 0$.

But what about Weierstrass's continuity—that is, preservation of closeness?
Here the two monsters differ. Monster 1 does not preserve closeness at the origin. If you pick an epsilon less than, say, 1/2, there will be no delta near zero
that will keep $f(x)$ within the value 1/2. The reason is that as x approaches zero,
$f(x)$ oscillates between 1 and –1 with indefinitely increasing frequency—and so
cannot be held to within the value of 1/2 when x is anywhere near zero. Since
preservation of closeness is what Weierstrass means by "continuity," Monster
1 is not Weierstrass-continuous. In this case, preservation of closeness matches
natural continuity: Both are violated by Monster 1.

Monster 2 is very different for Weierstrass. Because $f(x)$ gets progressively
smaller as x approaches zero, it does preserve closeness at zero. If you pick some
number epsilon much less than 1, then for every delta less than epsilon, the
value of $f(x)$ for Monster 2 will stay within epsilon. Since preservation of closeness is Weierstrass's metaphor for continuity, his "definition of continuity" des-
ignates Monster 2 as "continuous" by virtue of preserving closeness.

This does not mean that Monster 2 is naturally continuous while Monster 1
is not. Neither is naturally continuous. It means only that Monster 2 preserves
closeness while Monster 1 does not.

What are we to make of this? In the geometric paradigm, Monster 2 is not
continuous. In the Weierstrass paradigm, Monster 2 *is* "continuous." There is
no contradiction here, just different concepts that have confusingly been given
the same name. However, there are two very important theoretical morals here.

- Weierstrass "continuity" is not just a formalization of the "vague" concept of natural continuity. If one were just a formalization of the other,

it could not be the case that one clearly holds but the other clearly doesn't. The two concepts of "continuity" are simply very different concepts. They have different cognitive structure.

- Weierstrass "continuity" is not a generalization over the concept of natural continuity. It cannot be a generalization in this case, since Monster 2 *is* Weierstrass "continuous" but *is not* naturally continuous.

"Space-Filling" Monsters

We discussed so-called space-filling curves, like the Hilbert curve, in Chapter 12 (see Figure 12.4). The Hilbert curve—let us call it Monster 3—and other curves like it were high on the list of monsters.

Monster 3 violates properties 1, 2, 3, 4, 7, and 8 of the prototypical curve. (Properties 5 and 6 are inapplicable.) It cannot be generated by a moving point. The reason is that such a point must move in some direction at each point of the "curve," but in the Hilbert curve there is never a particular direction that the curve is going in. It changes direction at every point. Therefore, it can have no single direction of motion at any point.

Since the Hilbert curve cannot be generated by a moving point, it is not naturally continuous in Euler's sense. Moreover, it has no direction at any point, nor does it have a tangent at any point. Additionally, it has no specific finite length for any arc. Every arc is of indefinitely long length. Thus it fails properties 1 through 4.

Suppose you take the geometric paradigm to include only real numbers and not infinitesimals. Since the Hilbert curve maps onto all the real points in the square, it maps onto the points on a surface, in violation of property 7. For the same reason, it cannot be formed by an intersection of surfaces, in violation of property 8. This monster is therefore even more monstrous than Monsters 1 and 2. You can see why anyone brought up to think in terms of the geometric paradigm would find such cases "pathological."

As we have just seen, the Hilbert space-filling curve is not a naturally continuous one-dimensional curve; therefore, it is not "continuous" in the geometric paradigm. But what about the Weierstrass paradigm? Does the Hilbert curve preserve closeness?

The Hilbert curve does preserve closeness. Points close to one another on the unit interval get mapped onto points close to one another in the unit square. Again, a monster that is not continuous in the geometric paradigm *is* "continuous" in the Weierstrass paradigm, showing once more that the two notions of continuity are different concepts. Weierstrass continuity is neither a formalization nor a generalization of natural continuity.

More Monsters

Here are two well-known functions, which we will call Monsters 4 and 5.

$$\text{Monster 4: } f(x) = \begin{cases} 1 \text{ if } x \text{ is irrational} \\ 0 \text{ if } x \text{ is rational} \end{cases}$$

$$\text{Monster 5: } g(x) = \begin{cases} 1 \text{ if } x \text{ is irrational} \\ \text{undefined if } x \text{ is rational} \end{cases}$$

The geometric paradigm can make no sense of these functions, since they are nothing at all like a prototypical curve. As far as natural continuity is concerned, Monster 4 is nowhere naturally continuous. At every point p, there is a point infinitely close to p where the function "jumps." Monster 5, which is undefined over the rationals, is defined over a domain that is not naturally continuous—the irrational numbers with "holes" at the rational points. The graph of Monster 5 will therefore also have "holes" at rational points and not be naturally continuous.

In the Weierstrass paradigm, Monster 4 is not "continuous," because it does not preserve closeness. No matter what point x you pick on the real-number line, if you take an $\varepsilon = 1/2$ there will be no δ that will keep the value of f within 1/2.

But Monster 5 *is* "continuous" in the Weierstrass paradigm. The function g does preserve closeness. Even though the function is now defined over a naturally discontinuous domain, the values of the function do not "jump"; rather, they are always the same. Thus, for any ε, $g(x)$ will stay within ε of 1, no matter what δ you pick.

Again, this example shows that the two notions of "continuity" are very different concepts. Weierstrass "continuity" is shown, once more, to be neither a formalization nor a generalization of natural continuity.

Pierpont's Address

The discretization program is more than a century old. But whereas it is now largely accepted without question, it was very much an issue of debate a century ago. It is instructive to see how an influential American mathematician argued in favor of the Weierstrass paradigm a hundred years in the past.

In 1899, Professor Pierpont felt compelled to address the American Mathematical Society to try to convince his colleagues of the necessity of Weierstrass's arithmetization of calculus. The list we gave of the eight properties of curves is taken from Pierpont's address. Pierpont presented the list and went through his discussion of various monster functions, arguing, as we have done, that they cannot make sense as curves in the geometric paradigm. He also invoked the rigor myth again and again.

> The notions arising from our intuitions are vague and incomplete . . . The practice of intuitionists of supplementing their analytical reasoning at any moment by arguments drawn from intuition cannot therefore be justified. (Pierpont, 1899, p. 405)

He expressed the idea that only arithmetization is rigorous, while geometric intuitions are not suitable for "secure foundations."

> There are, however, a few standards which we shall all gladly recognize when it becomes desirable to place a great theory on the securest foundations possible . . . What can be proved should be proved. In attempting to carry out conscientiously this program, analysts have been forced to arithmetize their science. (p. 395)

The idea that numbers do not involve intuition comes through clearly.

> The quantities we deal with are numbers; their existence and laws rest on an arithmetic and not on an intuitional basis . . . and therefore, if we are endeavoring to secure the most perfect form of demonstrations, it must be wholly arithmetical. (p. 397)

What was interesting about Pierpont is that he knew better. He knew that ideas are necessary in mathematics and that one cannot, *within mathematics*, rigorously put ideas into symbols. The reason is that ideas are in our minds; even mathematical ideas are not entities within formal mathematics, and there is no branch of mathematics that concerns ideas. The link between mathematical formalisms using symbols and the ideas they are to represent is part of the study of the mind—part of cognitive science, not part of mathematics. Formalisms using symbols have to be understood, and the study of that understanding is outside mathematics per se. Pierpont understood this:

> From our intuition we have the notions of curves, surfaces, continuity, etc. . . . No one can show that the arithmetic formulations are *coextensive* with their corresponding intuitional concepts. (pp. 400–401; original emphasis)

As a result he felt tension between this wisdom and the appeal of the arithmetization of calculus.

Pierpont was torn. He understood that mathematics was irrevocably about ideas, but he could not resist the vision of total rigor offered by the arithmetization program:

> The mathematician of today, trained in the school of Weierstrass, is fond of speaking of his science as "die absolut klare Wissenschaft" [the absolutely clear science]. Any attempts to drag in metaphysical speculations are resented with indignant energy. With almost painful emotions, he looks back at the sorry mixture of metaphysics and mathematics which was so common in the last century and at the beginning of this. The analysis of today is indeed a transparent science built up on the simple notion of number, its truths are the most solidly established in the whole range of human knowledge. It is, however, not to be overlooked that the price paid for this clearness is appalling; it is total separation from the world of our senses. (p. 406)

This is a remarkable passage. Pierpont knows what is going to happen when mathematics comes to be conceived of mainly—or only—as being about "rigorous" formalism. Mathematical ideas—he uses the unfortunate term "intuition," which misleadingly suggests vagueness and lack of rigor—not only will be downplayed but will be seen as the enemy, a form of mathematical evil to be fought and overcome. He can't help himself. He has been converted and comes down on the side of "rigor," but he sees the cost and it is "appalling."

What Weierstrass Accomplished

Weierstrass's accomplishment was enormous. He achieved the remetaphorization of a major part of mathematics. With Dedekind, Cantor, and others, he played a major role in constructing the metaphorical worldview of most contemporary mathematicians. His work was pivotal in getting the following collection of metaphors accepted as the norm:

- Spaces Are Sets of Points.
- Points on a Line Are Numbers.
- Points in an n-dimensional Space Are n-tuples of Numbers
- Functions Are Ordered Pairs of Numbers.
- Continuity for a Line Is Numerical Gaplessness
- Continuity for a Function Is Preservation of Closeness

Where would contemporary mathematics be without these metaphors?

Weierstrass was also responsible for the demise of the respectability of the geometric paradigm. As Pierpont understood, that was a shame. The problem was not that the geometric paradigm was vague. The problem was that its limits were not properly understood. It is a wonderful tool for understanding naturally continuous, everywhere differentiable functions—functions that can be conceptualized as curves in a plane. Its advantage is that *for such functions* it permits visualization and spatial understanding to inform arithmetic and vice versa.

The existence of the monster functions shows that the geometric paradigm is not a fully general intellectual tool for studying *all* functions. It does not follow that it should be scrapped; all that follows is that it needs to be supplemented and its limits well understood.

One of the wonderful effects of the Weierstrass revolution was that it made mathematicians break down the relevant properties of functions and spaces and study them independently: continuity, differentiability, connectedness, compactness, and so on. Weierstrass's preservation of closeness concept (mistakenly called "continuity") is at the heart of modern topology. The grand opening up of twentieth-century mathematics had everything to do with Weierstrass's remetaphorization of analysis.

Continuity and Its Opposite, Discreteness

The German mathematician Hermann Weyl (1885–1955) noted in his classic work *The Continuum* (1987, p. 24):

> We must point out that, in spite of Dedekind, Cantor, and Weierstrass, the great task which has been facing us since the Pythagorean discovery of the irrationals remains today as unfinished as ever; that is, the continuity given to us immediately by intuition (in the flow of time and in motion) has yet to be grasped mathematically as a totality of discrete "stages" in accordance with that part of its content which can be conceptualized in an "exact" way.

Why should it, as Weyl says, be a "task" of mathematics to "grasp . . . the continuity given to us immediately by intuition (in the flow of time and motion) . . . as a totality of discrete 'stages'"? Why does mathematics have to understand the continuous in terms of the discrete?

Each attempt to understand the continuous in terms of the discrete is necessarily metaphorical—an attempt to understand one kind of thing in terms of another kind of thing. Indeed, it is an attempt to understand one kind of thing—the naturally continuous continuum—in terms of its very opposite—the discrete. We find it strange that it should be seen as a central task of mathematics to provide a metaphorical characterization of the continuum in terms of its opposite. Any such metaphor is bound to miss aspects of what the continuum is, and miss quite a bit.

If "the great task" is to provide absolute, literal foundations for mathematics, then the attempt to conceptualize the continuous in terms of the discrete is self-defeating. First, such foundations cannot be literal; they can only be metaphorical. Second, as Weyl himself says, only "part of its content" can be conceptualized discretely. The rest must be left out. If Weyl is right, the task cannot be accomplished.

We believe there is a greater task: understanding mathematical ideas.

Le trou normand:
A Classic Paradox of Infinity

IT IS TRADITIONAL IN FINE FRENCH CUISINE to offer the diner a *trou nor-mand*—usually a kind of *sorbet* with calvados—after a number of heavy courses. The sorbet refreshes the palate. It is a gastronomically light interlude before more courses. At this point in the book, we offer you the following *trou normand*.

There is a classic paradox that involves the following mathematical construction, as given in Figure T.1. Start at stage 1 with a semicircle of diameter 1, extending from 0 to 1 on the *X*-axis of the Cartesian plane. The perimeter of the semicircle is of length $\pi/2$. The center will be at $x = 1/2$, and the semicircle is above the *X*-axis.

At stage 2, divide the diameter in half and form two semicircles extending from 0 to 1/2 and 1/2 to 1. The two centers will be at $x = 1/4$ and $x = 3/4$ (see Figure T.2). The perimeter of each semicircle is $\pi/4$. The total length of both perimeters is $\pi/2$. The length of each diameter is 1/2. The total length of both diameters is 1.

At stage 3, divide the diameters in half again to form two more semicircles. There will now be four semicircles (see Figure T.3). The centers will be at $x = 1/8$, $x = 3/8$, $x = 5/8$, and $x = 7/8$. The perimeter of each semicircle will be $\pi/8$. The total length of all four perimeters is $\pi/2$. The length of each diameter is 1/4. The total length of all four diameters is 1.

Continue this process indefinitely. This is an infinite process without an end. At every stage *n*, there will be a bumpy curve made up of 2^{n-1} semicircles, whose total length is $\pi/2$, and where all the diameters taken together correspond to a segment of length 1. As *n* gets larger, the bumpy curve gets closer and closer to the diameter line, with the area between the bumpy curve and the diameter line getting smaller and smaller. But the length of the bumpy curve remains $\pi/2$ at all stages, while the length of the diameter line remains 1 at all stages. As *n* approaches infinity, the area between the bumpy curve and the diameter line ap-

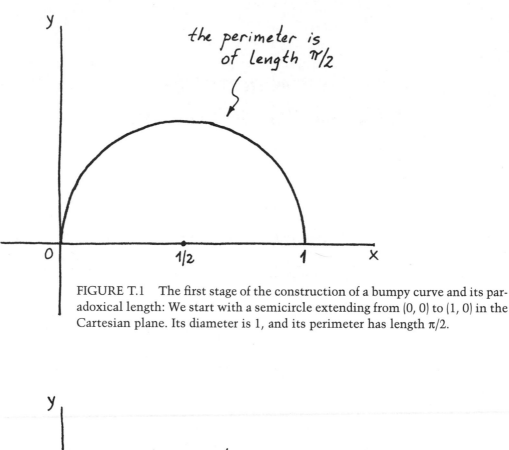

FIGURE T.1 The first stage of the construction of a bumpy curve and its paradoxical length: We start with a semicircle extending from (0, 0) to (1, 0) in the Cartesian plane. Its diameter is 1, and its perimeter has length π/2.

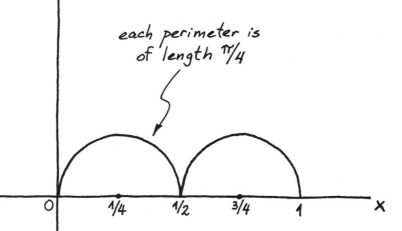

FIGURE T.2 The second stage of the construction: We divide the line segment from (0, 0) to (1, 0) in half, with each half as the diameter of a smaller semicircle. Each semicircle now has a perimeter of length π/4, and since there are two of them, the total perimeter is π/2, the same as the perimeter of the single semicircle in the previous stage (Figure T.1).

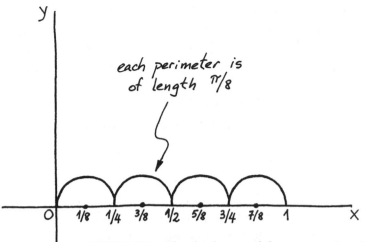

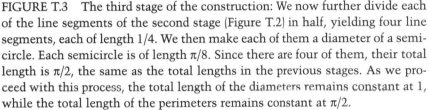

FIGURE T.3   The third stage of the construction: We now further divide each of the line segments of the second stage (Figure T.2) in half, yielding four line segments, each of length 1/4. We then make each of them a diameter of a semicircle. Each semicircle is of length π/8. Since there are four of them, their total length is π/2, the same as the total lengths in the previous stages. As we proceed with this process, the total length of the diameters remains constant at 1, while the total length of the perimeters remains constant at π/2.

proaches zero, while the lengths of the curve and the line remain constant at π/2 and 1 respectively.

What happens at $n = \infty$?

The Length Paradox

At $n = \infty$, there is no area between the bumpy curve and the diameter line. They occupy the same place in space. Yet the bumpy curve is still of length π/2 and the diameter line is still of length 1. How is this possible? The bumpy curve and the diameter line appear to have become the same line, but with two different lengths! And as we know, a single line should have only a single length.

A clearer statement of the problem will reveal why the apparent paradox arises. In the construction, there is an infinite sequence of curves approaching a limit. But sequences that have limits are sequences of *numbers*, as characterized by the BMI in Chapter 8. How can one get from limits of sequences of *numbers* to limits of sequences of *curves*?

To do so, we will have to use the metaphor we have been discussing in which naturally continuous curves and lines are conceptualized as sets. Given an appropriate sequence of functions $f_1(x)$, $f_2(x)$, . . . , we can conceptualize the ith bumpy curve as a set of ordered pairs of real numbers $(x, f_i(x))$ in the unit square in the Cartesian plane. The first semicircle will be represented by the set of ordered pairs of real numbers $\{(x, f_1(x))\}$. This metaphor allows us to replace the sequence of geometric curves with a sequence of sets of ordered pairs of real numbers. In short, we have gone, via metaphor, from the geometry of space to a different mathematical domain consisting of sets and numbers.

Now that spaces, curves, and points have been replaced metaphorically by sets, ordered pairs, and numbers, we can use the characterization of limits of sequences of numbers as given by the BMI. For each number x between 0 and 1, there will be a sequence of numbers y—$y_1, y_2, y_3, \ldots$—given by the values of y in the functions $f_1(x) = y_1, f_2(x) = y_2, f_3(x) = y_3, \ldots$. Each of these sequences of y-values defined for the number x will have a limit as the y's get smaller and smaller—namely, down to zero. Thus, for each real number x between 0 and 1, there will be a sequence of ordered pairs $(x, y_1), (x, y_2), (x, y_3), \ldots$ that converges to $(x, 0)$ (see Figure T.4).

But a subtle shift has occurred. We have replaced each bumpy curve by a bumpy-curve-set consisting of ordered pairs of numbers (x, y), with $y = f(x)$, where x ranges over all the real numbers between 0 and 1. But what converges to a limit is not this sequence of bumpy-curve-sets. Instead, we have an infinity of convergent sequences of y-values—one from each member of the sequence of bumpy-curve-sets—for each number x between 0 and 1. The limit of each such sequence is the pair $(x, 0)$. The set of all such limits is the set of ordered pairs of numbers $\{(x, 0)\}$, where x is a real number between 0 and 1. This set of ordered pairs of numbers corresponds, via the metaphors used, to the diameter line.

But this set is a *set of limits of sequences of ordered pairs* of numbers. What we wanted was the limit of a sequence of *curves*—that is, the *limit of a sequence of sets of ordered pairs* of numbers. Those are very different things conceptually.

To get what we want from what we have, we need a metaphor—one that we will call the Limit-Set metaphor: The Limit of a Sequence of Sets Is the Set of the Limits of Sequences.

Only via such a metaphor can we get the diameter line to be the limit of the sequence of bumpy-curve-sets. Incidentally, this is the same metaphor used to characterize the so-called space-filling curves discussed in Chapter 12.

We have used two conceptual metaphors:

- Curves (and Lines) Are Sets of Ordered Pairs of Numbers, and
- The Limit-Set metaphor.

If we accept these two metaphors, then the sequence of bumpy curves can be reconceptualized as a sequence of bumpy-curve-sets consisting of ordered pairs of numbers. That sequence will have as its limit the set of ordered pairs $\{(x, 0)\}$, where x is between 0 and 1. This is identical to the set of ordered pairs of num-

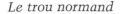
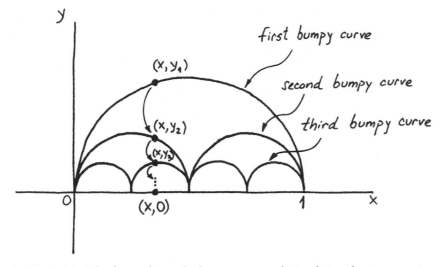

FIGURE T.4 The figure shows the bumpy curves obtained at each stage superimposed. Their maximal heights get smaller and smaller, reduced by a factor of 2 at each stage. In the process, the curves get closer and closer to the line segment from (0, 0) to (1, 0), and the area under the curve approaches zero as the sequence progresses. For every point on the line (x, 0), there is a sequence of points, one on each bumpy curve: (x, y_1), (x, y_2), (x, y_3), ... This sequence of points converges to (x, 0) as the sequence of numbers (y_1, y_2, y_3, ...) converges to 0.

Thus, for each point (x, 0) on the line segment between (0, 0) and (1, 0), there is a sequence of corresponding points (one on each bumpy curve) that converges to (x, 0). The set of all the limits of all the convergent sequences (x, y_1), (x, y_2), ... is thus identical to the set of all the points on the line between (0, 0) and (0, 1).

At each stage, the bumpy curve is a set of points with length $\pi/2$, while the line segment is a set of points with length 1. How can a sequence of curves of length $\pi/2$ converge point by point to a line segment of length 1?

bers representing the diameter line under the metaphor Curves (and Lines) Are Sets of Ordered Pairs of Numbers. Here we can see that the Limit-Set metaphor is one of the sources of the apparent paradox.

What Is the Length of a Set?

In order to characterize the limit of a sequence of curves, we have had to metaphorically reconceptualize each curve as a set—a set of ordered pairs of numbers. The reason is that limits of sequences are technically defined only for numbers, not for geometric curves. But now a problem arises. What is "length" for such a set?

In physical space as we experience it every day, there are natural lengths, like the length of your arm or your foot. Hence, we have units of measurements like "one foot." But when curves are replaced by sets, we no longer have natural lengths. Sets, literally, have no lengths. To characterize the "length" of such a set, we will need a function from the set to a number called its "length." In general, curves in the Cartesian plane have all sorts of numerical properties—the area under the curve, the curvature at each point, the tangent at each point, and so on. Once geometric curves are replaced by sets, then all those properties of the curves will have to be replaced by functions from the sets to numbers.

The Length Function

The length of a line segment $[a, b]$ along a number line is conceptualized metaphorically as the absolute value of the difference between the numbers, namely, $|b - a|$ (see Chapter 12). This is extended via the Pythagorean Theorem to any line segment oriented at any angle in the Cartesian Plane. Suppose its endpoints are (a_1, b_1), and (a_2, b_2). Its length is $\sqrt{(|a_2 - a_1|^2 + |b_2 - b_1|^2)}$.

What about the length of a curve? Choose a finite number of points along the curve (including the end points). Draw the sequence of straight lines connecting those points. Call it a *partition* of the curve. The length of the partition is the sum of the lengths of the straight lines in the partition. Define the length of the curve via the BMI as the *least upper bound of the set of the lengths of all partitions of the curve*. This gives us a *length function* for every curve.

Now think of the line segments as measuring sticks. As the measuring sticks get shorter and shorter, they measure the length of the curve more and more accurately. The length of the curve is *the limit of measurements as the length of the measuring sticks approaches zero*. Let us call this the Curve Length Metaphor.

The Sources of the Apparent Paradox

The appearance of a paradox comes from two sources:
    (1) A set of expectations about naturally continuous curves, and
    (2) The metaphors used to conceptualize curves in formal mathematics.

It should come as no surprise that our normal expectations are violated by the metaphors of formal mathematics.

Let us start with our normal expectations.

* Length, curvature at each point, and the tangent at each point are inherent properties of a naturally continuous curve.
* Identical curves should have identical properties.
* Nearly identical curves should have nearly identical properties.
* If a sequence of curves converges to a limit curve, the sequence of properties of those curves should converge to the properties of the limit curve.

The reason we have these expectations is that we metaphorically conceptualize curves as *objects in space* and properties that are inherent to a curve as *parts of the curve*. For example, we take the curvature at a point in a curve as being *part of the curve*. If we think of a curve as being traced out by a point in motion, we think of the direction of motion at each point (mathematicized as the tangent to the curve) as *part of the curve*. If we think of curves as objects and their inherent properties as parts of those objects, then as the curves get very close, their properties should get correspondingly close.

The metaphors that characterize formal mathematics, when taken together, violate these expectations. Here are the relevant metaphors.

- Functions Are Ordered Pairs of Real Numbers.
- Real Numbers Are Limits of Sequences of Rational Numbers (uses the BMI)
- Curves (and lines) Are Sets of Points.
- Points Are Ordered Pairs of Numbers (in the Cartesian Plane Blend)
- The Limit Metaphor (uses the BMI for limits of sequences of numbers)
- The Limit-Set Metaphor (defines the limit of a sequence of curves as the set of point-by-point limits, as in Figure T.4)
- Properties of Curves Are Functions
- Spatial Distance (between points a and b on a line) Is Numerical Difference ($|b - a|$)
- The Curve Length Metaphor (uses the BMI)
- Closeness (between two curves) Is a Number (defined by a metric, which assigns numbers to pairs of functions)

It should be clear why such metaphors violate the expectations discussed above. Curves are not physical objects, they are sets. Inherent properties are not parts, they are functions from one entity to a distinct entity. When two "curves" (sets) are "close" (have a small number assigned by a metric), there is no reason to think that their "properties" (numbers assigned to them by functions) should also be "close."

Moreover, the Limit-Set Metaphor that defines limits for curves says nothing about properties (like curvature, tangent, and length). From the perspective of the metaphors inherent in the formal mathematics, there is no reason to think that properties like curvature, tangent, and length *should* necessarily converge when the curves converge point by point.

Tangents and Length

Imagine measuring the length of a semi-circle on one of the bumpy curves, using measuring sticks that get shorter and shorter. If there were n semi-circles on that bumpy curve, the measurements of each semi-circle would approach $\pi/2n$ as a limit. The total length, n times $\pi/2n$, is always $\pi/2$.

As the measuring sticks get shorter, they change direction and eventually approach the orientations of tangents to the curve. The Curve Length Metaphor thus provides a link between lengths of curves and orientations of tangents, which in turn are characterized by the first derivative of the function defining the curve.

Compare the semi-circles with the diameter line. There the measuring sticks are always flat, with tangents at zero degrees. Correspondingly, the first derivative is zero at each point.

The Bumpy Curves in Function Space

A *function space* is defined by the metaphor that A Function Is a Point in a Space. The metaphor entails that there is a "distance" between the "points," that is, the functions. By itself, that metaphor does not tell us how "close" the "points" are to one another. For this, one needs a metric, a function from pairs of functions to numbers. The numbers are understood as metaphorically measuring the "distance" between the functions.

All sorts of metrics are possible, providing that they meet three conditions on distance d: $d(a, a) = 0$, $d(a, b) = d(b, a)$, and $d(a, b) + d(b, c) \leq d(a, c)$. In the field of *functional analysis*, metrics are defined so as to reflect properties of functions. To get an idea of how this works, imagine the bumpy curves and the diameter line as being points in a space. Imagine the metric over that space as being defined in the following way.

(1) The distance between any two functions $f(x)$ and $g(x)$ is defined as the sum of
    (i)   the maximum difference in the values of the functions, plus
    (ii)   the average difference in the values of the derivatives of the functions.

Formally, this is written:

$$d(f, g) = \sup_x \left| f(x) - g(x) \right| + \int_0^1 \left(\left| f'(x) - g'(x) \right| \right) dx$$

Via the metaphors Curves Are Sets of Points and Functions Are Ordered Pairs of Real Numbers, let $g(x)$ be the diameter line and let $f(x)$ vary over the bumpy curves. As the bumpy curves get closer to the diameter line, the maximum distance (the first term of the sum) between each bumpy curve $f_i(x)$ and the diameter line $g(x)$ approaches zero. The second term of the sum does not, however, approach zero. It represents the average difference between the values of the tangents at each value (x). In the diameter line $g(x)$ the tangents are always zero, so $g'(x) = 0$ for all x. Since the tangents on each bumpy curve go through the same range of values, the average of the absolute values of the tangents will be the same for each bumpy curve. Thus term (ii) will be a non-zero constant when $g(x)$ is the diameter line and $f(x)$ is any bumpy curve.

Here's what this means:

- Curves that are close in the Cartesian plane point-by-point, but *not* in their tangents, are not "close" in the function space defined by this metric.

In this function space, the metric given in (1) takes into account more than the difference between the values of the functions. It also considers the crucial factor that keeps the length of the bumpy curves from converging to the length of diameter line, namely, the difference in the behavior of the tangents. In this metaphorical function space, the sequence of "bumpy curve" points do *not* get close to the diameter-line point as n approaches infinity.

In Figures T.1–4, we represented the sequence of functions as curves — bumpy curves. This was a *metaphorical* representation of the functions, using the metaphors Points in the Plane Are Ordered Pairs of Numbers and Functions Are Sets of Ordered Pairs of Numbers, which are built into the Cartesian Plane Blend (see p. 385 below for details). This spatial representation of the function gave the illusion that, as n approached infinity, the bumpy curves "approached" (came indefinitely close to) the diameter line. But this metaphorical image ignored the derivatives (the tangents) of the functions, which are crucial to the question of length. In this sense, this particular metaphorical representation of these functions in the Cartesian Plane is degenerate: it leaves out crucial information. But in the function space defined by the metaphor Functions Are Points in Space and metric (1), this crucial information is included and it becomes clear that the bumpy curve functions do *not* come close to the diameter line function. There is not even the appearance of paradox here.

Under this metric, curves that are close *both* point-by-point *and* in their tangents will be represented by points that are close in this metaphorical function space.

Continuity over Function Spaces: The Role of Metrics

In the function space defined by this metric, the Length function will be "continuous" in Weierstrass' sense of "preserving closeness"— mapping close points onto close points (see Chapter 14). For example, suppose we pick any bumpy curve, and consider a partition of this curve into line segments. Each such partition is itself a curve, represented by a point in this function space. As the line segments in the partitions of the bumpy curve get smaller and smaller, the points in the function space representing the successive partitions get closer and closer to the point representing the bumpy curve. Correspondingly, the lengths of the partitions get closer and closer to the length of the curve. At the limit, the Length Function is continuous.

Note that the Length Function would not be "continuous" in Weierstrass' sense if the metric were changed to omit term (ii) — the average difference between derivatives, that is, slopes of tangents. Term (i) includes only the maximum difference between the values of the functions. In a function space with this metric, the bumpy curve points converge to the diameter line point. Weierstrass continuity fails for the Length Function in this space, since it does not preserve closeness in all cases. *Pairs of points that are close*, e.g., the point for a bumpy curve for a large n and the point for the diameter line *map via the Length Function onto a pair of numbers that are not close, $\pi/2$ and 1.*

Continuity as Normality

Here we see formal mathematics capturing a mathematical intuition: Because the *tangents* of the bumpy curves do not converge to the tangents of the diameter line, the *lengths* of the bumpy curves *should not* converge to the length of the diameter line. Not only is there no paradox, but the behavior of the Length Function is *normal*!

In other words, the Length Function is acting *normally* if it is *continuous* over a function space defined by a metric that takes into account *precisely* the factors relevant to characterizing relative length. Here those factors are not just how close the points in the curves get to one another, but how the tangents behave.

The Moral

The appearance of paradox arose for the following reasons:

(1) We took the functions as literally *being* the curves in the Cartesian Plane, when those curves were just metaphorical representations of the functions.

(2) Those representations distorted the role of a crucial property of the functions and so misled us into thinking that the bumpy curves got "close" in all respects to the diameter line.

(3) This was reinforced by our ordinary understanding of curves as objects with their properties as inherent parts of the objects.

(4) Mathematics provides an alternative conceptual metaphor — Functions Are Points in Space — in which these issues can be conceived more clearly from a different metaphorical perspective.

(5) The choice of conceptual metaphors matters. It is important to be aware of how metaphor choice affects our understanding of a problem.

Part V

Implications for the
Philosophy of Mathematics

15

The Theory of
Embodied Mathematics

A S COGNITIVE SCIENTISTS, we have been studying mathematics as a subject matter, asking certain basic questions:

1. What are mathematical ideas from the perspective of cognitive science? What commonplace cognitive mechanisms do they use?
2. Given that innate mathematics is minuscule—consisting of subitizing and a tiny bit of basic arithmetic—what cognitive mechanisms allow this tiny innate basis to be extended to generate all of advanced mathematics?
3. How are mathematical ideas grounded in our experience?
4. Which mathematical ideas are metaphorical and which are conceptual blends?

To answer these questions, we've had to look at details: the laws of arithmetic, Boolean classes, symbolic logic, group theory, set theory, and hyperset theory. We've also had to look at various forms of infinity: points at infinity, infinite sets, mathematical induction, infinite decimals, infinite sequences, infinite sums, least upper bounds, infinite intersections, transfinite cardinals, transfinite ordinals, infinitesimals, points, and continuity. And in the case study to follow, we will try to answer how $e^{\pi i}$ is conceptualized, which will lead us to analytic geometry, trigonometry, logarithms, exponentials, and complex numbers.

This hardly exhausts the details that cognitive science needs to look at. Though it is a bare beginning, we think it is a sufficiently rich range of topics to be representative of the kinds of questions that cognitive science will ultimately have to deal with. If any approach to cognitive science cannot answer questions 1 through 4 for at least these subject matters, then we would consider it inadequate as a theory of mind.

That is why we have taken the time and trouble to write this book.

A Short Advertisement for the Case Study of What $e^{\pi i} + 1 = 0$ Means

We realize that this is a long book and that you may not make it all the way through the case study in the last four chapters. But if you are at all interested in the power of mathematical idea analysis, then we heartily commend the case study to you. It is only in a study of that scope and length that we can show how really complex mathematical ideas get built up and how ideas are explicitly linked across subfields of mathematics.

The case study goes systematically through dozens of central ideas that use conceptual metaphors and blends. It shows how these fit together in complex networks—how one is embedded within another, and within another. What emerges from the case study is the beginning of what might be considered the Mathematical Idea Genome Project—a study in detail of the intricate structure of sophisticated mathematical ideas.

The purpose of mathematical idea analysis is to provide a new level of understanding in mathematics. It seeks to explain *why* theorems are true on the basis of what they mean. It asks what ideas—especially what metaphorical ideas—are built into axioms and definitions. It asks what ideas are implicit in equations and how *ideas* can be expressed by mere numbers. And finally it asks what is the ultimate grounding of each complex idea. That, as we shall see, may require some complicated analysis:

1. tracing through a complex mathematical idea network to see what the ultimate grounding metaphors in the network are;
2. isolating the linking metaphors to see how basic grounded ideas are linked together; and
3. figuring out how the immediate understanding provided by the individual grounding metaphors permits one to comprehend the complex idea as a whole.

Encounters with the Romance

In doing the research for this book, we have talked to a lot of mathematicians, mathematics students, and mathematics educators and analyzed a lot of books and articles. In the course of this research, we kept bumping up against a mythology—what we have called the *Romance of Mathematics*. As our research progressed, it became clear that our findings contradicted this mythology. This is not an unusual occurrence in cognitive science; it happens all the time when you study people's unconscious conceptual systems and value systems. People's conscious beliefs about time, causation, morality, and politics are typically inconsistent with their unconscious conceptual systems. It is also not unusual for people to get angry when told that their unconscious conceptual systems contradict their fondly held conscious beliefs, especially in sensitive areas like morality, religion, and politics. What we have found is that mathematics is one such sensitive area. Those who understand and use advanced mathematics tend to hold strong views about what mathematics is.

The discrepancies between the findings of cognitive science and the folk theories of the people whom cognitive scientists study is a particularly sensitive issue in the case of mathematics. The reason is that a great many of those who have a serious knowledge of mathematics not only tend to believe the mythology we call the Romance of Mathematics but tend to believe it fiercely. The Romance of Mathematics is part of their worldview, their very identity. Since our findings in general contradict the romance, we would not be surprised to find that this book infuriated such people.

We could just shrug our shoulders and say "Such is the lot of the empirical scientist" and ignore them. But we think it is wiser to confront the issue head-on, since, for the most part, these are people we deeply respect and like. In some cases, they are even heroes of ours. What we will do is state here again, at greater length than in the Preface, what the romance is and how our results contradict it, so that at least the points at issue will be clear.

THE ROMANCE OF MATHEMATICS

- Mathematics is an objective feature of the universe; mathematical objects are real; mathematical truth is universal, absolute, and certain.
- What human beings believe about mathematics therefore has no effect on what mathematics really is. Mathematics would be the same even if there were no human beings, or beings of any sort. Though mathematics is abstract and disembodied, it is real.
- Mathematicians are the ultimate scientists, discovering absolute truths not just about this physical universe but about any possible universe.

- Since logic itself can be formalized as mathematical logic, mathematics characterizes the very nature of rationality.
- Since rationality defines what is uniquely human, and since mathematics is the highest form of rationality, mathematical ability is the apex of human intellectual capacities. Mathematicians are therefore the ultimate experts on the nature of rationality itself.
- The mathematics of physics resides *in* physical phenomena themselves—there are ellipses in the elliptical orbits of the planets, fractals in the fractal shapes of leaves and branches, logarithms in the logarithmic spirals of snails. This means that "the book of nature is written in mathematics," which implies that the language of mathematics is the language of nature and that only those who know mathematics can truly understand nature.
- Mathematics is the queen of the sciences. It defines what precision is. The ability to make mathematical models and do mathematical calculations is what makes science what it is. As the highest science, mathematics applies to and takes precedence over all other sciences. Only mathematics itself can characterize the ultimate nature of mathematics.

Of course, not everyone who believes some of these statements believes them all. People vary in how much of the Romance they accept. But these themes are commonplace in popular writings about mathematics, and many of them are taken for granted in textbooks and mathematics courses, as well as in everyday discussions about mathematics. As you might guess, some mathematicians believe none of this and say so loudly and forcefully. Yet the Romance provides the standard folk theory of what mathematics is for our culture.

A Beautiful Story

The Romance of Mathematics makes a wonderful story. It is the premise of most popular books on mathematics and many a science-fiction movie. It has attracted generations of young people to mathematics. It perpetuates the mystique of the Mathematician, with a capital "M," as someone who is more than a mere mortal—more intelligent, more rational, more probing, deeper, visionary.

It is a story that many people *want* to be true. We want to know that amid the uncertainties and doubts of life, *something* is certain and absolutely true—that amid all the irrationalities around us, *some* people are supremely rational, some order is possible, and that at least in doing mathematics *we* can be rational, logical, and certain of our conclusions.

The Romance of Mathematics is sexy. Wouldn't *you* want to be the mathematician in the romance, the hero of the story? And if you're not the hero yourself, don't you want such heroes to look up to? Don't you want to live in a world where such heroes exist? It is a beautiful and inspiring story. We grew up with it, and it still reverberates within us. But sadly, for the most part, it is not a true story.

The Sad Consequences

The Romance of Mathematics is not a story with a wholly positive effect. It intimidates people. It makes mathematics seem beyond the reach of even excellent students with other primary interests and skills. It leads many students to give up on mathematics as simply beyond them.

The Romance serves the purposes of the mathematical community. It helps to maintain an elite and then justify it. It is part of a culture that rewards incomprehensibility, in which it is the norm to write only for an audience of the initiated—to write in symbols rather than clear exposition and in maximally accessible language. The inaccessibility of most mathematical writing tends to perpetuate the Romance and, with it, its ill effects: the alienation of other educated people from mathematics, and the inaccessibility of mathematics to people who are interested in it and could benefit from it. Socially, the inaccessibility of mathematics has contributed to the lack of adequate mathematical training in the populace in general. And that lack of adequate mathematical training contributes to an alarming trend—the division of our society into those who can function in an increasingly technical economy and those who cannot. We believe that the Romance of Mathematics is not an entirely harmless myth—that at least indirectly it is contributing to the social and economic stratification of society.

Because the Romance of Mathematics is scientifically untenable and doing social harm, it is time to reconsider it. Our goal is to give a more realistic picture of the nature of mathematical cognition, and in the process to make mathematics more accessible as well.

If you deeply believe the Romance of Mathematics, if it is part of your worldview and your identity, we ask you in the name of scientific understanding to put the Romance aside and be open to the possibility that it is false.

A Question of Faith:
Does Mathematics Exist Outside Us?

The central claim of the Romance, the one on which the entire myth depends, is that there is a *transcendent mathematics*—one that has an objective existence, external to human beings or any other beings. The unspoken assumption is this: Transcendent mathematics includes that portion of human mathematics that has been proved. This is a significant presupposition. The Romance would have no importance if it assumed that what has been proved by human mathematicians has no relation at all to any objectively true mathematics—that proofs by human mathematicians were not necessarily universal mathematical truths.

Proof therefore plays a central role in the romance. It is through proof that human mathematicians transcend the limitations of their humanity. Proofs link human mathematicians to truths of the universe. In the romance, proofs are discoveries of those truths.

Is there any scientific evidence that this is true—that what is proved in human mathematics is an objective universal truth, true of this physical universe or any possible universe, regardless of the existence of any beings? The answer is no. *There is no such evidence!* The argument is given in the Introduction to this book.

Moreover, as far as we can tell, there can be no such evidence, one way or the other. There is no way to tell empirically whether proofs proved by human mathematicians are objectively true, external to the existence of human beings or any other beings.

The Inherent Implausibility of Transcendent Mathematics

We should point out that even on the basis of mathematics itself—without any scientific evidence—the claim that transcendent mathematics exists appears to be untenable. One important reason is that mathematical entities such as numbers are characterized in mathematics in ontologically inconsistent ways. Here are some of the ways that natural numbers are characterized within modern mathematics:

1. On the number line, all numbers are points on a line. The number line is as central to mathematics as any mathematical concept could be. On the number line, points are zero-dimensional geometric objects. If numbers are literally and objectively points on a line, then they, too—as abstract transcendent entities—must be zero-dimensional geometric objects!

2. In set theory, numbers are sets. Zero is the empty set. One is the set containing the empty set. Two is the set containing (a) the empty set and (b) the set that contains the empty set. And so on. In set theory, numbers are therefore not zero-dimensional geometric entities.

3. In combinatorial game theory (Berlekamp, Conway, & Guy, 1982), numbers are values of positions in combinatorial games. Combinatorial game theory is a normal, respectable (though not widely known) branch of mathematics. On the transcendent mathematics position, the entities of combinatorial game theory and their properties should be real. Numbers, therefore, should really be values of positions in combinatorial games, and not points on a line or sets.

Since transcendent mathematics takes each branch of mathematics to be literally and objectively true, it inherently claims that it is literally true of the number line that numbers are points, literally true of set theory that numbers are sets, and literally true of combinatorial game theory that numbers are values of positions. None of these branches of mathematics has a branch-neutral account of numbers.

But, again according to transcendent mathematics, there should be a single kind of thing that numbers are; that is, there should be a unique ontology of numbers. Numbers should either be zero-dimensional geometric entities lined up in an order, or they should be sets of the appropriate structure, or they should be values of positions in games, or something else, or just numbers—but only one kind! That is an inherent problem in the philosophical paradigm of transcendent mathematics.

But human mathematics is richer than the transcendent mathematics view allows. It simply has many ontologically distinct and incompatible notions of number. Scientific considerations aside, the transcendent mathematics view does not make sense within mathematics itself. To make sense of the ontology of number in mathematics, one has to give up on the idea that mathematics has an objective existence outside human beings and that human beings can have correct knowledge (via proof) of transcendent mathematics.

For reasons both external and internal to mathematics, the transcendent mathematics position cannot work.

Is Mathematics in the Physical World?

There is another popular view close to transcendent mathematics but not identical—the idea that mathematics is part of the physical world. This view is based on the enormous success that mathematics has had in physics, where mathematics is used to describe phenomena and make correct predictions with great accuracy. In many cases, physical regularities can be stated briefly and succinctly in terms of "laws" formulated in mathematical terms. This fact has often been misstated as "The universe runs according to mathematical laws," as if the laws came first and the physical universe "obeyed" the laws. Accordingly, the "truth" of physical laws formulated in mathematical terms is taken as indicating that the mathematics used in stating the physical laws is actually there in the physical universe. Since the regularities of the physical universe exist external to human beings, so mathematics itself must exist external to human beings as part of the physical universe.

There is a great deal that is wrong with this argument. First, no one observes laws of the universe as such; what are observed empirically are *regularities* in the universe. Regularities in the universe exist independent of us. *Laws* are mathematical statements made up by human beings to attempt to characterize those *regularities* experienced in the physical universe.

Physicists, having physical bodies and brains themselves, can comprehend regularities in the world *only by using the conceptual systems that the body and brain afford.* Similarly, they understand mathematics *using the conceptual systems that the body and brain afford.*

What they do in formulating "laws" is fit their human conceptualization of the physical regularities to their prior human conceptualization of some form of mathematics. There is no unmediated fit between mathematics and physical regularities in the world. All the "fitting" between mathematics and the regularities of the physical world is done *within the minds of physicists who comprehend both.* The mathematics is in the mind of the mathematically trained observer, not in the regularities of the physical universe.

Moreover, there is no way to make sense of what it would mean for the mathematics to literally be out there in the physical universe. Consider a few cases.

1. Take the use of Cartesian coordinates in stating physical laws. There are no physical Cartesian coordinates literally extending out there into space, with an origin (zero-point) at some point in space and positive numbers extending off in one direction and negative numbers extending off in the opposite direction. The idea of Cartesian coordinates is a human idea that human physicists impose upon space in order to use mathematics to do calculations. They are not *physically* out there.

2. Consider the fact that baseballs move in parabolic trajectories, which can be calculated with numbers by the use of mathematical equations. Are the numbers that are used in the calculation actually out there in

space lined up along the parabola? When Mark McGwire hits a home run, are the numbers that would be used to calculate the trajectory literally out there in the flight path of the ball, number by number? Obviously not. The ball is out there and people observe it. There is a regularity about how balls move in a gravitational field. The numbers are used by people to calculate the path of motion. The numbers are in people's minds, not out in space.

3. The mathematics used to comprehend and calculate how electromagnetism works is complex analysis—calculus using complex numbers. Electromagnetic forces are periodic (they recur) and they systematically wax and wane; that is, they get stronger and weaker. As we shall see in Case Study 3, the mathematics of complex numbers is, among other things, the mathematics of periodicity, of recurrent systematic increases and decreases. The physical laws governing electromagnetism are therefore formulated using complex numbers, which have so-called real and imaginary parts. In those laws, the value of the "real" part of the complex numbers is used to indicate the degree of electromagnetic force. The imaginary part of those complex numbers is necessary in the calculation. Does this mean that the imaginary numbers used in the calculations are literally out there in the physical universe? Should we believe that there is an additional—imaginary—dimension to the physical world because imaginary numbers are used in the calculation of the periodic waxing and waning of electromagnetic forces? There is no *physical* reason to believe this.

The nature of the physical regularity is clear: Electromagnetic forces periodically wax and wane in a systematic way. Human beings have come up with a mathematics that can be used to calculate such periodic variations. Human beings use the mathematics to calculate the physical regularities. The mathematics is in the minds of the physicists, not in the physical regularities themselves. The imaginary numbers, in this case, are in the minds of the physicists, not in the universe. The same is true of the "real" numbers used in the calculations.

In summary,

1. There are regularities in the universe independent of us.
2. We human beings have invented consistent, stable forms of mathematics (usually with unique right answers).
3. Sometimes human physicists are successful in fitting human mathematics as they conceptualize it to their human conceptualization of the

regularities they observe in the physical world. But the human mathematical concepts are not out there in the physical world.

In addition, it should be observed that most of the mathematics that has been done in the history of the discipline has no physical correlates at all. There are no transfinite numbers in physical laws, no parallel lines that physically meet at infinity, no quaternions, no empty sets, no Diophantine equations, no Sierpinski gaskets, no space-filling curves. For this reason alone, mathematics cannot be said to be external to human beings simply on the grounds that mathematics can be used as an effective tool for making accurate calculations and predictions in physics.

Why the Only Mathematics
Is Embodied Mathematics

If the Romance of Mathematics were true, if mathematics had an objective existence outside of human minds and brains, then the cognitive science of mathematics would be relatively innocuous and uninteresting. The argument runs like this:

- The study of mathematical cognition may tell us about how we human beings conceptualize and understand mathematics, how mathematics might be realized in the human mind and brain, how it might be learned, or how we make mathematical discoveries, but it cannot tell us anything about mathematics itself!
- After all, if mathematics exists independent of human beings, studying the cognitive processes and neural structure of the human mind and brain won't get us any closer to the true nature of mathematics.

But the only access that human beings have to any mathematics at all, either transcendent or otherwise, is through concepts in our minds that are shaped by our bodies and brains and realized physically in our neural systems. For human beings—or any other embodied beings—mathematics *is* embodied mathematics. The only mathematics we can know is the mathematics that our bodies and brains allow us to know. For this reason, the *theory of embodied mathematics* we have been describing throughout the book is anything but innocuous. As a theory of the only mathematics we know or can know, it is a theory of what mathematics *is*—what it really is!

Because it is an empirical theory about the embodied mind, the theory of embodied mathematics is framed within the study of embodied cognition. The elements of embodied cognition are not axioms and proofs but image schemas, aspectual concepts, basic-level concepts, semantic frames, conceptual metaphors, conceptual blends, and so on. Because mathematics does not study the mind, it cannot study itself as a product of mind. The methods and apparatus of embodied cognitive science are necessary.

Weak and Strong Requirements for a Theory of Embodied Mathematics

All conceptualization, knowledge, and thought makes use of the physical neural structure of our brains. In the weak sense, embodiment means that every concept we have must somehow be characterized in the neural structure of our brains, and that every bit of thinking we do must be carried out by neural mechanisms of exactly the right structure to carry out that form of thought. Moreover, everything we learn can be learned only through a neural learning mechanism capable, by virtue of its structure, of learning that kind of thing.

Everything we know either must be learned or must be built into the innate neural wiring of our brains. As we saw in Chapter 1, it appears that the most fundamental aspects of arithmetic—the apprehension of small numbers of things—is indeed innate. But most of mathematics is not. The concept of forty-nine is not innate, nor is the concept of zero, nor is the concept of infinity, nor are square roots, logarithms, convergent infinite series, the empty set, cosines, complex numbers, transfinite numbers, and so on for most of the concepts of advanced mathematics. The fact that these have been created and learned means that, from the perspective of cognitive science, there must ultimately be a biologically based account of the mechanisms by which they are created, learned, represented, and used. When someone presents you with an idea, the appropriate brain mechanism must be in place for you to understand it and learn it. Language and the meanings conveyed by language do not come out of thin air. Every bit of the meaning of language must also be accounted for by neural and cognitive mechanisms.

These are the *weak* requirements imposed on the cognitive science of mathematics by what we know about the embodiment of mind. But there are even stronger requirements (for a discussion, see Núñez, 1999). Cognitive science must explain how abstract reason is possible and how it is possible to have abstract concepts and to *understand* them. This is anything but a trivial enterprise. The reason is that abstract concepts cannot be perceived by the senses. You cannot see or hear or smell or touch the concepts of justice, responsibility, and honor, much less the concepts of ecological danger, evolution, due process, or entropy. It is a challenge for cognitive science to explain—in terms of our bodies and brains—exactly how we can comprehend such concepts and think using them.

The same is true of every concept in advanced mathematics: continuity, limits, fractals, open sets, infinite intersections, transfinite numbers, hypersets, logarithms, infinitesimals. The cognitive science of mathematics must do no less than cognitive science in general. The cognitive scientist must ultimately be able to account, in terms of neural and cognitive mechanisms, for how we understand abstract concepts, both concepts outside mathematics, like a healthy marriage and marginal utility, and concepts within mathematics, like the complex plane or function spaces.

The theory of embodied mathematics is not, and cannot be, a theory within mathematics. Rather, it is a special case of the theory of the embodied mind in cognitive science, where the subject matter happens to be mathematics. The answer to the question *How do we understand complex numbers at all?* cannot be a set of definitions, axioms, theorems, and proofs. That just pushes the question back one step further: What are the cognitive mechanisms involved in conceptualizing and learning those definitions, axioms, theorems, and proofs and all of the concepts used in them?

Mathematics cannot adequately characterize itself! That is, it is not the kind of enterprise that is equipped to answer scientific questions about mathematical cognition. Cognitive scientists, of course, use mathematical tools, just as other scientists do. But that is different from trying to characterize the automatic and unconscious aspects of mathematical cognition using axiom systems and proofs, which is hopeless.

The answer to questions about mathematical cognition must be given in terms of the kind of cognitive and neural mechanisms found in the automatic, unconscious, human conceptual system. As we saw in Chapter 2, a lot is known about the nature of human concepts and how they are embodied. Presumably the same mechanisms are at work in mathematical cognition as are at work in the rest of human cognition. Indeed, we have made that argument, chapter after chapter, throughout this book. There is not one analysis we have given that uses mechanisms that are specialized to mathematics alone. Every analysis in every chapter starting from basic arithmetic makes use of cognitive mechanisms found throughout the rest of our conceptual system: image schemas, frames, aspectual schemas, conceptual metaphors, conceptual blends, and so on.

Strong Constraints

What is known about the embodiment of mind additionally imposes strong constraints on any would-be theory of embodied mathematics. First, it must make sense of variation and change within mathematics. Mathematics has changed enormously over time, and forms of mathematics often vary from community to community across the mathematical world. Mathematicians differ in their interpretations of mathematical results; for instance, what does the independence of the Continuum hypothesis really imply? Fashions come and go in mathematics. The Bourbaki program of deriving everything from the most abstract formal structures was all the rage thirty or forty years ago. Today there is much more interest in studying specific details of specific subject matters: the relationship between modular forms and elliptic curves, the topology of four-dimensional spaces, nonlinear dynamics and complexity, and questions of computability relevant to computer science. A cognitively adequate account of mathematical cognition must be flexible enough to characterize historical change, different forms of mathematics (e.g., standard and nonstandard analysis, well-founded and non-well-founded set theory), alternative understandings of results, and the characterization of all the fashions. And of course, much of mathematics remains stable and that must be accounted for too. The Pythagorean theorem hasn't changed in twenty-five hundred years and, we think, won't in the future.

Next, a theory of embodied cognition applied to mathematics must answer the question of how, with pretty much the same kinds of brains and bodies, we are able to do all this. Given a very small amount of innate arithmetic, what capacities, used both within and outside of mathematics, allow us to go so far beyond the few mathematical capacities we are all born with?

The Challenge for an Embodied Mathematics

Human mathematics is not a reflection of a mathematics existing external to human beings; it is neither transcendent nor part of the physical universe. But there are excellent reasons why so many people, including professional mathematicians, think that mathematics *does* have an independent, objective, external existence. The properties of mathematics are, in many ways, properties that one would expect from our folk theories of external objects. The reason is that they are metaphorically based on our experience of external objects and experiences: containers, continuous paths of motion, discrete objects, numerosity for subitizable numbers, collections of objects, size, and on and on. Here are some of the basic properties of external objects as we experience them in everyday life that also apply to mathematics.

SOME PROPERTIES OF EXTERNAL OBJECTS THAT ARE CHARACTERISTIC OF MATHEMATICS

- *Universality:* Just as external objects tend to be the same for everyone, so basic mathematics is, by and large, the same across cultures. Two plus two is always four, regardless of culture.
- *Precision:* In the world of physical subitizable objects, two objects are two objects, not three or one. As an extension of this, given a sack of gold coins there is a precise answer to the question of how many there are in the sack.
- *Consistency* for any given subject matter: The physical world as we normally experience it is consistent. A given book is not both on the desk and not on the desk at a given time.
- *Stability:* Basic physical facts—that is, particular occurrences at a given time and place—don't change. They are stable over time. If there was a book on your desk at 10 A.M. this morning, it will always be the case that on this day in history there was a book on your desk at 10 A.M.
- *Generalizability:* There are basic properties of trees that generalize to new trees we have never encountered, properties of birds that generalize to birds yet unborn, and so on.
- *Discoverability:* Facts about objects in the world can be discovered. If there is an apple on the tree in the backyard, you can discover that the apple is there.

Mathematics is a mental creation that evolved to study objects in the world. Given that objects in the world have these properties, it is no surprise that mathematical entities should inherit them. Thus, mathematics, too, is *universal, precise, consistent* within each subject matter, *stable* over time, *generalizable,* and *discoverable.* The view that mathematics is a product of embodied cognition—mind as it arises through interaction with the world—explains why mathematics has these properties.

The Properties of Embodied Mathematics

Recent research in neuroscience, cognitive science, and the history of mathematics points, we believe, in the direction of an embodied mathematics. The theory of embodied mathematics makes the following claims.

1. Mathematics is a product of human beings. It uses the very limited and constrained resources of human biology and is shaped by the nature of our brains, our bodies, our conceptual systems, and the concerns of human societies and cultures.

2. The parts of human cognition that generate advanced mathematics as an enterprise are normal adult cognitive capacities—for example, the capacity for conceptual metaphor. Such cognitive capacities are common to all human beings. As such, the potential for mathematics, even advanced mathematics, is a human *universal*.

3. Simple numeration is built into human brains. Like many other animals, human beings can "subitize"—that is, instantly and accurately perceive the numbers of entities in a very small collection. This is clearly an embodied capacity.

4. The subject matters of mathematics—arithmetic, geometry, probability, calculus, set theory, combinatorics, game theory, topology, and so on—arise from human concerns and activities: for example, counting and measuring, architecture, gambling, motion and other change, grouping, manipulating written symbols, playing games, stretching and bending objects. In other words, mathematics is fundamentally a human enterprise arising from basic human activities.

5. The mathematical aspect of these concerns is *precision*—precise sums, measurements, angles, estimates, rates of change, categorizations, operations, and so on. Precision is made possible because human beings can make very clear and accurate distinctions among objects and categories under certain circumstances and can fix in their minds and consistently recall abstract entities like numbers and shapes.

6. Precision is greatly enhanced by the human capacity to *symbolize*. Symbols can be devised to stand for mathematical ideas, entities, operations, and relations. Symbols also permit precise and repeatable calculation.

7. Conceptual metaphor is a neurally embodied fundamental cognitive mechanism that allows us to use the inferential structure of one domain to reason about another. It allows mathematicians to bring to one

domain of mathematics the ideas and the methods of precise calculation of another domain.

8. Once established firmly within a community of mathematicians, mathematical inferences and calculations for a given subject matter tend not to change over time or space or culture. The stability of embodied mathematics is a consequence of the fact that normal human beings all share the same relevant aspects of brain and body structure and the same relevant relations to their environment that enter into mathematics.

9. Mathematics is not monolithic in its general subject matter. There is no such thing as *the* geometry or *the* set theory or *the* formal logic. Rather, there are mutually inconsistent versions of geometry, set theory, logic, and so on. Each version forms a distinct and internally consistent subject matter.

10. Mathematics is effective in characterizing and making predictions about certain aspects of the real world as we experience it. We have evolved so that everyday cognition can, by and large, fit the world as we experience it. Mathematics is a systematic extension of the mechanisms of everyday cognition. Any fit between mathematics and the world is mediated by, and made possible by, human cognitive capacities. Any such "fit" occurs in the human mind, where we cognize both the world and mathematics.

Given these properties, the theory of embodied mathematics is anything but radical. It yields a new understanding of why mathematics is *universal* (at least as far as human beings are concerned); *consistent* within each subject matter; *able to generalize* beyond humanly possible experience; *precise; symbolizable; stable* in its inferences and calculations across people, cultures, and time; and *effective* for describing major aspects of the natural world. These are properties of mathematics that anyone in the community of practicing scientists and mathematicians will recognize.

Several of these properties are not obvious and need further clarification. Exactly what makes inference and calculation stable in a view of mathematics as embodied? How can mathematics be consistent if there is no One True Geometry or no One True Set Theory or no One True Logic? If mathematics can be shaped by the concerns of culture, how can mathematical results be stable across cultures?

The Stability of Inference and
the Appearance of Timelessness

Mathematical inference tends to be stable across individuals, time, and culture because it uses basic cognitive mechanisms—mechanisms like category formation, spatial-relations concepts, conceptual metaphors, subitizing, and so on. Some of those mechanisms are innate, some have developed in all of us during childhood, and some develop only with special training of basic cognitive capacities. Once we learn a basic cognitive mechanism, it is stable in each of us (barring injury and pathology). For example, once we develop a system of categories, we keep on reasoning with that system consistently throughout our lives.

Another source of stability are spatial-relations concepts. The primitive spatial-relations schemas used in mathematics are universal across human languages—for example, the concept of containment, a path, or a center. What counts as a bounded region of space or a path in the spatial-relations system of any language in any culture is the same. Moreover, conceptual metaphors, such as Numbers Are Points on a Line, have been shown to have a property extremely important for mathematics: namely, that they preserve inferences; that is, the inferential structure of one domain (say, geometry) can be used by another (say, arithmetic). Once a metaphorical mapping is established for a mathematical community, the inferences are the same for anyone in that community, no matter what culture they come from. If that mathematical community extends over generations or longer, the inferences are stable over those generations.

Important consequences for mathematics follow from the fact that certain aspects of human conceptual systems are universal and that they have inference-preserving mechanisms such as conceptual metaphor.

1. *Inferential stability.* The first important consequence is that mathematical proof and computations are cognitively stable. Proofs made using inference-preserving cognitive mechanisms remain valid. Correct computations remain correct.
2. *The possibility of discovery.* Once mathematical concepts and assumptions are established within a mathematical community, it is possible to make discoveries by reasoning alone—that is, without recourse to empirical evidence.

3. *Abstraction.* Mathematics has general conceptual categories, such as integer, prime, circle, line, torus, parabola, and so on. Proofs about such categories can hold for all members of those categories, whether they have ever been, or could ever be, thought of by any real human beings or not. Proofs about all prime numbers, for example, are proofs about all members of a category; they hold even for prime numbers so large we could never conceptualize them as individual numbers.

4. *Stable, natural connections among the branches of mathematics.* As we have seen, conceptual metaphors such as Numbers Are Sets can be formulated precisely within a mathematical community. Such metaphors allow mathematicians to model one branch of mathematics using another. In the case of the metaphor Numbers Are Sets, arithmetic is modeled using set theory. Such mathematical models across branches of mathematics, once established, can remain stable. Since conceptual metaphors preserve inference, the consequences of such a metaphorical model can be drawn out systematically by mathematicians over the course of generations.

5. *The systematic evolution of mathematics over time.* Generation after generation of mathematicians draw out the consequences of assumptions and models established by previous mathematicians. Once results are established, they are stable and take on a seemingly "timeless" quality.

Equally Valid, Mutually Inconsistent Subject Matters in Mathematics

The general subject matters of mathematics (arithmetic, geometry, probability, topology) arise from general human activities. However, differing versions of these subject matters can be created by mathematicians using normal conceptual mechanisms. Thus, a specialist in geometry can choose to study one or another form of geometry defined by whatever axioms he or she is interested in (e.g., Euclidean or non-Euclidean, projective, inversive, differential, and so on).

Each type of geometry is a separate subject matter, *a human creation that is the product of human imagination, which in turn uses ordinary human cognitive mechanisms.* Each subject matter is internally consistent, though there are inconsistencies across the subject matters. In projective geometry, for example, all parallel lines meet at infinity. In Euclidean geometry, parallel lines never meet. In spherical geometry, there are no two lines (great circles) that are parallel. These are not contradictory theories of a single subject matter but different subject matters defined by different assumptions. In some cases, they are metaphorical assumptions, as when, in projective geometry, we conceptualize parallel lines as meeting at infinity.

The same is true of set theory. There are lots and lots of set theories, each defined by different axioms. You can construct a set theory in which the Continuum hypothesis is true and a set theory in which it is false. You can construct a set theory in which sets cannot be members of themselves and a set theory in which sets can be members of themselves. It is just a matter of which axioms you choose, and each collection of axioms defines a different subject matter. Yet each such subject matter is itself a viable and self-consistent form of mathematics.

There are thus no monolithic subject matters in mathematics like *the* geometry or *the* set theory or *the* formal logic. For example, the theorems of non-Euclidean geometries are not absolutely true or false, as if there were only one true geometry, but only true or false relative to choices of axioms, which amounts to a choice among the kinds of geometry you want to talk about. The same is true of set theory. There is no one true set theory. Whether sets can be members of themselves depends on whether you want to talk about well-founded set theories (where sets *cannot* be members of themselves) or non-well-founded set theories (where sets *can* be members of themselves).

It is important to remember that such alternative forms of mathematics—like well-founded versus non-well-founded set theory or Euclidean versus non-Euclidean geometry—evolve over time. A mere thirty years before this book was written, there was no non-well-founded set theory. It was invented by mathematicians responding to problems in modeling recursive mechanisms—problems that became important because of the expansion of computer science over those years.

The moral here is that subject matters in mathematics tend to have multiple versions for historical reasons—at some point in our future history it may become important to certain mathematicians to have still another alternative version of set theory or of geometry or of calculus. Mathematicians, being a creative lot, will invent them, and there is no way to predict in advance just what new forms of mathematics mathematicians will invent.

How Culture Gives Rise to Alternative Versions of Mathematics

In the Romance of Mathematics, culture is assumed to be irrelevant. If mathematics is an objective feature of this or any other universe, mere culture could not have any effect on it.

Is mathematics really independent of culture? There is one sense in which mathematics is not culture-dependent and another in which it *is* culture-dependent.

- Mathematics is independent of culture in the following very important sense: Once mathematical ideas are established in a worldwide mathematical community, their consequences are the same for everyone regardless of culture. (However, their establishment in a worldwide community in the first place may very well be a matter of culture.)
- Mathematics is culture-dependent in another very important sense, a sense recognized by mathematicians such as Wilder (1952) and Hersh (1997). Historically important, culturally specific ideas from outside mathematics often find their way into the very fabric of mathematics itself. Culturally specific ideas can permanently change the actual content of mathematics forever.

Let us consider some examples where the *permanent content* of mathematics has been shaped by culture. Here are some important cultural ideas from outside mathematics that arose at particular times in history and changed the content of mathematics itself.

1. *The idea of essence.* Everything in the universe has an essence that defines what it is and is the causal source of its natural behavior. This is also true of subject matters—for example, geometry. What is the essence of geometry? The answer since Euclid has been that the essence can be given by a small number of obviously true postulates. Part of the idea of essence is that it is the causal source of the natural behavior of an entity. Euclid interpreted this idea as follows for geometry: If the postulates constitute the essence of geometry, all truths of geometry should follow as logical consequences from the postulates.

If you think about it, this is not a necessary truth. In fact, as a consequence of what Kurt Gödel proved, it isn't even true of geometry. Yet the idea of a small set of axioms characterizing all the truths of a subject matter took hold throughout Europe, just as the idea of essence did. Both ideas are still with us today, Gödel notwithstanding.

2. *The idea that all human reason is a form of mathematical calculation called logic.* Mathematical reasoning must therefore also be a version of logic, which can itself be mathematicized.

Again, this is neither a necessary truth nor even a truth. Cognitive science has shown that people do not reason using mathematical logic. Metaphorical reasoning and spatial reasoning are obvious counterexamples.

3. *The idea of foundations for a subject matter.* Every respectable subject matter is to be conceptualized metaphorically as if it were a physical structure like a building, which has to have secure, solid, permanent foundations if it is not to "collapse" or "fall apart." This includes mathematics.

About this idea, too, there is nothing necessarily true nor even true: Biology and geology do not have such "foundations," but they are not likely to "collapse."

Each of these notions is a case of a *general* idea—an idea that originated outside mathematics, in a specific culture at a specific time in history. Each was then applied to mathematics as a special case. None of these was purely an idea about mathematics. When these ideas were applied to mathematics at particular historical moments, mathematics was changed forever.

As noted in Chapter 5, the idea of essence arose in Greek philosophy in pre-Socratic times. It was at the heart of just about all Greek philosophy, and it was applied to mathematics as a special case before Euclid by the Pythagoreans, who believed that the Essence of all Being was number (See Lakoff & Johnson, 1999, ch. 16). For example, the essence of a triangle was taken to be the number 3, while that of a square was the number 4. Similarly it was believed that every shape could be described in terms of some mathematics that characterized its essence. That idea has come down to us today in the folk belief that ellipses are in the orbits of the planets, that Fibonacci series are in flowers, that logarithms are in snails shaped like logarithmic spirals, and that π is in everything with a spherical shape, from bubbles to stars.

The second idea—that thought is mathematical calculation—also arose with the Greeks and has come down to us through Enlightenment philosophers as diverse as Descartes, Hobbes, and Leibniz. Hobbes believed that all thought was a form of "reckoning." He metaphorically thought of ideas as numbers and reasoning as being like arithmetic operations such as addition and subtraction. This metaphor persists to the current day and can be seen in expressions about ideas in general, like "He put two and two together," "It just doesn't add up," and "What's the bottom line?" The idea that all human reasoning is logic, which can be precisely formulated in mathematical terms, came to fruition in the nineteenth century, with the invention of symbolic logic by George Boole and Gottlob Frege. Mathematical logic has been seen by philosophers from Bertrand Russell to Richard Montague as characterizing human reason. Even one of the present authors—George Lakoff—once held that view (but that was back in the 1960s, before cognitive science developed).

The third idea—that theories, like buildings, must have secure, solid, permanent foundations on which all else is built—is at least as old as Aristotle.

Throughout the Western tradition, it has been the governing metaphor behind all philosophical theories that pretend to give an account of certain and absolute knowledge. In mathematics, the tradition goes back at least to Euclid. It, too, reached its full flowering in nineteenth-century Europe, with Frege's attempt to provide foundations for mathematics in formal logic.

Indeed, the Foundations of Mathematics movement of the early twentieth century brought all three of these ideas together:

1. The essence of a subject matter is to be given by a small set of axioms.
2. Mathematical reasoning is a form of mathematical calculation, which allows all mathematical truths to be calculated (using mathematical logic).
3. All respectable subject matters can and must have secure foundations on which everything in the subject matter is built.

The Foundations movement itself collapsed. None of these ideas has stood the test of time within mathematics. All three have been found to be mathematically untenable. But they all shaped the structure of mathematics itself— not just how it was done but its very content. One cannot even imagine contemporary mathematics without these ideas.

Where did these ideas come from? They are not part of the structure of the universe. They are not built into the brain structure of all human beings, as basic numeration is. And they are not cognitive universals. They are, rather, products of human culture and human history. They were not part of ancient Babylonian, Egyptian, Indian, Mayan, or Chinese mathematics. They are special cases of important general ideas that are part of European cultural history.

Thus, the very idea that there must be absolute "foundations" for mathematics is itself a culture-dependent feature of mathematics! The irony in this is worthy of mention: Those who most vocally argue against the idea that culture plays a role in mathematics are the contemporary supporters of the Foundations movement, which is itself a cultural feature of the discipline.

The moral is clear: Many of the most important ideas in mathematics have come not out of mathematics itself, but arise from more general aspects of culture. The reason is obvious. Mathematics always occurs in a cultural setting. General cultural worldviews will naturally apply to mathematics as a special case. In some cases, the result will be a major change in the content of mathematics itself.

The content of mathematics is not given in advance. Mathematics evolves. It is grounded in the human body and brain, in human cognitive capacities, and in common human activities and concerns. Mathematics also has a cultural dimension, which, from the perspective of embodied mathematics, is entirely natural. Since mathematical ideas are products of human beings with normal human cognitive capacities living in a culture, it is perfectly natural that general cultural ideas should be applied to many special cases, including mathematics.

The Historical Dimension of Embodied Mathematics

From the perspective of embodied mathematics, we would expect other effects of history and culture on mathematics. Since mathematicians all live at some specific time and base their work on that of previous mathematicians, we would expect mathematics to evolve over time, and it does. Even ideas in mathematics as basic as what a number is have changed over time. Numbers were first restricted to the positive integers, then extended to include the rational numbers. Zero was added afterward. Later came the negative numbers, the irrationals, the transcendentals, the complex numbers, the quaternions, the transfinite numbers, the hyperreal numbers, the surreal numbers, and so on.

This development is not a matter of linear progress. There are inconsistencies between, say, the quaternions and the complex numbers, and between the transfinite and the hyperreal numbers. Thus, not only has our idea of number changed over time, but we now have equally mathematically valid but mutually inconsistent versions of what numbers are. This should not be surprising. Such a nonlinear mathematical evolution is a natural by-product of the historical dimension of the embodiment of mathematics.

Similarly, the embodied nature of mathematics leads one to expect that there should be competing schools of mathematics, and there are. New mathematical ideas are invented by mathematicians, most of whom live in an academic culture (a culture which has always been highly politicized); and there are equally valid mathematical ideas that are inconsistent with each other.

Moreover, we would expect some forms of mathematics to be generally accepted by "mainstream" mathematicians and other forms not to be. For example, in the Newton-Leibniz dispute over whether calculus should be formulated in terms of Leibniz's infinitesimals, Newton won out. Mainstream mathematicians do not accept infinitesimals, though nowadays there is a perfectly fine mathematics of infinitesimals formalized by Abraham Robinson (see Chapter 11) and a small but active group of mathematicians working with them. Do infinitesimal numbers exist? Your answer depends on which school of thought you are in. Since both schools practice perfectly valid forms of mathematics, there are different and equally valid answers to the question. This, too, is natural and to be expected, given that mathematics is embodied.

We would also expect, from the embodiment of mathematics, that major historical events and trends outside mathematics proper should have an important effect on the very content of mathematics itself. The recent development of computer science is a case in point. Computers have led to the creation of huge new branches of mathematics that would not otherwise have existed.

Floating-Point Arithmetic: A Case of the Relevance of History in Mathematics

Take a simple case. Consider versions of floating-point arithmetic, the form of arithmetic used in all computers. Many versions were developed for different computers—although through agreements across companies and countries, an international standard has finally been adopted by means of a political process. The standard version of floating-point arithmetic now used is not used because it is objectively true, or even truer than other forms. The version that is now standard was chosen for pragmatic reasons. It is now *the* form of arithmetic most used in the world by far.

Floating-point arithmetic is a branch of mathematics on its own. As new computers develop, updated versions meeting the international standard will have to be invented and adopted. There is a whole field that studies the properties of different floating-point arithmetics, which are invented regularly.

Each version of floating-point arithmetic gives pretty much the same answers as ordinary arithmetic for numbers as large as most people are likely to care about, and with an accuracy that will satisfy most computer users. However, over the full range of real numbers, each floating-point arithmetic is somewhat different and gives different answers.

Not all theorems of ordinary arithmetic hold for floating-point arithmetics, which can have two zeroes, one positive and one negative, and have $+\infty$ and $-\infty$ as numbers. Yet each is a form of the real thing: floating-point arithmetic, completely precise and rigorous within itself.

Note, incidentally, that it makes no sense to ask if there is *really* one zero or two in arithmetic in general. It depends on which kind of arithmetic one chooses, normal or one of the floating-point versions. One might decide to use the arithmetic most commonly used by scientists throughout the world to answer such questions. If one adopts that pragmatic criterion, then we must admit that there are two zeroes, +0 and –0, in arithmetic. The reason is that most of the arithmetic done in the history of the world has been done on computers within the past fifty years. All of that arithmetic used floating-point, not standard arithmetic.

If one chooses one's arithmetic according to the version most used by scientists, floating-point arithmetic is the clear winner. A consequence of that would be the loss of the existence of real numbers. Infinite decimals cannot exist in floating-point arithmetic, since each number must be representable by a small finite number of bits. Infinite decimals take an infinite number of bits.

There is an intriguing consequence of choosing to believe in the real existence of the most commonly used arithmetic around, floating-point arithmetic: π, as a precise and unique infinite decimal, does not exist in floating-point arithmetic, which is finite.

Now consider a common claim made about mathematics by true believers in the Romance of Mathematics:

> We know that mathematics is true in the physical world because it works for science. It follows that π therefore exists in the physical world.

Since floating-point arithmetic is the form of arithmetic most used in scientific calculation, and since π (as an infinite decimal) does not exist in floating-point arithmetic, it follows that even if one accepted arguments of that form, it would not follow that the infinite polynomial π has an objective existence in the world!

When thinking about the claim that mathematics exists independent of all beings and all culture and peculiarities of history, think about floating-point arithmetics. The field of floating-point arithmetic would not have existed without the explosive development of computers, which was made possible by the development of transistors and microchips. What better example could one have of a historical event outside mathematics proper, making possible the development of a form of mathematics that would not otherwise exist?

There are, of course, other examples. The serious development of such fields of mathematics as complexity theory and fractal geometry was made possible only through the development of computers. And computers have even raised the question of what counts as a proof. Consider a proof done by a computer in millions of steps—a proof that no human being could possibly follow and that provides no human insight. Is this to be taken as a proof or not? Mathematicians disagree (Kleiner & Movshovitz-Hadar, 1997). There is no ultimate answer, as you would expect from the embodiment of mathematics. The reason is that the very concept of a proof is a human concept involving evaluation, and such evaluative concepts are what cognitive linguists call "essentially contested concepts"—concepts that necessarily have different versions, since people have different values. In this case, the bifurcation of the very concept of a proof is the result of a grand historical event, the development of computers.

Why Embodied Mathematics Is Not a Form of Postmodernism

As we have just noted, a significant part of mathematics itself is a product of historical moments, peculiarities of history, culture, and economics. This is simply a fact. In recognizing the facts for what they are, we are *not* adopting a postmodernist philosophy that says that mathematics is *merely a cultural artifact.* We have gone to great lengths to argue against such a view.

The theory of embodied mathematics recognizes alternative forms of mathematics (like well-founded and non-well-founded set theories) as equally valid but about different subject matters. Although it recognizes the profound effects of history and culture upon the content of mathematics, it strongly rejects radical cultural relativism on empirical grounds.

In recognizing all the ways that mathematics makes use of cognitive universals and universal aspects of experience, the theory of embodied mathematics *explicitly rejects any possible claim that mathematics is arbitrarily shaped by history and culture alone.*

Indeed, the embodiment of mathematics accounts for real properties of mathematics that a radical cultural relativism would deny or ignore: conceptual stability, stability of inference, precision, consistency, generalizability, discoverability, calculability, and real utility in describing the world.

This distinguishes an embodied view of mathematics from a radical relativist perspective. The broad forms of postmodernism recognize the effects of culture and history. But they do not recognize those effects of embodiment that are *not* arbitrary. It is the nonarbitrariness arising from embodiment that takes mathematics out of the purview of postmodernism.

Moreover, the embodiment of mind in general has been scientifically established by means of convergent evidence within cognitive science. Here, too, embodied mathematics diverges from a radically relativistic view of science as just historically and culturally contingent. We believe that a science based on convergent evidence can make real progress in understanding the world.

Why Transcendent Mathematics Is Just
As Antiscientific as Radical Postmodernism

Consider the radical form of postmodernism which claims that mathematics is purely historically and culturally contingent and fundamentally subjective. This is an a priori philosophical view. Scientific evidence is seen as irrelevant. No evidence from any science, including cognitive science, is given priority over that a priori view. Arguments based on empirical evidence have no weight for someone who comes to the discussion with such an a priori view.

Now consider the radical version of the transcendent mathematics position. It adopts the Romance as a fundamental, unshakable truth. It accepts as an a priori philosophical position that mathematical entities have a real, objective existence and that mathematics is objectively true, independent of any beings with minds.

This, too, is an a priori philosophical view. Scientific evidence is seen as irrelevant. No evidence from any science, including cognitive science, is given priority over that a priori view. Arguments based on empirical evidence have no weight for someone who comes to the discussion with such an a priori view.

16

The Philosophy of
Embodied Mathematics

The Embodied Nature of Mathematics

The cognitive science of mathematics has an immediate consequence for the philosophy of mathematics. It provides a new answer for perhaps the most basic philosophical question of all regarding mathematics: What is the nature of mathematics? This answer can be summarized as follows:

- Mathematics, as we know it or can know it, exists by virtue of the embodied mind.
- All mathematical content resides in embodied mathematical ideas.
- A large number of the most basic, as well as the most sophisticated, mathematical ideas are metaphorical in nature.

The way that mathematics is embodied, as we saw in Chapter 3, accounts for the basic properties of mathematics; namely, it is *stable, precise, generalizable, symbolizable, calculable, consistent within each of its subject matters, universally available,* and *effective for precisely conceptualizing a large number of aspects of the world as we experience it.* These are, of course, the properties of mathematics that any philosophy of mathematics must account for.

The Romance Disconfirmed

One immediate consequence is the disconfirmation of the Romance of Mathematics.

- From a scientific perspective, there is *no* way to know whether there are objectively existing, external, mathematical entities or mathematical truths.
- Human mathematics is embodied; it is grounded in bodily experience in the world.
- Human mathematics is *not* about objectively existing, external mathematical entities or mathematical truths.
- Human mathematics is primarily a matter of mathematical ideas, which are significantly metaphorical in nature.
- Mathematics is not purely literal; it is an imaginative, profoundly metaphorical enterprise.
- There is no mathematics out there in the physical world that mathematical scientific theories describe.

Radical Social Constructivism Disconfirmed

A second immediate consequence is a disconfirmation of the radical social constructivist theory of mathematics.

Mathematics, being embodied, uses general mechanisms of embodied cognition and is grounded in experience in the world. Therefore, it is *not arbitrary!* That means:

- Mathematics is not purely subjective.
- Mathematics is not a matter of mere social agreement.
- Mathematics is not *purely* historically and culturally contingent.

This is not to say that historical and cultural factors don't enter into mathematics. They do, in a very important way, as we discussed in the previous chapter.

Ontology and Truth

For well over two millennia, ontology and truth have been concerns of Western philosophy. Consequently, two of the central questions of the philosophy of mathematics have been:

- What are mathematical objects?
- What is mathematical truth?

The most common contemporary answers to these questions are those of the adherents of both the Romance and the Foundations movement:

- Mathematical objects are real, objectively existent entities.
- Mathematical truths are objective truths of the universe.

But as we have seen, such answers are ruled out by the cognitive science of mathematics in general and mathematical idea analysis in particular. From this perspective there are new answers:

- Mathematical objects are embodied concepts—that is, ideas that are ultimately grounded in human experience and put together via normal human conceptual mechanisms, such as image schemas, conceptual metaphors, and conceptual blends.
- Mathematical truth is like any other truth. A statement is true if our embodied understanding of the statement accords with our embodied understanding of the subject matter and the situation at hand. Truth, including mathematical truth, is thus dependent on embodied human cognition (see Lakoff & Johnson, 1999, chs. 6–8; Núñez, 1995).

Since human mathematics is all that we have access to, mathematical entities can only be conceptual in nature, arising from our embodied conceptual systems. A mathematical statement can be true only if the way we understand that statement fits the way we understand the subject matter that the statement is about. Conceptual metaphors often enter into those understandings.

Some Examples

To make these very abstract views clearer, let us consider some examples.

Example 1
Entity: Zero
Truth: $n + 0 = n$

Since human mathematics is all we have any scientific access to, we have to ask what zero is in human mathematics. Zero is not a subitized number and it is not a part of innate mathematics. Indeed, it took a long time in the history of mathematics for zero to be recognized as a number at all.

Zero has become a number through the metaphorical extension of the natural numbers (see Chapters 3 and 4). Thus, in the Arithmetic Is Object Collection metaphor, zero is the empty collection; in the Arithmetic Is Object Construction metaphor, it is the lack of an object; in the measuring stick metaphor, it is the initial point of the measurement; and in the Arithmetic Is Motion metaphor, zero is the origin of motion. It is an entailment of each of these metaphors that $n + 0 = n$. In the object-collection metaphor, adding a collection of n objects to an empty collection yields a collection with n objects. In the moving metaphor, taking n steps from the origin and then taking no steps leaves you at the same place as just taking n steps from the origin.

Given our understanding of zero and our understanding of the operation of addition and its result, we can see why we must take $n + 0 = n$ to be true.

It should be recalled that in many cultures around the world, there is no number zero and no such "truth." This was the case in Western mathematics for many centuries.

Example 2

Entity: The empty class
Truth: The empty class is a subclass of every class.

The empty class (see Chapter 6) is a metaphorical creation of Boole's. It exists only by virtue of a mapping from zero in the source domain of Boole's metaphor.

"The empty class is a subclass of every class" is a truth of Boole's algebra of classes, arising from Boole's metaphor. It is the metaphorical projection of $n \cdot 0 = 0$—namely, $A \cap \varnothing = \varnothing$. In Boole's framework, B is a subclass of A if and only if $A \cap B = B$. When $B = \varnothing$, this is true. Hence, the empty class is a subclass of every class.

Thus the truth of this statement is a consequence of the ideas used in Boole's formulation of his algebra of classes. Those ideas include the metaphors characterizing zero in arithmetic, the grounding metaphor Classes Are Containers, and Boole's metaphor, which creates the empty set from zero.

Of course, Boole's metaphor does not exist for most people. In most people's ordinary nonmathematical understanding of classes, there is no empty class and no such "truth."

Example 3

Entity: The infinite sum $\sum_{n=1}^{\infty} \frac{1}{2^n}$
Truth: $\sum_{n=1}^{\infty} \frac{1}{2^n} = 1$

This infinite sum "exists" by virtue of the Basic Metaphor of Infinity. The truth is "true" only by virtue of the Basic Metaphor of Infinity. (See Chapters 8 and 9.)

The existence of this entity and this truth depends upon the acceptance of the idea of actual (as opposed to potential) infinity. If one rejects such a notion, as many mathematicians have, then there is no such entity and no such truth.

Example 4

Entity: $\aleph_0$

Truth: $\aleph_0 + \aleph_0 = \aleph_0$

The entity $\aleph_0$ exists only by virtue of Cantor's metaphor plus the Basic Metaphor of Infinity. $\aleph_0 + \aleph_0 = \aleph_0$ is true only by virtue of accepting those metaphors. (See Chapter 10.)

Example 5

Entity: A "line" L in projective geometry connecting points "at infinity."

Truth: L is unique and every point at infinity lies on L.

Points "at infinity" exist only by virtue of the Basic Metaphor of Infinity. The entity, line L, connecting two such points at infinity, thus also exists only by virtue of the BMI. And, correspondingly, the truth that every point at infinity lies on L is "true" only by virtue of the BMI.

Without such conceptual metaphors, those entities do not exist and those truths do not hold. For example, zero does not exist conceptually without the grounding metaphors for arithmetic, and the "truth" $n + 0 = n$ is not "true" if there is no zero. The empty class, as a unique object that is a subclass of every class, does not exist without Boole's metaphor mapping zero onto the domain of classes to create the empty class. The "truth" that the empty class is a subclass of every class is not true if there is no Boolean empty class for it to be true of.

Similarly, the infinite sum $\sum_{n=1}^{\infty} \frac{1}{2^n}$, the transfinite cardinal $\aleph_0$, and the "line at infinity" L in projective geometry do not exist as mathematical entities without the Basic Metaphor of Infinity, nor could the "truths" about them be "true" without that conceptual metaphor.

The Point

In each of the above cases, the entity in question is a metaphorical entity—that is, an entity existing conceptually as a result of a metaphorical idea. In each case, the truth is "true" only relative to that metaphorical idea. Each such mathematical entity exists only in the minds of beings with those metaphorical ideas, and each mathematical truth is true only in the minds of beings with such mathematical ideas.

As we shall see in the case study, the same holds for the entity $e^{\pi i}$ and the truth $e^{\pi i} + 1 = 0$.

All of these amount to new philosophical claims about mathematical entities and mathematical truths.

These views are, of course, very different from the views held by mathematicians who accept the Romance of Mathematics. Unfortunately, aspects of the Romance are often confused with an important approach to doing mathematics—namely, formalist mathematics. Formalist mathematics is a remarkable intellectual program that rests upon a fundamental metaphor, what we will call the *Formal Reduction Metaphor*.

The Formal Reduction Metaphor

This remarkable conceptual metaphor provides a prescription for (1) conceptualizing any mathematical subject matter in terms of sets and (2) symbolizing any such subject matter in a uniform way that meshes with formal logic. Accordingly, what are called mathematical "proofs" can be symbolized and made mechanical.

The target domain of this metaphor consists of mathematical concepts in general. The source domain consists only of sets, structures within set theory, symbols, and strings of symbols.

The metaphor provides a way to project a set-theoretical structure onto any domain of mathematics at all—provided that this domain of mathematics has already been discretized. As we discussed in Chapters 12 through 14, metaphors like Spaces Are Sets of Points are necessary to conceptualize geometry, topology, and calculus in discrete terms. Given such a metaphorical discretization of those subject matters—a discretization that essentially changes the concepts involved—the Formal Reduction Metaphor can come into play.

According to this metaphor, any conceptual structure in any branch of mathematics can be mapped onto a set-theoretical structure in the following way:

- Mathematical elements are mapped onto sets, elements of sets, and structures of sets defined within set theory.
- n-place relations are mapped onto ordered n-tuples.

- Functions and operations are mapped onto suitably constrained sets of ordered pairs.
- The meaningful symbolizations of a mathematical subject matter are mapped onto uninterpreted symbol strings in a way that fits formal logic, with predicates and arguments, operators, quantifiers, and so on.
- What we have called "symbolization mappings" (which assign symbols to meaningful ideas) are mapped onto formal "interpretations"—mappings from symbols to set-theoretical structures (e.g., sets of ordered pairs).
- Conceptual axioms are mapped onto constraints on the set-theoretical structures.
- Symbolizations of conceptual axioms are mapped onto symbol strings.
- Symbolization mappings (relating symbols to concepts) are mapped onto interpretations—mappings from symbol strings to set-theoretical structures.

Let us take a simple illustrative example. The axiom of arithmetic that says "$a + b = b + a$" would be expressed in a set-theoretical model as a constraint on the operation "+." The operation would be represented as an infinite set of ordered pairs, each of which consists of an ordered pair of numbers paired with a single number: $\{((1, 3), 4), ((3, 1), 4), \dots\}$. Here the first element represents "$1 + 3 = 4$" and the second represents "$3 + 1 = 4$." The axioms would be expressed as a constraint on this set-theoretical structure so that whenever $((a, b), c)$ is in the set, then $((b, a), c)$ is in the set. The formal symbol string "$a + b = b + a$" (which in itself means nothing) would be mapped by an interpretive function onto this constraint.

Of course, the "numbers" in this structure would also be sets, as given by the Natural Numbers Are Sets metaphor. For example, the number 1 would be represented as the set $\{\emptyset\}$ and the number 3 would be represented by the set $\{\emptyset, \{\emptyset\}, \{\emptyset, \{\emptyset\}\}\}$, where $\emptyset$ is the empty set. Every ordered pair (a, b) would be represented via the Ordered Pairs Are Sets metaphor as the set $\{a, \{a, b\}\}$. Thus, the ordered pair $(1, 3)$ would be represented, via both metaphors, as the set $\{\{\emptyset\}, \{\{\emptyset\}, \{\emptyset, \{\emptyset\}, \{\emptyset, \{\emptyset\}\}\}\}\}$, while the ordered pair $(3, 1)$ would be represented as the set $\{\{\emptyset, \{\emptyset\}, \{\emptyset, \{\emptyset\}\}\}, \{\{\emptyset, \{\emptyset\}, \{\emptyset, \{\emptyset\}\}\}, \{\emptyset\}\}\}$. In short, many of the metaphors we have discussed throughout this work are used in the reduction of various branches of mathematics to set theory. Those metaphors work together with the Formal Reduction Metaphor to accomplish complete reductions.

Of course, there will be different reductions depending on the version of set theory used. For example, if hyperset theory (discussed in Chapter 7) is used, the "sets" will be hypersets, which are conceptualized as graphs rather than as containers.

The Formal Reduction Metaphor is actually a metaphorical schema, since it generalizes over all branches of mathematics. Each particular reduction for a branch of mathematics would have a somewhat different special case of the schema.

THE FORMAL REDUCTION METAPHOR

Source Domain SETS AND SYMBOLS		Target Domain MATHEMATICAL IDEAS
A set-theoretical entity (e.g., a set, a member, a set-theoretical structure)	→	A mathematical concept
An ordered *n*-tuple	→	An *n*-place relation among mathematical concepts
A set of ordered pairs (suitably constrained)	→	A function or operator
Constraints on a set-theoretical structure	→	Conceptual axioms: ideas characterizing the essence of the subject matter
Inherently meaningless symbol strings combined under certain rules	→	The symbolization of ideas in the mathematical subject matter
Inherently meaningless symbol strings called "axioms"	→	The symbolization of the conceptual axioms—the ideas characterizing the essence of the subject matter
A mapping (called an "interpretation") from the inherently meaningless symbol strings to the set-theoretical structure	→	The symbolization relation between symbols and the ideas they symbolize.

Mathematics, in its totality, has enormously rich and interesting ideas—ideas like magnitude, space, change, rotation, curves, spheres, spirals, probabilities, knots, equations, roots, recurrence, and on and on. Under the Formal Reduction Metaphor, all this conceptual opulence is assigned a relatively diminished structure in terms of the relative conceptual poverty of symbol strings and the ideas of set theory: elements, sets, *n*-tuples, sets of *n*-tuples, a membership relation, a subset relation, unions, intersections, complements, and so on.

There are two interpretations of this conceptual metaphor. The one that makes sense to us is what we call the cognitive interpretation. On this interpretation, most practicing mathematicians who make use of a formal reduction of a subject matter unconsciously make use of a metaphorical blend of both the mathematical subject matter (the target domain) and the set-and-symbol structure (the source domain). For example, suppose the subject matter (the target domain) is addition in arithmetic and the set-and-symbol structure is the one we used earlier in discussing "$a + b = b + a$." Any practicing mathematician using such a formal reduction would also be thinking of it in terms of what it means—that is, in terms of our ordinary ideas of numbers and addition. No human mathematician thinks of numbers and addition only in terms of set-theoretical structures like $\{\{\varnothing, \{\varnothing\}, \{\varnothing, \{\varnothing\}\}\}, \{\{\varnothing, \{\varnothing\}, \{\varnothing, \{\varnothing\}\}\}, \{\varnothing\}\}\}$, which represents the ordered pair (3, 1). Real mathematicians use both at once, thinking of arithmetic as structured by the set-and-symbols.

On this interpretation, what matters is cognitive reality—a reality that can be studied scientifically. What is cognitively real for practicing mathematicians using the reduction is that they use the ideas of number and arithmetic operations as paired by the Formal Reduction Metaphor with the ideas of set theory and symbol systems.

This interpretation shows the utility of formal reduction: Different branches of mathematics with different ideas can be compared under the reduction, which imposes structures with only one domain of ideas: sets and symbols. Moreover, the Formal Reduction Metaphor inherently makes an interesting claim: The relatively impoverished conceptual structure of set theory and formal logic is sufficient to characterize the structure (though not the cognitive content) of every mathematical proof in every branch of mathematics.

There is, however, a second interpretation, which we call the formal foundations interpretation of the Formal Reduction Metaphor. In this interpretation, the metaphor is taken as literally true: All branches of mathematics reduce ontologically to set theory and mathematical logic. On this interpretation, the ordered pair (3, 1) really is the set $\{\{\varnothing, \{\varnothing\}, \{\varnothing, \{\varnothing\}\}\}, \{\{\varnothing, \{\varnothing\}, \{\varnothing, \{\varnothing\}\}\}, \{\varnothing\}\}\}$. A line really *is* a set, and the points on the line really *are* the members of the set.

Under the first interpretation, formal mathematics is just one branch of mathematics, structured by human conceptual systems. Under the second interpretation, formal mathematics is *all* of mathematics!

The second interpretation defines a philosophy of mathematics that began in the late nineteenth century—namely, the Formal Foundations program, which proclaims: Mathematics *is* formal logic plus set theory! This is a historically-contingent philosophical program. Its goal has been to provide unique, well-understood, formally characterizable, certain, and "secure" foundations for the entire edifice of mathematics, which is itself taken as having these properties.

The philosophical approach to mathematics defined by this interpretation of the Formal Reduction Metaphor fits the Romance of Mathematics: The sets are objectively existing entities. The symbol strings of the formal language are objectively existing entities. The mappings between them are objectively existing entities. Since this is all mathematics is, mathematics has an objective existence.

On this interpretation of the Formal Reduction Metaphor, mathematics is *mind-free*. It has nothing to do with minds, human or otherwise. It has nothing to do with ideas or concepts of any sort, metaphorical or otherwise. It has nothing to do with bodies, human or otherwise.

On the Formal Foundations interpretation, Numbers Are Sets is not a metaphor but a literal statement. Since set theory is all there is, numbers would have to be certain set-theoretical entities. Similarly, Spaces Are Sets is not a metaphor but a literal truth. So-called points would have to really *be* members of such sets. Points could not be locations in spaces, because sets do not literally have locations in them.

Mathematics Without Ideas: A Scientific Fallacy

Under the Formal Foundations interpretation of the Formal Reduction Metaphor, mathematics *has no ideas!* The very notion of mathematical ideas as constitutive of mathematics must appear as nonsense from this perspective. Mathematical idea analysis might, from this perspective, be *at best* a study external to mathematics itself of how human beings happen to understand real, objectively existing mathematics. From this perspective, mathematical idea analysis can have nothing to do with mathematics per se.

This is, of course, nonsense from the perspective of cognitive science. In any mathematics that is humanly comprehensible, classes and sets have a conceptual structure, and even a metaphorical structure. That metaphorical structure makes use of the Classes Are Containers metaphor, the 4Gs (needed for zero), Boole's metaphor, Cantor's metaphor, and the Basic Metaphor of Infinity.

Choosing the First Interpretation

The first interpretation of the Formal Reduction Metaphor is the only one that makes sense to us. We see the use of the Formal Reduction Metaphor as defining an important and enormously interesting branch of mathematics. Under this interpretation, all the results of formal mathematics stand as mathematical results within one branch of mathematics. This interpretation also makes sense of what you learn when you learn formal mathematics: You learn how to apply the Formal Reduction Metaphor.

We find the second interpretation both scientifically irresponsible and incoherent. It is scientifically irresponsible because it ignores relevant contemporary findings about the nature of the human mind, which are in turn relevant to understanding the issue of whether human mathematics could be an instance of some external, objectively existing mathematics.

We find it incoherent because it does not allow one to take into account alternative versions of set theory (e.g., well-founded and non-well-founded) and alternative versions of logic (there are thousands—two-valued, multivalued, types of modal logic, and so on).

On this use of the second interpretation, mathematics cannot be unique and yet *defined* only by set theory and logic. Set theories and logics have lots and lots of versions—all of them within mathematics. Which is supposed to define the "unique mathematics"? Any choice of a single set theory or logic rules out the mathematical viability of alternative set theories and logics.

Thus we advocate accepting the first interpretation of the Formal Reduction Metaphor. This interpretation allows one to glory in the full richness of mathematical concepts, to accept all the real mathematical results of formal mathematics, and to accept the cognitive science of mathematics and mathematical idea analysis. On this interpretation, there is no conflict whatever between formal mathematics as a discipline within mathematics and the content of this book.

This is a sensible interpretation. We can accept numbers as being numbers. We don't have to eliminate them in favor of sets alone. And we can, when appropriate, understand them as sets, as points on a line, as strategic advantages in combinatorial games, and so on.

Moreover, we don't have to give up what Pierpont lamented having to give up—the geometric understanding of functions (see Chapter 14). One can see that, under the Formal Reduction Metaphor and the Spaces Are Sets metaphor, functions are conceptualized one way; under the geometric metaphors, they are conceptualized another way. Each mode of metaphorical understanding has different uses. And each is precise on its own terms. Each defines continuity differently—natural continuity versus Weierstrass continuity (again, see Chapter 14). But you do not have to choose! As long as you can keep your metaphors straight, you can use whichever is most useful for a given purpose.

Mathematical Ideas and "Intuitions"

Formalist mathematicians often inveigh against "intuition," on the grounds that it is vague and often mathematically incorrect. It is important to contrast mathematical ideas, as we have been discussing them, with "vague intuitions."

Cognitive science has begun to make possible the scientific study of the precise structure of mathematical ideas. This book exploits the techniques developed to date. Mathematical ideas, as we have seen and will see in the case study, have a precise structure that can be discovered. This book is our first step in such an exploration.

What we are studying in the book are not vague intuitions. We are studying the structure and grounding of many of the most central ideas in all mathematics. The precise characterizations given of metaphorical mappings, blends, and special cases reveal real, stable, and precise conceptual structure.

What we are doing is making as precise as we can at this stage in cognitive science what was left vague in the *practice* of formal mathematics. A "formalization" of a subject matter in terms of set theory often hides the conceptual structure of that subject matter. Mathematical idea analysis begins to make that conceptual structure precise. What we are doing is making explicit what is implicit in the practice of formal mathematics: We are characterizing in precise cognitive terms the mathematical ideas in the cognitive unconscious that go unformalized and undescribed when a formalization of conscious mathematical ideas is done. This includes the bodily grounding of mathematical ideas and a characterization of the metaphors and blends that give mathematical thought an abstract character.

What this implies is that formalization using the Formal Reduction Metaphor does not do what many people claim for it. It does *not* formulate otherwise vague ideas in a rigorous fashion. Instead, it uses metaphor to *replace* certain ideas with other ideas—ideas for which there is a well-understood symbolization and method of calculation. It is important to bear in mind what does and does not occur when the Formal Reduction Metaphor is used to "formalize" some domain of mathematics:

1. The original ideas are not kept; they are replaced metaphorically by other ideas.
2. The formalization is *not* an abstract generalization over the original ideas.
3. The original ideas receive no mathematical idea analysis under the formalization.

4. The ideas of set theory receive no mathematical idea analysis under the formalization.

As long as these points are understood, as they are under the first interpretation, there is no problem whatever with formal mathematics. The problem is only with the second interpretation—that of the Formal Foundations program.

What Might "Foundations" Be?

If there are "foundations" for mathematics, they are *conceptual foundations—mind-based* foundations. They would consist of a thorough mathematical idea analysis that worked out in detail the conceptual structure of each mathematical domain, showing how those concepts are ultimately grounded in bodily experience and just what the network of ideas across mathematical disciplines looks like.

That would be a major intellectual undertaking. We consider this book an early step in that direction.

What Is Equality?

The general point of this chapter can be illustrated with what is usually taken as the simplest and most obvious of mathematical concepts—equality. When a child is taught mathematical notation, one of the first symbols that she learns is "=," as in "1 + 1 = 2." From a contemporary perspective, the symbol "=" seems absolutely basic and indispensable. Yet it is a relatively recent innovation in mathematics, no more than a few hundred years old (see Cajori, 1928/1993). It may be hard to believe, but for two millennia—up to the sixteenth century—mathematicians got by without a symbol for equality. It wasn't because they had no symbols at all. They had symbols for numbers and operations—just none for equality.

What mathematicians used instead of "=" was language to express what they meant—statements like "This number, when added to 3, *yields* 5" or "By solving the simultaneous equations, we *get* 17" or "Which numbers *can be decomposed into* a sum of two squares?" In short, they used words in various languages meaning "yields," "gives," "produces," "can be decomposed into," "can be factored into," "results in," and so on. These expressions all have meanings that are appropriate to specific mathematical contexts.

For example, "3 + 2 = 5" is usually understood to mean that the operation of adding 3 to 2 yields 5 as a result. Here 5 is the result of a process of addition and "=" establishes the relationship between the process and the result. But "5 = 3 + 2" is usually understood in a very different way—namely, that the number 5 can be decomposed into the sum of 3 plus 2. And as we saw in Chapter 4, the equation "$3 + 2 = 4 + 1$" is usually understood in terms of an equal result frame stating, in this case, that the result of the process of adding 3 and 2 yields the same number as the result of the process of adding 4 and 1. From a cognitive perspective, "$=$" in these instances represents three different concepts. The use of the same symbol for these different concepts hides the cognitive nuances.

We have also seen uses of "$=$" that depend on the Basic Metaphor of Infinity in cases such as $0.999\ldots = 1$, the infinite sum $\sum_{n=1}^{\infty}\frac{1}{2^n} = 1$, the granular equation $H = 1/\partial$, and the transfinite equation $\aleph_0 + \aleph_0 = \aleph_0$. Each of these uses of "$=$" has a meaning that depends on some special case of the BMI. Mathematics counts on such metaphorical uses of "$=$" as $e^{\pi i} + 1 = 0$, which we will discuss in detail in the case study to follow.

Even an idea as apparently simple as equality involves considerable cognitive complexity. From a cognitive perspective, there is no single meaning of "$=$" that covers all these cases. An understanding of what "$=$" means requires a cognitive analysis of the mathematical ideas involved. Moreover, none of these meanings of "$=$" is abstract and disembodied. Every such meaning is ultimately embodied via our everyday experience plus the metaphors and blends we use to extend that experience.

A Portrait of Mathematics

Here is the view of mathematics that emerges from this book.

- Mathematics is a natural part of being human. It arises from our bodies, our brains, and our everyday experiences in the world. Cultures everywhere have some form of mathematics.
- There is nothing mysterious, mystical, magical, or transcendent about mathematics. It is an important subject matter for scientific study. It is a consequence of human evolutionary history, neurobiology, cognitive capacities, and culture.
- Mathematics is one of the greatest products of the collective human imagination. It has been constructed jointly by millions of dedicated people over more than two thousand years, and is maintained by hundreds of thousands of scholars, teachers, and people who use it every day.
- Mathematics is a system of human concepts that makes extraordinary use of the ordinary tools of human cognition. It is special in that it is stable, precise, generalizable, symbolizable, calculable, consistent

within each of its subject matters, universally available, and effective for precisely conceptualizing a large number of aspects of the world as we experience it.

- The effectiveness of mathematics in the world is a tribute to evolution and to culture. Evolution has shaped our bodies and brains so that we have inherited neural capacities for the basics of number and for primitive spatial relations. Culture has made it possible for millions of astute observers of nature, through millennia of trial and error, to develop and pass on more and more sophisticated mathematical tools—tools shaped to describe what they have observed. There is no mystery about the effectiveness of mathematics for characterizing the world as we experience it: That effectiveness results from a combination of mathematical knowledge and connectedness to the world. The connection between mathematical ideas and the world as human beings experience it occurs within human minds. It is human beings who have created logarithmic spirals and fractals and who can "see" logarithmic spirals in snails and fractals in palm leaves.

- In the minds of those millions who have developed and sustained mathematics, conceptions of mathematics have been devised to fit the world as perceived and conceptualized. This is possible because concepts like change, proportion, size, rotation, probability, recurrence, iteration, and hundreds of others are both everyday ideas and ideas that have been mathematicized. The mathematization of ordinary human ideas is an ordinary human enterprise.

- Through the development of writing systems over millennia, culture has made possible the notational systems of mathematics. Because human conceptual systems are capable of conceptual precision and symbolization, mathematics has been able to develop systems of precise calculation and proof. Through the use of discretization metaphors, more and more mathematical ideas become precisely symbolizable and calculable. It is the human capacity for conceptual metaphor that makes possible the precise mathematization and sometimes even the arithmetization of everyday concepts—concepts like collections, dimensions, symmetry, causal dependence and independence, and many more.

- Everything in mathematics is comprehensible—at least in principle. Since it makes use of general human conceptual capacities, its conceptual structure can be analyzed and taught in meaningful terms.

- We have learned from the study of the mind that human intelligence is multifaceted and that many forms of intelligence are vital to human culture. Mathematical intelligence is one of them—not greater or lesser than musical intelligence, artistic intelligence, literary intelligence, emotional and interpersonal intelligence, and so on.

- Mathematics is creative and open-ended. By virtue of the use of conceptual metaphors and conceptual blends, present mathematics can be extended to create new forms by importing structure from one branch to another and by fusing mathematical ideas from different branches.

- Human conceptual systems are not monolithic. They allow alternative versions of concepts and multiple metaphorical perspectives of many (though by no means all!) important aspects of our lives. Mathematics is every bit as conceptually rich as any other part of the human conceptual system. Moreover, mathematics allows for alternative visions and versions of concepts. There is not one notion of infinity but many, not one formal logic but tens of thousands, not one concept of number but a rich variety of alternatives, not one set theory or geometry or statistics but a wide range of them—all mathematics!

- Mathematics is a magnificent example of the beauty, richness, complexity, diversity, and importance of human ideas. It is a marvelous testament to what the ordinary embodied human mind is capable of—when multiplied by the creative efforts of millions over millennia.

- Human beings have been responsible for the creation of mathematics, and we remain responsible for maintaining and extending it.

The portrait of mathematics has a human face.

Part VI

$$e^{\pi i} + 1 = 0$$

A Case Study of the Cognitive Structure
of Classical Mathematics

THIS EQUATION OF LEONHARD EULER'S is one of the deepest in classical mathematics. It brings together in one equation the most important numbers in mathematics: e, π, i, 1, and 0. In doing so, this one equation implicitly states a relationship among the branches of classical mathematics—arithmetic, algebra, geometry, analytic geometry, trigonometry, calculus, and complex variables. It also embodies the relationship among many of the most central mathematical ideas—trigonometric functions, logarithms and exponentials, power series, derivatives, imaginary numbers, and the complex plane. To understand Euler's equation is to understand many of the deep connections among these ideas and among the branches of classical mathematics.

The equation itself appears mysterious—some would say mystical. What does it mean? We are taught that exponentiation is the multiplication of a number by itself, that 2^3 *means* 2 times 2 times 2. e is a real number, the infinite decimal 2.71812818284. . . . π is an infinite decimal 3.14159. . . . i is an imaginary number, $\sqrt{-1}$. π times i is an imaginary number with an infinite number of decimal places. What could it mean to multiply a real number, e, by itself an imaginary number of times?

Moreover, how could an equation among *numbers* relate deep and important *ideas?* Numbers in themselves are just numbers. 4,783.978 is just a number. But e, π, and i are more than just numbers. They express ideas. But how can a mere *number* express an *idea?* And how can an equation be a crystallization of the cognitive relationships among these ideas?

The four chapters of this case study are dedicated not only to answering such questions but to providing a model of how such questions can be approached through mathematical idea analysis.

Case Study 1

Analytic Geometry
and Trigonometry

Benjamin Peirce, one of Harvard's leading mathematicians in the nineteenth century (and the father of Charles Sanders Peirce), lectured in one of his classes on Euler's proof that $e^{\pi i} = -1$. He found it somewhat shocking. In teaching the equation and its proof, he remarked,

> Gentlemen, that is surely true, it is absolutely paradoxical; we cannot understand it, and we don't know what it means. But we have proved it, and therefore we know it must be truth. (cited in Maor, 1994, p. 160)

Peirce was not the last mathematics teacher to fail to understand what $e^{\pi i} = -1$ means. Relatively few mathematics teachers understand it even today, and fewer students do. Yet generation after generation of mathematics teachers and students continue to go uncomprehendingly through one version or another of Euler's proof, understanding only the regularity in the manipulations of the symbols.

They are much like Mr. M, Laurent Cohen and Stanislas Dehaene's patient discussed in Chapter 1, who knows that "three times nine is twenty-seven" but not what it means (Dehaene, 1997, p. 189).

Mr. M, being brain damaged, has no choice. Benjamin Peirce was born too soon. But in the age of cognitive science one can at least try to do a bit better.

The Problem

Exactly what does $e^{\pi i}$ mean and why does it equal -1?

From a cognitive perspective, the answer is not a matter of axioms, definitions, theorems, and proofs. That kind of answer is given in trigonometry or elementary calculus classes, where the proof is taught. But, as Benjamin Peirce said eloquently, knowing how to prove something does not necessarily mean that you understand the deep meaning of what you've proved.

It is not our job here to teach Euler's proof as it is done in a calculus class. Our job is to give an account in cognitive terms of the meaning and significance of the equation, by characterizing the *concepts* involved and the relevant relationships among the concepts. What is needed is a conceptual analysis, a mathematical idea analysis, framed in terms of cognitive mechanisms, of what is required to understand—really understand—$e^{\pi i}$.

The answer cannot be that raising a number to a power is just multiplying that number by itself. $e^{\pi i}$ makes no sense as e multiplied by itself π times (since π, too, is an infinite decimal) and then i times, where $i = \sqrt{-1}$, an imaginary number. What could "$\sqrt{-1}$ times" possibly mean? A different answer is required, one that follows from the conceptual content of the ideas involved in e, π, i, and -1. Purely formal definitions, axioms, and proofs do not capture conceptual content. Formal proofs are simply inadequate for the enterprise of characterizing meaning and providing understanding.

Other questions must be answered as well. Why *should* $e^{\pi i}$ equal, of all things, -1? $e^{\pi i}$ has an imaginary number in it; wouldn't you therefore expect the result to be imaginary, not real? e is about differentiation, about change, and π is about circles. What do the ideas involved in change and in circles have to do with the answer? e and π are both transcendental numbers—numbers that are not roots of any algebraic equation. If you operate on one transcendental number with another and then operate on the result with an imaginary number, why *should* you get a simple integer like -1? The answer requires an overview of the conceptual structure of a lot of classical mathematics. Moreover, it requires understanding how different branches of mathematics are related to one another: how algebra is related to geometry, and how both are related to trigonometry, to calculus, and to the study of complex numbers.

To answer this, we must look closely at the cognitive mechanisms that connect these different branches of mathematics. Not surprisingly, linking metaphors will be involved.

Our study begins with an examination of analytic geometry, where arithmetic, algebra, and geometry come together.

Analytic Geometry

The field of analytic geometry, which unites geometry and algebra and on which so much of mathematics depends, was invented by René Descartes and rests on the concept of the Cartesian plane. The Cartesian plane is a conceptual blend of (a) two number lines and (b) the Euclidean plane, with two lines perpendicular to each other.

Since Descartes and Euler preceded the discretization program described in Chapter 12, we will be assuming the earlier geometric paradigm, where space is naturally continuous space.

The Cartesian Plane Blend

Conceptual Domain 1 Number Lines		Conceptual Domain 2 The Euclidean Plane with Line X Perpendicular to Line Y
Number line x	↔	Line X
Number line y	↔	Line Y
Number m on number line x	↔	Line M parallel to line Y
Number n on number line y	↔	Line N parallel to line X
The ordered pair of numbers (m,n)	↔	The point where M intersects N
The ordered pair of numbers $(0,0)$	↔	The point where X intersects Y
A function $y = f(x)$; that is, a set of ordered pairs (x,y)	↔	A curve with each point being the intersection of two lines, one parallel to X and one parallel to Y
An equation linking x and y; that is, a set of ordered pairs (x,y)	↔	A figure with each point being the intersection of two lines, one parallel to X and one parallel to Y
The solutions to two simultaneous equations in variables x and y	↔	The intersection points of two figures in the plane

This conceptual blend systematically links geometry with arithmetic and algebra. In the blend, one can geometrically visualize functions and equations in geometric terms, and also conceptualize geometric curves and figures in algebraic terms.

- The simple equation $x = y$ corresponds in this conceptual blend to the diagonal line through the origin from the lower left to upper right, and conversely, the diagonal line corresponds to the equation.
- A line in general maps onto an equation of the form $y = ax + b$, where a is the slope of the line and b is the number on the y-axis corresponding to the point at which the line intersects the axis, and conversely.
- The equation $x^2 + y^2 = 4$ corresponds to a circle of radius 2 with its center at the origin, and conversely.

In general, curves in space can be conceptualized as being traced by motions of points. Those motions can in turn be conceptualized as having components of motion along the *x*- and *y*-axes of the Cartesian plane. Descartes observed that such motions in the plane correspond to continuous variations of numbers in algebraic equations. Continuous changes of numerical variables in algebraic equations thus became correlated with the motion of points along the *x*- and *y*-axes. In the Cartesian Plane blend, continuous functions are correlated with curves formed by the motion of a point over the plane.

The Cartesian Plane blend is, of course, generalizable to spaces of any dimension. This can be done by adding the following correspondences to the blend given above.

The *n*-tuple of numbers $(n_1, n_2, \ldots)$	↔	A point in *n*-dimensional space
The collection of all *n*-tuples of numbers	↔	An *n*-dimensional space

This allows us to visualize spaces of many dimensions in terms of *n*-tuples of numbers, and equations in many variables in terms of figures in multidimensional spaces.

The Functions Are Numbers Metaphor

Literally, functions are not numbers. Addition and multiplication tables do not include functions. But once we conceptualize functions as ordered pairs of points in the Cartesian plane, we can create an extremely useful metaphor. The operations of arithmetic can be metaphorically extended from numbers to functions, so that functions can be metaphorically added, subtracted, multiplied, and divided in a way that is consistent with arithmetic. Here is the metaphor:

FUNCTIONS ARE NUMBERS

Source Domain NUMBERS (VALUES OF FUNCTIONS)		*Target Domain* FUNCTIONS
Any arithmetic operation on the *y*-values of two functions $y = f(x)$ and $y = g(x)$	→	The corresponding arithmetic operation on the two functions $f(x)$ and $g(x)$

For example, the sum $(f + g)(x)$ of two functions $f(x)$ and $g(x)$ is the sum of the values of those functions at *x*: $f(x) + g(x)$. Note that the "+" in "$f(x) + g(x)$" is the ordinary literal "+" that is used in the addition of two numbers, while the "+" in "$f + g$" is metaphorical, a product of this metaphor. Similarly, all the following are special cases of this metaphor:

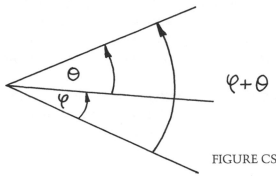

FIGURE CS 1.1   The intuitive idea of adding angles.

- $(f - g)(x)$ Is $f(x) - g(x)$
- $(f \cdot g)(\mathbf{x})$ Is $f(x) \cdot g(x)$
- $(f \div g)(x)$ Is $f(x) \div g(x)$, where $g(x) \neq 0$

In the Cartesian plane, simple arithmetic functions of one variable can be conceptualized metaphorically as curves. The Functions Are Numbers metaphor allows us to conceptualize what it means to add, subtract, multiply, and divide such curves. Conceptualizing curves as elements subject to arithmetic operations is a central idea in classical mathematics.

The Basic Metaphor of Trigonometry

Many of the linking metaphors in mathematics are arithmetization metaphors—metaphors that conceptualize some domain of mathematics in terms of arithmetic, so that arithmetic calculations can be applied to it. We have just seen how points in space can be conceptualized in terms of numbers, how spaces of any number of dimensions can be conceptualized in terms of numbers, and how functions themselves can be conceptualized as numbers.

The fundamental metaphor that defines the field of trigonometry conceptualizes angles as numbers. Consider the following question: How can we add angles? The obvious intuitive answer is to place their vertices together, with a leg in common, and take the total angle as the "sum" (see Figure CS 1.1).

But that's not precise enough. It doesn't give you a way to calculate sums of angles, much less to do subtraction, multiplication, division, and raising to powers for angles. Angles in themselves are nonnumerical. In plane geometry, there are angles but no numbers. If we are going to do arithmetic calculation on angles, we have to get angles to *be* numbers. That is a job for a metaphor—what we will call the Trigonometry metaphor.

The post-Cartesian version of the Trigonometry metaphor begins with a conceptual blend we will call the Unit Circle blend. We will look in great detail at the structure of this blend from a cognitive perspective. Many of the details we will discuss are implicit in mathematics texts. Our job here is to make explicit what is implicit.

The Unit Circle blend has four corresponding domains, and we will build it up in stages. At the first stage, there is a blend of two elements in two domains: a circle in the Euclidean plane blended with the Cartesian plane to yield a unit circle—a circle whose center is at the origin (0, 0) of the Cartesian plane and whose radius is 1 (see Figure CS 1.2).

When the circle is blended with the Cartesian plane in this way, numbers are assigned to lengths of lines, because the Number-Line blend is inherent in the Cartesian plane. In particular, the radius of the circle has the length 1 in the unit circle.

THE UNIT CIRCLE BLEND (STAGE 1)

Domain 1 A CIRCLE IN THE EUCLIDEAN PLANE WITH CENTER AND RADIUS		*Domain 2* THE CARTESIAN PLANE, WITH X-AXIS, Y-AXIS, ORIGIN AT (0,0)
Euclidean plane	↔	Cartesian plane
Center	↔	Origin
Radius	↔	Distance 1 from origin

At stage 2, we add the angle ϕ which is ultimately to be conceptualized metaphorically as a number. We place the vertex of ϕ at the center of the circle with one leg of ϕ on the x-axis (see Figure CS 1.3).

Note that at this stage there is a correlation between ϕ and the length of the arc subtended by ϕ. Here is the blend at Stage 2. (For the sake of simplicity, when a blend is composed of three or more domains we will display the table without the double arrows.)

THE UNIT CIRCLE BLEND (STAGE 2)

Domain 1 A CIRCLE IN THE EUCLIDEAN PLANE WITH CENTER AND RADIUS	*Domain 2* THE CARTESIAN PLANE, WITH X-AXIS, Y-AXIS, ORIGIN AT (0,0)	*Domain 3* AN ANGLE ϕ IN THE EUCLIDEAN PLANE WITH LEG 1 AND LEG 2
Euclidean plane	Cartesian plane	Euclidean plane
Center	Origin	Vertex of ϕ
Radius	Distance 1 from origin	——
——	The interval [0,1] on the X-axis	Leg 1

| Line from center to point P on the circle | Line from origin to point (x_1, y_1) | Leg 2 |
| Arc subtended by ϕ | Arc length | —— |

There are several things of interest in the blend at this stage. First, Leg 2 of the angle ϕ intersects the circle. We have called the intersection point on the circle P, and the corresponding point in the Cartesian plane (x_1, y_1). Second, there is now an arc subtended by angle ϕ, and in the Cartesian plane it has a length that is a number; this is indicated in the last row. Third, leg 1 of angle ϕ now corresponds to the line segment [0, 1] in the Cartesian plane, as indicated in the fourth row. Moreover, several of the correspondences do not extend to all three domains of the blend; for example, in the fourth row there is nothing in the Euclidean circle to correspond to leg 1 of the angle or line segment [0, 1] in the Cartesian plane.

Finally, we reach Stage 3 of the blend, where the right triangle of figure CS 1.4 is added. The vertex of angle A corresponds to the center of the circle, the origin in the Cartesian plane and the vertex of angle ϕ. The result is the full Unit Circle blend, which is conceptually quite complex (see Figure CS 1.4).

THE UNIT CIRCLE BLEND			
Domain 1 A CIRCLE IN THE EUCLIDEAN PLANE WITH CENTER AND RADIUS	*Domain 2* THE CARTESIAN PLANE, WITH X-AXIS, Y-AXIS, ORIGIN AT (0,0)	*Domain 3* AN ANGLE ϕ IN THE EUCLIDEAN PLANE WITH LEG 1 AND LEG 2	*Domain 4* A RIGHT TRIANGLE, WITH HYPOTENUSE, RIGHT ANGLE, ∠A, ∠P, SIDE a AND SIDE b
Euclidean plane	Cartesian plane	Euclidean plane	Euclidean plane
Center	Origin	Vertex of ϕ	Vertex of $\angle A$
Radius	Distance 1 from origin	——	Length of hypotenuse
——	The interval [0,1] on the X-axis	Leg 1	——
Line from center to point P on the circle	Line from origin to point (x_1, y_1)	Leg 2	Hypotenuse
Arc subtended by ϕ	Arc length	Size of ϕ	Size of $\angle A$
——	A segment from $(0, 0)$ to $(x_1, 0)$	——	Side a
——	A segment from (x_1, y_1) to $(x_1, 0)$	——	Side b

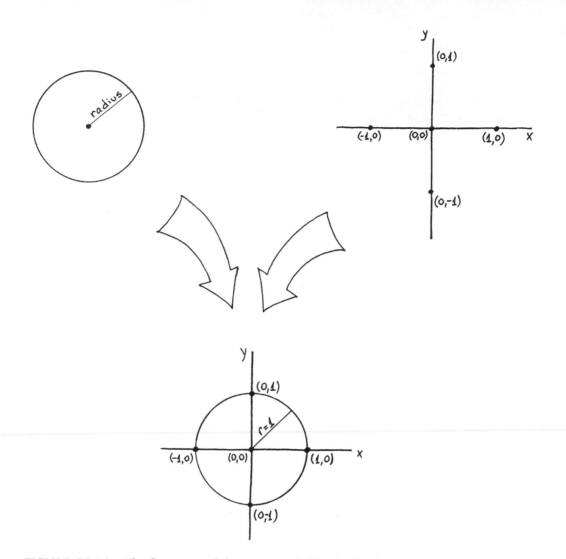

FIGURE CS 1.2   The first stage of the Unit Circle blend. The domain of a circle with center and radius (upper left) is conceptually combined with the Cartesian Coordinate domain (upper right) to yield a conceptual blend (bottom), in which the center of the circle is identical to the origin (0, 0) in the Cartesian plane and the length of the radius of the circle equals 1.

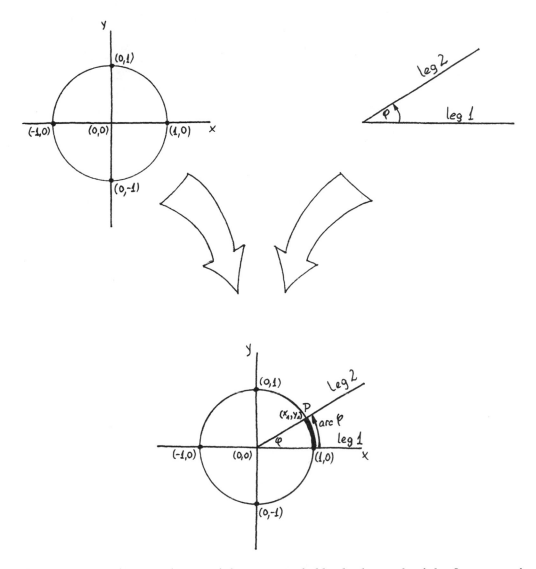

FIGURE CS 1.3 The second stage of the Unit Circle blend. The result of the first stage of the Unit Circle blend (upper left) combines conceptually with an angle φ (upper right) to produce a conceptual blend (bottom), in which the vertex of the angle is identical to the origin-center and leg 1 of the angle is identical to the nonnegative portion of the x-axis. Note that a new entity emerges in the blend: the arc on the circle subtended by the angle φ.

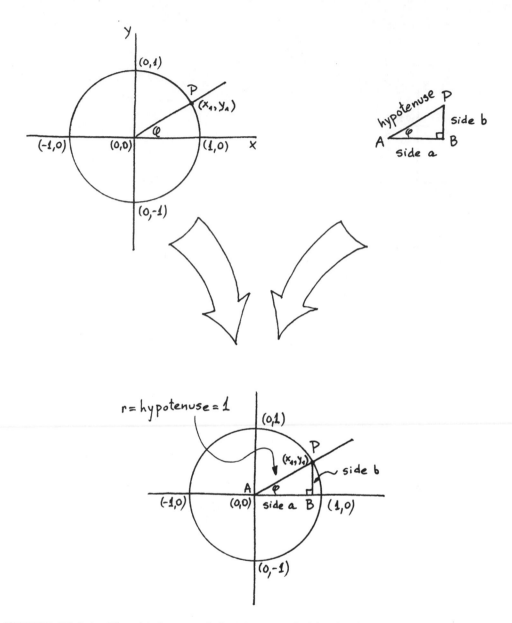

FIGURE CS 1.4 The third stage of the Unit Circle blend. The result of the second stage (upper left) is conceptually combined with a right triangle (upper right) to form a new conceptual blend (bottom), in which vertex *A* is identical to the center-origin and the hypotenuse is identical to the radius. Note that new entities and entailments emerge in the blend. It is entailed that the hypotenuse of the right triangle = 1. Sides *a* and *b* of the right triangle are now interdefined with the other elements of the blend in such a way that new mathematical functions emerge—namely, the cosine of angle ϕ (whose value is the *x*-coordinate of *P*) and the sine of angle ϕ (whose value is the *y*-coordinate of *P*).

Note that important correspondences have been added. The hypotenuse of the right triangle now corresponds to the line from the center to point P. It is an entailment of the blend that that line has the length of the radius—namely, 1. Additionally, new lengths have been introduced into the blend: the lengths of sides a and b of the triangle, which correspond to numbers in the Cartesian plane. These lengths are also entailments of the blend.

The Trigonometry Metaphor

Given the Unit Circle blend, a blend of remarkable complexity, it is now possible to state the Trigonometry metaphor with great simplicity.

THE TRIGONOMETRY METAPHOR

Source Domain THE UNIT CIRCLE BLEND		Target Domain TRIGONOMETRIC FUNCTIONS
The length of the arc subtended by angle ϕ	$\rightarrow$	The number assigned to angle ϕ
The length of side a	$\rightarrow$	The value of the function $\cos \phi$
The length of side b	$\rightarrow$	The value of the function $\sin \phi$

Entailments

The unit circle: $x^2 + y^2 = 1$	$\rightarrow$	$\sin^2 \phi + \cos^2 \phi = 1$
ϕ is a right angle	$\rightarrow$	$\phi = \pi/2$
ϕ spans half of the circle	$\rightarrow$	$\phi = \pi$
ϕ spans the whole circle	$\rightarrow$	$\phi = 2\pi$
ϕ is zero	$\rightarrow$	$\sin 0$ (which equals $\sin \pi$) $= 0$
ϕ is two right angles (180°)	$\rightarrow$	$\cos \pi = -1$

The first entailment of this metaphor is a straightforward consequence of the Pythagorean theorem applied to the right triangle in the blend, whose hypotenuse is of length 1.

The next three entailments are consequences of

a. the fact that the circumference of a circle is π times the diameter,
b. the assignment of 1 as the radius of the unit circle, and
c. the fact that via the Trigonometry metaphor, ϕ is measured in units of length on the unit circle.

Since the diameter of the unit circle is 2, its circumference is 2π. Thus, if ϕ spans the whole circle, the number assigned to it by the Trigonometry metaphor will be 2π. If ϕ spans half the circle, it follows that the number assigned to it will be π.

Given the metaphorical definition of the functions sine and cosine, it follows that $\sin \pi = 0$, $\cos \pi = -1$, and $\cos 2\pi = 1$. Here we have a functional correlation between basic arithmetic elements 0 and 1 on the one hand and on the other hand π, the ratio of the circumference to the diameter of the circle.

In Case Study 4, where we take up Euler's famous equation $e^{\pi i} + 1 = 0$, we will see that the relationship among π, 0, and 1 in that equation is exactly the relationship given by the Trigonometry metaphor. The relationship between π and -1 in Euler's equation arises from the fact that in the blend, $\cos \pi = -1$.

The remaining entailments of the Trigonometry metaphor are as follows. Recall that in the Cartesian plane a curve is conceptualized as if it were traced by the motion of a point. Thus, a circle is conceptualized as the curve traced by a point that moves in a circle. The point completes traversing the circle once when it moves a length 2π. However, the point can be conceptualized as moving further—that is, as moving around the circle k times (where k is a natural number), tracing a line of length $2k\pi$. Thus, the angle in the unit circle can be conceptualized either as ϕ (where ϕ is less than 2π) or as $\phi + 2k\pi$, if the point has moved around the circle k times before moving a length ϕ.

In the blend, an angle and the arc it subtends determine the lengths of sides a and b. But the converse is not true, since the size of the angle and the arc may be $\phi + 2k\pi$, if the point tracing the arc has moved around the circle k times. No matter how many times k the point has moved around the circle, the length of legs a and b are the same at angle ϕ. This means that the values of the functions $\sin \phi$ and $\cos \phi$ are the same as the values of $\sin (\phi + 2k\pi)$ and $\cos (\phi + 2k\pi)$.

The Recurrence Is Circularity Metaphor

The Trigonometry metaphor produces the sine and cosine functions. Descartes's metaphor conceptualizes functions as curves in the Cartesian plane. What do the sine and cosine functions look like as such curves?

Sin ϕ and cos ϕ are functions of the angle ϕ, which is now a number, via the Trigonometry metaphor. Since ϕ varies over numbers, it can vary over the numbers on a number line and therefore over the numbers on an axis in the Cartesian plane. Let the x-axis specify values of ϕ, and the y-axis, values of $\sin \phi$ and $\cos \phi$.

Recall that $\sin \phi$ and $\cos \phi$ vary only between -1 and 1, because the extreme values they can take are limited by the radius of the circle in the blend, which is 1. The curves are therefore confined to a long horizontal strip of the plane

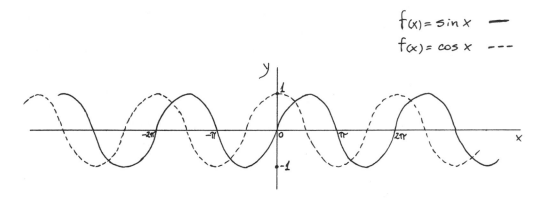

FIGURE CS 1.5   The sine and cosine functions showing the period 2π, which results from the fact that the functions are defined relative to the Unit Circle blend. Such functions give a characterization of the concept of recurrence according to the metaphor Recurrence Is Circularity.

where y varies between -1 and 1 (see Figure CS 1.5). Moreover, since the values of $\sin\phi$ and $\cos\phi$ are the same for ϕ and $\phi + 2k\pi$, the values recur at intervals along the x-axis of 2π. The sine and cosine curves thus repeat themselves over and over indefinitely—as a point traces around the circle in the Unit Circle blend. This property of the curves, in which values recur after a fixed "period," is called periodicity. In this case the period is 2π.

There is a deep and inherent link between the periodicity—the recurrence of values—of the functions and the circularity of the Unit Circle blend in terms of which the functions are defined. In English and many other languages, the Recurrence Is Circularity metaphor occurs in the conceptual system of the language in general, completely outside mathematics.

For example, a recurrent event like the Christmas holiday is spoken of as "coming *around* again." We speak of recurrent events as happening "over and over" (as in the image of going over the top of a circle again and again). Cultures in which history is believed to recur are spoken of as having "*circular* concepts of time." The Aymara in the Andes, for example, believe that events recur—that they happen again and again. They draw recurring events using a circle. Thus, the relationship between the unit circle and functions that have recurrent values is a metaphorically natural one.

Because of this conceptual connection between recurrence and circularity, the concept of recurrent values is expressed mathematically in terms of sine and cosine functions, which are themselves conceptualized via the Unit Circle blend.

The Polar Coordinate Metaphor

Given the Trigonometry metaphor, we can now form a further metaphorical blend from that metaphor:

THE TRIGONOMETRY BLEND

Source Domain THE UNIT CIRCLE BLEND		*Target Domain* TRIGONOMETRIC FUNCTIONS
The length of the arc subtended by angle ϕ	$\leftrightarrow$	The number assigned to angle ϕ
The length of side a	$\leftrightarrow$	The function $\cos \phi$
The length of side b	$\leftrightarrow$	The function $\sin \phi$

This allows us to use the Trigonometry blend to conceptualize points in the Cartesian plane using sine and cosine functions. The results are polar coordinates.

THE POLAR-TRIGONOMETRY METAPHOR

Source Domain THE TRIGONOMETRY BLEND		*Target Domain* THE CARTESIAN PLANE BLEND
r	$\rightarrow$	the distance from the origin O to point $P = (a, b)$
θ	$\rightarrow$	the angle formed by a counter clockwise rotation from the x-axis to OP, where $P = (a, b)$
The ordered pair (r, θ)	$\rightarrow$	The point (a, b)
$r \cos \theta$	$\rightarrow$	a
$r \sin \theta$	$\rightarrow$	b

As a result, ordinary Cartesian coordinates come to be conceptualized in trigonometric terms. Each point in the Cartesian plane (a, b) is uniquely characterized by the ordered pair of numbers (r, θ), where r is a number (the distance of (a, b) from the origin) and θ is a number by virtue of the Trigonometry metaphor (see Figure CS 1.6).

What makes polar coordinates metaphorical? Recall that the Cartesian plane in itself has no trigonometry—no angles-as-numbers, no sines and cosines. Through this metaphor, the Cartesian plane becomes conceptualized in terms of trigonometry, with each point (a, b) being conceptualized as the pair of numbers (r, θ).

As we shall see, this metaphor is crucial in understanding $e^{\pi i} + 1 = 0$.

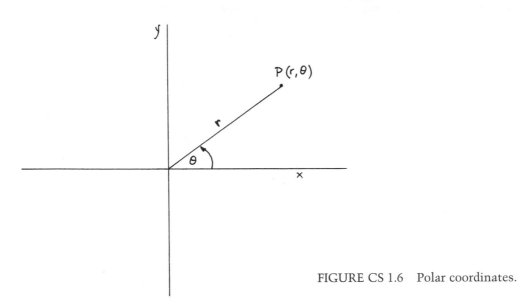

FIGURE CS 1.6   Polar coordinates.

The metaphors discussed above constitute a remarkable achievement in the history of mathematical thought. They link together arithmetic, algebra, and geometry to create two whole new branches of mathematics—analytic geometry and trigonometry. Because metaphors and metaphorical blends allow the inferential structure of one domain to be used in another domain, these metaphors and blends permit arithmetic calculations, algebraic equations, and the methods of analytic geometry to be used for calculations on angles. They permit the visualization in spatial terms of functions and equations. And they make possible a mathematics of recurrent phenomena, with the number π taking on a new meaning. Where π was previously only the ratio of the circumference of a circle to its diameter, 2π now becomes a measure of periodicity for recurrent phenomena, with π as the measure of half a period. This is a new *idea*: a new *meaning* for π.

The Meaning of Numbers

Here is a place where the cognitive science of mathematics is needed to make systematic sense of mathematics as a discipline. Formal mathematics—with symbolic proofs and calculations—does not itself contain the *meanings* of those symbols, proofs, and calculations. Moreover, the meaning of a symbol like π is not merely its numerical value. The meaning of π in a particular formula, proof, or calculation is conceptual: a ratio of circumference to diameter or a standard of recurrence. And the links among the multiple meanings of π are given by conceptual metaphors.

The metaphors and blends used in the conceptualization of functions like sine and cosine and of numbers like π are anything but superfluous in mathematical thought. Those metaphors play a crucial role in understanding what numbers and symbols mean. The "recurrence" sense of π always evokes the Trigonometry blend and the Recurrence Is Circularity metaphor. Understanding a use of π in a discussion of recurrence necessarily involves understanding that blend and that metaphor. It is by means of such cognitive mechanisms that numbers have meaning for us.

Have you ever wondered what $e = 2.718281828459045\ldots$ means?

Case Study 2

What Is *e*?

THE EXPRESSION 2^5, two raised to the fifth power, is usually taken to mean $2 \cdot 2 \cdot 2 \cdot 2 \cdot 2$—two multiplied by itself repeatedly. But the corresponding expression $e^{\pi i}$, the number $e = 2.718281828459045 \ldots$ raised to the power $\pi \cdot i$ $(3.1415926 \ldots \cdot \sqrt{-1})$, cannot mean e multiplied by itself π times and the result then multiplied by itself $\sqrt{-1}$ number of times. In general, raising a number to a power cannot be understood as self-multiplication. What then *is* raising a number to a power?

We can see the answer most easily if we look first at *logarithms*. Think for a moment of how a slide rule works. If you want to multiply two numbers, you *add* their lengths on the slide rule. Slide rules allow us to do multiplication by doing addition. The reason that they work is that they use a logarithmic scale: Multiplying is adding the logarithms. By adding the logarithms of the numbers, you multiply the numbers.

What is a logarithm, then? A logarithm is a mapping that allows you to multiply by adding. Logarithms were invented (by John Napier in the late sixteenth century) because the multiplication algorithm is cognitively more complicated than the addition algorithm: It is simply harder to multiply two large numbers than to add them. In the days before computers and mechanical calculating machines, multiplication was done by hand. Napier sought to make the process easier by finding a way to accomplish multiplication by doing simple addition.

Notice that $q \cdot q \cdot q \cdot q \cdot q = (q \cdot q) \cdot (q \cdot q \cdot q) = q^2 \cdot q^3 = q^{2+3} = q^5$. For example, $100{,}000 = 100 \cdot 1{,}000 = 10^2 \cdot 10^3 = 10^{2+3} = 10^5$. Suppose you had a table of the form:

1	—	0
10	—	1
100	—	2
1,000	—	3
10,000	—	4
100,000	—	5

You could multiply 100 times 1,000 by going to numbers on the right—namely, 2 and 3; then adding 2 plus 3 to get 5; and then going back to the column on the left to get 100,000.

The table specifies values of a mapping, called the logarithm, or log for short. Let us look at some of its properties. It maps real numbers onto real numbers and products onto sums. That is, it maps a product $a \cdot b$ onto a sum $a' + b'$ by mapping numbers a and b onto a' and b' and by mapping the operation of multiplication onto the operation of addition.

Suppose you want to multiply numbers a and b. You first find their images a' and b' under the mapping log. You might want to construct a table listing pairs (a, a'), (b, b') such that $\log(a) = a'$ and $\log(b) = b'$. To multiply a times b, you add a' and b', get the sum, and take the inverse of log to get the product of a and b. In other words, $\log^{-1}(a' + b') = a \cdot b$. In order to get a unique result, log needs to be a one-to-one mapping. In order to multiply by adding, you need to first apply log, then add $a' + b'$, then take the inverse mapping $\log^{-1}$ of the sum.

Log is a mapping from the group of positive real numbers with the operation of multiplication to the group of all real numbers with the operation of addition. As such, log maps the multiplicative identity, 1, onto the additive identity, 0. That is, $\log(1) = 0$. Thus, $\log(a \cdot 1)$ is mapped onto $\log(a) + \log(1) = a' + 0 = a'$, which is $\log(a)$. Log also maps multiplicative inverses $(1/a)$ onto additive inverses $(-a')$. Thus, the log $(a \cdot 1/a)$ is mapped onto $\log(a) + \log(1/a) = a' + (-a') = 0 = \log(1)$. We can see this by extending the table, as follows:

Number		Logarithm
1/10,000	—	−4
1/1,000	—	−3
1/100	—	−2
1/10	—	−1
1	—	0
10	—	1
100	—	2
1,000	—	3
10,000	—	4
100,000	—	5

For example, $100{,}000 \cdot (1/100) = 10^5 \cdot 10^{-2} = 10^{5-2} = 10^3 = 1{,}000$.

To multiply 100,000 times 1/100, you take the logarithms 5 and –2, add 5 + (–2), get 3, and take the inverse logarithm to get 1,000. Note that $1{,}000 \cdot (1/1{,}000) = 10^3 \cdot 10^{-3} = 10^{3-3} = 10^0 = 1$. Via logarithms, you take the logarithms of 1,000 and 1/1,000—namely, 3 and –3; add them to get 0; and take the inverse logarithm of 0 to get 1.

The table we have so far is very simpleminded. To be of any real use, it has to be filled in with a huge number of values. Each line in the required table must have a number n paired with a power of 10, n', such that $n = 10^{n'}$. Then $n' = \log(n)$. Napier's achievement was to construct such a huge table in a consistent manner. It took many years.

As an abstract mathematical function, log maps *every* positive real number onto a corresponding real number, and maps *every* product of positive real numbers onto a sum of real numbers. Of course, there can be no table for such a mapping, because it would be infinitely long. But abstractly, such a mapping can be characterized as outlined here. These constraints completely and uniquely determine every possible value of the mapping. But the constraints do not in themselves provide an algorithm for computing such mappings for all the real numbers. Approximations to values for real numbers can be made to any degree of accuracy required by doing arithmetic operations on rational numbers.

The general constraints completely defining the mapping are:

- 1 is mapped onto 0.
- 10 is mapped onto 1.
- 10^n is mapped onto n, for all real numbers n.
- Products are mapped onto sums.

Notice that for every real number in the range of the mapping, there is a positive real number mapped onto it. The negative numbers and zero in the domain of the mapping do not get mapped onto any real number. Figure CS 2.1 illustrates the constraints on the logarithm mapping.

Bases

The account we have just given for the logarithm mapping fits the choice of 10 as a base. The base is the number that is mapped onto 1. We chose 10 as an illustration because our system of numerals happens to use 10 as a base; that is, each real number is numerically represented as the sum of products of 10 to some integral power (e.g., 3.14 is 3 times 10 to the 0 power, plus 1 times 10 to the –1 power, plus 4 times 10 to the –2 power).

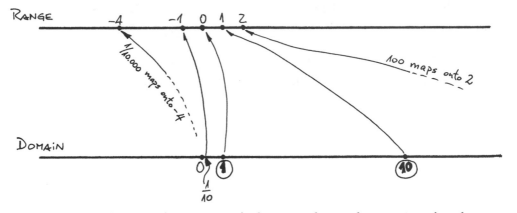

FIGURE CS 2.1   The logarithm mapping for base 10—that is, the mapping of products onto sums with base 10. 1 is mapped onto 0, the base 10 is mapped to 1, and, in general, 10^n is mapped onto n. Thus, the logarithm to base 10 of a number in the domain is the corresponding number in the range. To find the product of two numbers in the domain, map them both to the range (this is called "taking their logarithms"), add those numbers in the range, and find the pre-image of the sum in the domain of the mapping.

However, this is an arbitrary choice. Numbers can be written as sums of products of this sort with any base at all greater than 1. Similarly, there can be different versions of the logarithm function depending on the choice of base, for any base greater than 1. The following table gives the number n as the logarithm with the base 2 for a corresponding number 2^n.

n	-2	-1	0	1	2	3	4	5	6	7	8	9	10
	↑	↑	↑	↑	↑	↑	↑	↑	↑	↑	↑	↑	↑
2^n	1/4	1/2	1	2	4	8	16	32	64	128	256	512	1024

Thus, to multiply 16 times 32, you add their logarithms in base 2—namely, 4 plus 5—to get 9. The inverse logarithm of 9 is 512, which is the product of 16 and 32. For logarithms with base 2, the constraints that define the function are:

- 1 is mapped onto 0.
- 2 is mapped onto 1.
- 2^n is mapped onto n, for all real numbers n.
- Products are mapped onto sums.

Since a logarithm mapping can be defined with any base b greater than 1, the general constraints are:

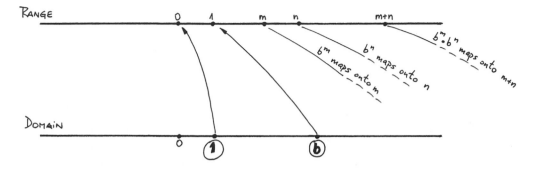

FIGURE CS 2.2   The logarithm mapping for an arbitrary base *b*—that is, the mapping of products onto sums with a base *b*. 1 is mapped onto 0, the base *b* is mapped onto 1, and b^n is mapped onto *n*. The product $b^m \cdot b^n$ is mapped onto the sum $m + n$. The logarithm to base *b* of a number in the domain is the corresponding number in the range.

- 1 is mapped onto 0.
- *b* is mapped onto 1, for *b* greater than 1.
- b^n is mapped onto *n*, for all real numbers *n*.
- Products are mapped onto sums.

Figure CS 2.2 displays the structure of such a mapping.

Exponentials

As we have just seen, the logarithm mapping with a base *b*, greater than 1, maps a positive real number with the structure b^n onto a real number *n*. Since the logarithm mapping with base *b* is one-to-one, there will be an inverse mapping that maps a real number *n* onto a positive real number b^n. This is the *exponential mapping*. It is the inverse of the logarithm mapping, and its general constraints are:

- 0 is mapped onto 1.
- 1 is mapped onto *b*, for *b* greater than 1.
- *n* is mapped onto b^n, for all real numbers *n*.
- Sums are mapped onto products.

This mapping is defined for *all* real numbers *n* and for *any* real number *b* greater than 1.

When *b* = 2, we have the inverse of the mapping just given for the logarithm to base 2.

n	–2	–1	0	1	2	3	4	5	6	7	8	9	10
↓	↓	↓	↓	↓	↓	↓	↓	↓	↓	↓	↓	↓	↓
2^n	1/4	1/2	1	2	4	8	16	32	64	128	256	512	1024

For the special cases given in the tables, it so happens that they can be computed via self-multiplication. For example, 2^6 can be computed by multiplying $2 \cdot 2 \cdot 2 \cdot 2 \cdot 2 \cdot 2 = 64$. But the exponential mapping applies to an infinity of cases where this is not true. Consider, for example, $2^{2/3}$. This is a well-defined instance of the exponential mapping with base 2—namely, $\sqrt[3]{4}$—but it is not an instance of 2 multiplied by itself 2/3 of a time, whatever that could mean. Similarly, 2^π is also a well-defined instance of the mapping, but it, too, is not an instance of 2 multiplied by itself, 3.14159. . . (to infinity) times—whatever *that* could mean! The exponential mapping for a base b is defined as specified above, but it is not self-multiplication. Figure CS 2.3 indicates the general form of the mapping.

It is important to understand the difference between self-multiplication and exponentiation. When we write $32 = 2 \cdot 2 \cdot 2 \cdot 2 \cdot 2$, we are implicitly making use of the grounding metaphor that Arithmetic Is Object Construction (see Chapter 3); that is, we understand 32 as a whole that is made up of five instances of the same part, 2, put together by multiplication. The symbolization 2^5 as usually understood is, from a conceptual point of view, ambiguous. It could mean "$2 \cdot 2 \cdot 2 \cdot 2 \cdot 2$" or it could mean "the result of the exponential mapping with base 2 applied to 5." An alternative way of writing the latter might be $\text{Exp}_2(5)$, where 5 is the input to the exponential mapping and 2 indicates what 1 maps onto in this version of the exponential mapping. 2^π, however, can mean only $\text{Exp}_2(\pi)$; that is, "π is the input to the exponential mapping and 2 indicates what 1 maps onto in this version of the exponential mapping."

Prior to the choice of a fixed base, logarithms and exponentials are each families of functions. To choose a base—say, 2 or 10—is to choose a member from the family, a unique logarithmic or exponential function. Suppose we were to choose the base $b = 2$, for example. The result of that choice would be a single function, $\log_2(x)$ (where x is a positive real number) and another single function 2^x (where x is a real number).

What Is b^0?

The exponential function, as we have just seen, is not characterized in terms of self-multiplication; rather, the exponential is a function from the additive group of real numbers to the multiplicative group of positive real numbers. That function maps the additive identity 0 onto the multiplicative identity 1, for any

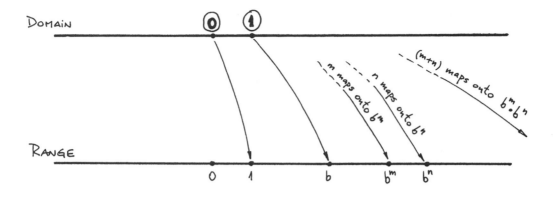

FIGURE CS 2.3   The exponential mapping for an arbitrary base b—that is, the mapping of sums onto products with a base b. 0 is mapped onto 1, 1 is mapped onto the base b, and n is mapped onto b^n. The sum $m + n$ is mapped onto the product $b^m \cdot b^n$.

choice of base b. The expression "b^0" does not mean "b multiplied by itself zero times," which makes no sense. Instead, it means: "Look at the exponential function that maps 1 into b. That function also maps 0 into 1." Since this is true for any base b, it is always true that $b^0 = 1$.

The way that the exponential function is written—"b^x"—is misleading. It is done that way for the purpose of ease of calculation, for those cases where self-multiplication will happen to allow you to calculate the value of the function. But the notation hides the relationships between b and x when you understand the exponential function as mapping from the domain of real numbers under addition to the range of positive real numbers under multiplication.

In such a mapping, the "x" in "b^x" is in the *domain* of the exponential function, whereas the "b" is in the *range* of the exponential function. b is the number that 1 is mapped onto—the number that picks out which particular exponential mapping is being used. b thus constrains what x is mapped onto.

For example, take 2^3. The "2" in this notation first picks out the exponential function that maps 1 onto 2. This choice of what 1 maps onto constrains the rest of the function. "$2^3 = 8$" means that when the exponential mapping is constrained so that 1 maps onto 2, then 3 maps onto 8. Compare this with "$10^3 = 1,000$." This equation means that when the exponential mapping is constrained so that 1 maps onto 10, then 3 maps onto 1,000.

From a cognitive perspective, the exponential function can be seen as a conceptual metaphor—namely, Multiplication Is Addition—where products are conceptualized in terms of sums and calculated by addition. The exponential function is an arithmetization of this metaphor.

The one-to-one correspondence defining exponentials in one direction and logarithms in the other is the conceptual blend characterized by this metaphor.

THE EXPONENTIAL BLEND

n	–2	–1	0	1	2	3	4	5	6	7	8	9	10
	↕	↕	↕	↕	↕	↕	↕	↕	↕	↕	↕	↕	↕
2^n	1/4	1/2	1	2	4	8	16	32	64	128	256	512	1024

We are not yet quite able to say what $e^{\pi i}$ means. First, we have to explain what e means. To do that, there are still a few steps to go.

- We need to understand the conceptual metaphors used in mathematicizing change.
- Then we need to see how the meaning of e makes use of those metaphors.
- Then we need to see why the number 2.718281828459045. . . has that meaning.
- And finally we need to get an overview of how the number 2.718281828459045. . . is systematically linked to its meaning through a system of conceptual metaphors and blends.

The Mathematical Metaphor for Change

In our everyday conceptual systems, change is understood metaphorically in terms of motion. Changes are expressed by the language of motion: We can *go into* or *come out of* a state of euphoria, prices can *rise* or *fall*, and a project where there is no change has *come to a standstill*.

There is a disparity between our everyday concept of motion and the complex metaphorical concept of motion used in mathematics. Human beings perceive and control motion directly, via motion detectors in the brain's visual system and motor-control mechanisms in the brain. But in the mathematical modeling of human concepts for the sake of precise calculation, it takes a lot of metaphorical apparatus to quantify aspects of motion such as speed, acceleration, distance traveled, and so on, using only the tools of mathematics.

Take, for example, the concept of the average speed of motion over a time interval—or, more precisely, speed at a given instant. It was a great achievement of seventeenth-century mathematical thought to conceptualize these ideas using the Cartesian plane, which in turn uses the Numbers Are Points on a Line metaphor and the Cartesian Plane blend (see Case Study 1). The *concepts* of average speed and instantaneous speed must be turned into *numbers*. Since concepts and numbers are different kinds of things, this is a job for metaphor. It is accomplished in several conceptual steps.

The Mathematicization of the Concept "Speed"

1. First, time must be metaphorically conceptualized in terms of distance. This is accomplished naturally using our everyday metaphor in which the passage of time is linear motion over a landscape (see Lakoff & Johnson, 1999, ch. 10; see also Núñez, 1999). An amount of time is metaphorically a distance traveled, as in sentences like *We're coming up on Christmas, That's still down the road,* and *There's still quite a long way to go to the end of the project.*

2. Motion must be conceptualized in terms suitable for mathematicization. Motion is directly perceived and performed by human beings. Certain very limited aspects of motion, like how far you have traveled in a given time, can be conceptualized in terms of other concepts—in particular, in terms of two independent dimensions, distance and time. Time is itself conceived of as metaphorical distance. This metaphorical conceptualization of time in terms of distance allows us to reason about time by using what we know about distance. Thus the concept of distance traveled over time can be further mathematicized as the ratio of two distances. Distance can be represented using the Cartesian Plane blend.

3. These two independent spatial dimensions must be mathematicized. This requires another metaphor: Logical Independence Is Geometrical Orthogonality in the Cartesian Plane. Thus, each spatial dimension can be mathematicized as an axis on the Cartesian plane. Distances are represented in terms of the two Number-Line blends that form the axes in the Cartesian plane.

4. The moment-to-moment process of motion must also be mathematicized. This is accomplished by metaphorically conceptualizing the process of motion in terms of a function from real numbers to real numbers. The domain of the function consists of the numbers in the *X*-axis of the Number-Line blend, where the line in the blend represents time, metaphorically conceptualized as distance. The range of the function consists of the numbers in the *Y*-axis of the Number-Line blend, where the line in the blend represents distance. The motion itself does not literally exist here; it is metaphorically conceptualized as the set of pairs of numbers (t, l) representing particular discrete times and locations. "Continuity" for motion is not directly represented, either, but is represented metaphorically via Weierstrass's discretized preservation-of-closeness notion of continuity (see Chapter 14).

5. Average speed over an interval of time, which is a concept, must be mathematicized so that it is conceptualized as a *number*: the ratio of the number representing the distance covered divided by the number representing the amount of time (conceptualized also as a distance covered). Average speed, of course, does not occur in the world. The concept of average speed more accurately fits the real speed of real objects in motion as the time interval considered is made shorter and shorter. To characterize real speed of motion, you need a concept of speed *at an instant*.

6. The conceptualization of average speed does not provide for instantaneous speed without an additional metaphor:

   Instantaneous Speed Is Average Speed over an Interval of Infinitely Small Size.

   An infinitely small interval does not literally exist. It is understood by still another metaphor, the BMI (see Chapters 8 and 14).

7. To get the general concept *rate of change* from the specific concept *speed of motion*, we need to use the Change Is Motion metaphor, applied to the concept of *speed of motion at an instant* and yielding the more general concept of *change at an instant*.

This complex conceptual apparatus is needed in order to get the mathematics of differential calculus to be the mathematics of the concept of change.

It is worth stopping for a moment to glance in awe at all the conceptual metaphors used to get differential calculus to be the mathematics of change. By the time you are taught this in a calculus course, most of the metaphorical apparatus has already been internalized over the course of many years of schooling.

It is also worth taking a look at what happens when you don't notice these metaphors as metaphors. Here is a quote from Bertrand Russell's *Principles of Mathematics*.

> Motion consists *merely* in the occupation of different places at different times, subject to continuity. . . . There is no transition from place to place, no consecutive moment or consecutive position, no such thing as velocity except in the sense of a real number, which is the limit of a certain set of quotients. (Russell, 1903, p. 473)

Russell believed that Weierstrass's metaphorical discretizing of natural continuity was literally true. His view that thought was mathematical logic blinded him to the metaphorical mechanisms of mind needed to go from the normal human concept of change to a conceptualization of change in Weierstrass's mathematics. The result is conceptually bizarre. Russell really believed that there was no such thing as a motion or speed but only sequences of pairs of numbers (called "states") and limits of such sequences (called "rates of change"). Russell's pathological literalness is revelatory. When you encounter beliefs this peculiar, you become more acutely aware of the elaborate cognitive mechanisms involved in mathematicizing concepts.

Now that we have seen the metaphors used in mathematicizing change, we are only three steps away from understanding *e:*

- First, we need to see how the meaning of *"e"* follows from the metaphors for mathematicizing the concept of change.
- Then, we need to understand why the number 2.718281828459045... has the meaning we have ascribed to the symbol *e*.
- And finally we need to get an overview of how the number 2.718281828459045... is systematically linked to its meaning through a system of conceptual metaphors and blends.

What Is e?

We are now in a position to understand *e* as a concept involved intimately in the mathematicization of change.

We will ask how the exponential function, b^x, "changes," using the concept of change mathematicized as indicated in 1 through 7 above. As we shall see, the meaning of *e* depends crucially on cognitive mechanisms used to mathematicize change.

We begin with the concept of the derivative as discussed in Chapter 14. There we used the notation $f'(x)$ for the derivative of $f(x)$. Here it will be more convenient to use the notation $\frac{d}{dx}f(x)$ for the derivative of $f(x)$:

$$\frac{d}{dx}f(x) = \lim_{\Delta x \to 0}\frac{f(x + \Delta x) - f(x)}{\Delta x}.$$

Suppose $f(x) = b^x$. Then the derivative $\frac{d}{dx}b^x$ becomes

$$\frac{d}{dx}b^x = \lim_{\Delta x \to 0}\frac{b^{x + \Delta x} - b^x}{\Delta x}.$$

Because the exponential function is a function mapping from sums to products, the following equality holds: $b^{x+\Delta x} = b^x \cdot b^{\Delta x}$. This equality, as we shall soon see, is of paramount importance. It is through this equality that the function that maps from sums to products is intimately linked with the concept of change.

Given this equality, the following logical steps become possible—steps that would otherwise be impossible:

$$\frac{d}{dx}b^x \quad = \lim_{\Delta x \to 0} \frac{b^{x+\Delta x} - b^x}{\Delta x}$$

$$= \lim_{\Delta x \to 0} \frac{(b^x \cdot b^{\Delta x}) - b^x}{\Delta x}$$

That was the crucial step. The rest is automatic.

By grouping the common factors b^x together, we get

$$= \lim_{\Delta x \to 0} \frac{b^x(b^{\Delta x} - 1)}{\Delta x}$$

Since b^x contains no Δx term, it can be taken outside the limit sign:

$$= b^x \lim_{\Delta x \to 0} \frac{b^{\Delta x} - 1}{\Delta x}$$

We know for other reasons (too long and irrelevant to go into here) that the limit $\lim_{\Delta x \to 0} \frac{b^{\Delta x} - 1}{\Delta x}$ exists. Calling it a number, C, which is a constant, we get

$$\frac{d}{dx}b^x = C \cdot b^x.$$

In other words, the derivative of an exponential function with base b—namely, b^x—is a constant times b^x itself, which is another way of saying that *an exponential function has a rate of change proportional to its size*. C is the factor stating that proportion.

From this discussion, we can see exactly why the function that maps sums onto products also has the property of being proportional to its size. It is because $b^{x+\Delta x} = b^x \cdot b^{\Delta x}$, which sanctions the crucial step in the proof. This is true because b^x maps sums onto products.

When the constant $C = 1$, then the rate of change of b^x is b^x itself; that is, b^x has a rate of change that is *identical* to its size. In mathematical terms, b^x is its own derivative. This happens when

$$\lim_{\Delta x \to 0} \frac{b^{\Delta x} - 1}{\Delta x} = 1$$

That limit constrains the value of the base b.

"e" is the name we give to the base b when that limit is 1. What does this mean? It means that e^x is a function whose rate of change is identical to itself.

Since exponential functions map sums onto products, we can now see what *e* means:

THE CONCEPT OF *e*

"e" is that number
which is the base of a function
that maps sums onto products and
whose rate of change is identical to
itself.

We have now shown how the meaning of *"e"* follows from the metaphors for mathematicizing the concept of change. At this point, we need to answer two remaining questions:

- Why does *e*, which at this point is a concept, have the numerical value it does? Why should *e*, meaning what it does, be the particular number 2.718281828459045. . . ?
- What system of metaphors and blends occurring throughout classical mathematics determines that the concept *e* is the unique real number 2.718281828459045. . . ?

Why 2.718281828459045. . . ?

We have just characterized the meaning of *e*. We have also seen that that meaning imposes a mathematical constraint on what the unique numerical value of the concept of *e* must be. The numerical value of *e* is determined by the following limit:

$$\lim_{\Delta x \to 0} \frac{e^{\Delta x} - 1}{\Delta x} = 1$$

We now ask:

Given this limit expression, what is the numerical value of *e*? Euler found a way to calculate it. The answer is 2.718281828459045. . . . Why that number? We are about to give two answers—one in the text and one in an appendix to this chapter.

The Power-Series Answer

Euler's most interesting answer to the question of the numerical value of *e* came through his investigation of power series. The reasoning involved implicitly makes use of an extremely complex conceptual blend. Our job is to make that blend explicit, show what its structure is, and show how it is used in the power-series answer.

The rational structure of the answer starts with the concept of e: "e is that number such that e^x is its own derivative." The blend links our understanding of power series with our understanding of derivatives, beginning with the following three well-known properties of derivatives:

1. The derivative of a constant equals 0. In symbols, $\frac{d}{dx}C = 0$. The reason is that constants, by definition, do not change, so their rate of change is zero.
2. If n is a positive integer, then $\frac{d}{dx}x^n = n \cdot x^{n-1}$. This is the general case of the example we just looked at: $\frac{d}{dx}x^2 = 2x$.
3. If $u = f(x)$ and $v = g(x)$, then $\frac{d}{dx}(u + v) = \frac{d}{dx}(u) + \frac{d}{dx}(v)$. That is, the derivative of the sum of two functions is the sum of the derivatives.

These properties will allow us to get derivatives of power series.

Implicit in the use of the Power-Series blend are three common conceptual metaphors used throughout classical mathematics:

a. Numbers Are Wholes That Are Sums of Their Parts. This allows us to conceptualize numbers as sums of other numbers (see Chapter 3).
b. Functions Are Numbers, with respect to the operations of arithmetic. This allows us to add functions (see Case Study 1).
c. Infinite Sums Are Limits of Infinite Sequences of Partial Sums (a special case of the Basic Metaphor of Infinity for limits of sequences; see Chapter 9).

These three metaphors taken together allow us to conceptualize a function as an infinite sum of other functions.

$$\sum_{n=1}^{\infty} f(n) = \lim_{n \to \infty} \sum_{k=1}^{n} f(k)$$

From the basic properties of derivatives, we know how to calculate the derivative for any power series—that is, any series of the form

$$\sum_{n=0}^{\infty} (a_n \cdot x^n).$$

Here is the blend that makes an understanding of power series possible:

			THE POWER-SERIES BLEND		
1 WHOLES AND PARTS	*2* NUMBERS	*3* FUNCTIONS	*4* INFINITE SERIES (VIA BMI)	*5* POWER SERIES (VIA BMI)	*6* CALCULUS APPLIED TO POWER SERIES
Whole	Resulting number c	Composite function $h(x)$	Infinite sum $S = \sum_{k=1}^{\infty} a_k$	Function at infinity $F(x) = \sum_{k=0}^{\infty} a_k x^k$	Derivative at infinity $F'(x) = \sum_{k=0}^{\infty} k a_k x^{k-1}$
Parts	Operand numbers a, b	Component functions $f(x), g(x)$	Finite terms a_k	Finite terms $a_k x^k$	Derivatives of finite terms $k a_k x^{k-1}$
Part-whole relation	Addition relation $a + b = c$	Addition relation $f(x)+g(x) = h(x)$	Addition relation $a_{k-1}+a_k$	Addition relation $a_{k-1}x^{k-1}+a_k x^k$	Addition relation $(k-1)a_{k-1}x^{k-2} + k a_k x^{k-1}$
—	—	—	Partial sum of functions through n terms $S = \sum_{k=1}^{n} a_k$	Partial sum through n terms $\sum_{k=0}^{n} a_k x^k$	Partial sum of derivatives through n terms $F^n(x) = \sum_{k=0}^{n} k a_k x^{k-1}$

This complex blend is made up of the following parts:

- Columns 1 and 2 constitute the metaphoric blend characterized by the object-construction metaphor (see Chapter 3). It allows numbers to be seen as wholes made up of parts that are added together. As we will see, the number e is calculated by adding together parts—the terms of the power series.
- Columns 2 and 3 constitute the metaphoric blend characterized by the metaphor Functions Are Numbers (see Case Study 1). This allows us to see the function e^x as being numberlike and hence able to be seen as a whole made up of parts that are added together.
- Column 4 is linked to column 2 (with its finite sums of numbers) via the use of the BMI that extends finite sums of numbers to infinite sums of numbers.
- Column 4 (which contains infinite sums of numbers) is linked to column 5 (which contains infinite sums of functions) via the metaphor Functions Are Numbers. This allows us to see the function e^x not just as a sum of functions but as an *infinite* sum of functions.

- Column 6 is an application of the idea of a derivative of a sum of functions to column 5, where there is an infinite sum of functions.

This conceptual structure is what is taken for granted in applying the basic properties of derivatives to power series. Reasoning in terms of this blend, we can find a power series that is its own derivative.

The essential thing to notice is this:

$$\text{The derivative of } a_n \cdot x^n = n \cdot a_n \cdot x^{n-1}.$$

This follows from the properties of derivatives given earlier.

When this condition holds, each term of the series will be the derivative of the term that follows it. To get the whole series, then, all we need is the first term.

- The first term of a power series contains no multiple of x, because x^0 is always 1. Therefore, the first term is always the constant a_0.

All other terms of a power series contain a multiple of x. When $x = 0$, all those other terms equal zero and the entire sum is a_0. Moreover, in this case when $x = 0$, $e^x = 1$. Therefore the first term of the power series is the constant $a_0 = 1$.

Starting from 1 as the first term and knowing that each term is the derivative of the following term, we can deduce the whole series.

- The second term must be "x," because $\frac{d}{dx}x = 1$, the first term.
- The third term must be $(1/2) \cdot x^2$, because $\frac{d}{dx}((1/2) \cdot x^2) = x$, the second term.
- The fourth term, $a_3 x^3$, must be $(1/2) \cdot (1/3) \cdot x^3$, because $\frac{d}{dx}((1/2) \cdot (1/3) \cdot x^3)$ $= (1/2) \cdot x^2$, the third term. And so on.
- The nth term, must be $(1/2) \cdot \ldots \cdot (1/n-1) \cdot x^{n-1}$ because $\frac{d}{dx}((1/2) \cdot \ldots$ $\cdot (1/n-1) x^{n-1}) = (1/2) \cdot \ldots \cdot (1/n-2) \cdot x^{n-2}$, the previous term.

In other words, the series is

$$e^x = 1 + x + x^2/2! + x^3/3! + \ldots$$

This gives us an easy way to compute approximations to the value of e, because when $x = 1$, $e^1 = e$.

Setting $x = 1$ in the power series, we get

$$e = 1 + 1 + 1/2! + 1/3! + \dots$$

The value of this series converges to 2.718281828459045. . . .

We have now answered all the questions we started out with. We have seen along the way just why *e* means what it does and has the numerical value that it does. And we have seen what the function e^x means and why it has the properties it has. Perhaps most important, we have seen an excellent example of how it is possible for a number to have meaning.

As we make our way toward an understanding of $e^{\pi i} + 1 = 0$, we should keep the following in mind.

WHAT IS THE FUNCTION e^x?

It is the function that:

- maps sums onto products,
- maps 2.718281828459045. . . onto 1,
- is its own derivative, and
- changes in exact proportion to its size.

WHAT IS *e*?

- *e* is the real number 2.718281828459045. . . .
- *e* is the base of an exponential function that has a rate of change exactly equal to its size.
- Thus, *e* is a real number that is the base of a function that maps sums onto products and whose rate of change is exactly the same as its size.

Appendix

When we look at textbooks, we are often told that *e* is a number whose value is determined by a limit, $\lim_{n \to \infty} (1 + \frac{1}{n})^n$. Indeed, the limit of this expression converges to the numeric value 2.718281828459045. . . . This is the same value we obtained above through the use of the power series. The relationship between the limit and the power series is discussed well in many places—for example, Simmons (1985, p. 372) and Maor (1994, p. 35).

What such texts do not do is answer the most basic of questions:

Why *should* this particular limit, $\lim_{n \to \infty}(1 + \frac{1}{n})^n$, be the base of the exponential function that is its own derivative?

We can see *why* by looking at the concepts that go to make up the concept of "*e*," for example, rate of change, length, function, tangent, curve, limit, and so on.

We know that in any exponential function, 1 maps to the base b; that is, the value of the function at $x = 1$ is b. In more familiar terms, $b^1 = b$. If b^x is its own derivative, b^x, at every value x, then the value of that derivative at $x = 1$ is b, since $b^1 = b$.

This is represented geometrically in Figure CS 2.4. There the curve is the exponential function that is its own derivative. The derivative at $x = 1$ is the slope of the tangent of the curve at the point $(1, b)$ in the Cartesian plane. The slope of that tangent is b.

Now consider a value close to 1 on the x-axis—say, $1 + q$ (see Figure CS 2.5). The exponential function maps the number $(1 + q)$ onto the number b^{1+q}. The average change between 1 and $1 + q$ is given by the ratio $\frac{b^{1+q} - b}{q}$.

Using the limit metaphor applied to this expression as q approaches zero, we can see what happens when q gets infinitely small. In such a case, what we have is the limit

$$\lim_{q \to 0} \frac{b^{1+q} - b}{q}$$

But this limit corresponds exactly to the value of the derivative of the function at 1, which we know is the number b, the base of the exponential function. So we have

$$\lim_{q \to 0} \frac{b^{1+q} - b}{q} = b$$

We have gotten this far starting with the meaning of *e*—namely, *the base of the function that maps sums onto products and that is identical to its own rate of change.* We have reached these formulas via the metaphors mathematicizing the concept of change. What we have done so far is mostly conceptual work, using metaphors and blends to link concepts to formulas.

Suppose we now take the expression "$(b^{1+q} - b)/q = b$" in isolation (without the limit metaphor) and do some simple algebraic manipulations on it.

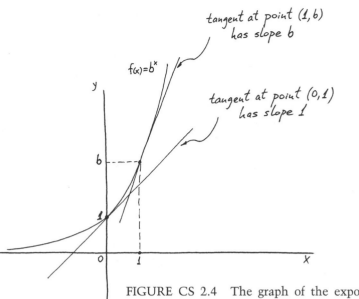

tangent at point (1,b)
has slope b

$f(x)=b^x$

tangent at point (0,1)
has slope 1

FIGURE CS 2.4   The graph of the exponential function $f(x) = b^x$. Since the derivative of the function is the function itself, the tangent at point $(1, b)$ has slope b and the tangent at point $(0, 1)$ has slope 1.

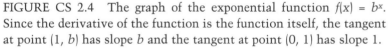

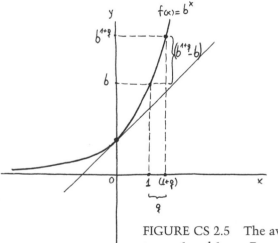

FIGURE CS 2.5   The average change in the function $f(x) = b^x$ between 1 and $1 + q$. From the diagram we can see that this average change is $((b^{1+q} - b) / q)$. When q goes to 0, the value of this ratio becomes b.

1. Given the exponential function, which maps sums onto products, we know that $b^{1+q} = b^1 \cdot b^q$. Substituting in the above equation, we get

$$(bb^q - b)/q = b.$$

2. Factoring out b yields

$$b(b^q - 1)/q = b.$$

3. Dividing both sides by b we get

$$(b^q - 1)/q = 1.$$

Note that from page 410 we know that this limit is 1 precisely when the rate of change of b^x is b^x itself, as it is in our example.

4. Now, when we solve for b in $(b^q - 1)/q = 1$, we obtain

$$(b^q - 1) = q$$

and then,
$$b^q = 1 + q$$

and after taking the qth-root at both sides we get,

$$b = (1 + q)^{1/q}.$$

This means that $b = (1 + q)^{1/q}$ is true if and only if the expression we started with, $(b^{1+q} - b)/q = b$, is true. And *that* means that the same numerical relation that holds between b and q in $(b^{1+q} - b)/q = b$ also holds in $b = (1 + q)^{1/q}$. In the latter expression, since we have solved for b, the relationship between b and q is expressed more directly.

5. We are interested in the value that b takes as q, under the limit metaphor, takes on smaller and smaller values, approaching zero. In other words,

$$b = \lim_{q \to 0} (1 + q)^{1/q}.$$

6. Now if we set q to be $1/n$, then we have

$$b = \lim_{n \to \infty} (1 + 1/n)^n.$$

With this formulation, *b* can be computed to as close an approximation as one wants, by letting *n* vary over bigger and bigger values.

What we have shown is that this limit is exactly the base *b* of the exponential function whose derivative is itself. But that is the definition of the concept "*e*." Thus, *b* = *e*, and

$$e = \lim_{n \to \infty} (1 + 1/n)^n.$$

There is nothing mysterious about this result. It is a consequence of four things:

1. What *e* means.
2. The metaphors for mathematicizing change.
3. The limit metaphor, which is a special case of the BMI.
4. Some simple algebraic manipulations.

There is a moral here. A system of conceptual metaphors and blends characterizes systematic relationships between concepts and formulas. We have to understand the nature of that system if we are to understand the mathematical ideas embodied by the formulas.

Case Study 3

What Is *i*?

$\sqrt{-1}$ IS TAKEN FOR GRANTED TODAY. No serious mathematician would deny that it is a number. Yet it took centuries for $\sqrt{-1}$ to be officially admitted to the pantheon of numbers. For almost three centuries, it was controversial; mathematicians didn't know what to make of it; many of them worked with it successfully without admitting its existence. Why?

Primarily for cognitive reasons. Mathematicians simply could not make it fit their *idea* of what a number was supposed to be. A number was supposed to be a magnitude. $\sqrt{-1}$ is not a magnitude comparable to the magnitudes of real numbers. No tree can be $\sqrt{-1}$ units high. You cannot owe someone $\sqrt{-1}$ dollars. Numbers were supposed to be linearly ordered. $\sqrt{-1}$ is not linearly ordered with respect to other numbers. Is it less than -1? Is it more than -1? The answer is that it is neither, as we shall see in detail below.

Why was it used at all? Because it was useful for solving equations and doing computations that needed to be done; that is, it came into existence for the most human of reasons. For an excellent history of $\sqrt{-1}$, see Nahin (1998).

The Complex Numbers and Ordinary Cognition

Our goal in this chapter is to show how the imaginary and complex numbers, the complex plane, and complex arithmetic arise from, and are understood in terms of, the mechanisms of ordinary cognition. This is counter to the impression many students often get when they are taught the subject by means of formal definitions and proofs.

From a formal perspective, much about complex numbers and arithmetic seems arbitrary. From a purely algebraic point of view, *i* arises as a solution to the equation $x^2 + 1 = 0$. There is nothing geometric about this—no complex *plane* at all. Yet in the complex plane, the *i*-axis is 90° from the *x*-axis. Why? Complex numbers in the complex plane add like vectors. Why? Complex numbers have a weird rule of multiplication:

$$(a + bi) \cdot (c + di) = (ac - bd) + (ad + bc)i.$$

Why? Is this an arbitrary invention of mathematicians?

We have just represented a complex number in the usual way as a sum "*a + bi*." Is this "sum" the same kind of sum that we see in $2 + 3 = 5$? Mathematicians write that $i = \sqrt{-1}$. Does "square root" here mean the same thing as in "$5 = \sqrt{25}$"?

These are the kinds of questions we seek to answer. Our answers will be given in terms of human cognition. As such, they will be very different from the answers one finds in mathematics books. We will argue that there is a deep *cognitive* reason why the *i*-axis is 90° from the *x*-axis and that metaphor plays a central role in complex arithmetic. The cognitive perspective, as we shall see, makes sense of things that otherwise look arbitrary.

Equations and Closure

In elementary algebra, the variable x is used in the activity of solving problems through computation. It is a number, but you don't yet know which number. It is a number whose identity you are trying to pin down. For example, suppose you are asked, "Find a number such that, when you multiply it by 3 and add 5 to the result, you get 32." You learn to write this as an equation, with "x" as the number you are asked to identify: $3x + 5 = 32$. As long as you know that x is a number, then the laws of arithmetic will apply and allow you to do ordinary calculations. You can subtract 5 from both sides, getting $3x = 27$ and then divide both sides by 3, giving $x = 9$. This "identifies" x as the number 9. The cognitive mechanism employed here is the Fundamental Metonymy of Algebra (see Chapter 3).

All this assumes a principle: The ordinary laws of arithmetic hold for x when you take x to be a number, and the laws of arithmetic hold for all "numbers." This is equivalent to the cognitive principle of closure, which we discussed in Chapter 4: Any arithmetic operations on "numbers" will result in a "number."

It is important to recall where the principle of closure comes from. It is an entailment of the four basic grounding metaphors for arithmetic for the basic operations of addition and multiplication, just as were the laws of arithmetic (see Chapter 3). Closure and the laws arise naturally as entailments of the grounding metaphors, when just addition and multiplication are considered. But in technical mathematics, beginning with the Greeks, the subject matter of the branches of mathematics was assumed to be defined by essences—principles that characterize precisely what is and isn't in a category of mathematical objects. The properties of numbers from a nontechnical perspective (e.g., that of closure and the laws) are reanalyzed as essences—part of what *defines* what a "number" is. It follows that in equations, when "x" is taken to be a "number," closure and the laws must be operating. Closure, of course, is defined relative to whatever operations we take to be legitimate in arithmetic. The closure concept was motivated by addition and multiplication but was extended metaphorically to subtraction and division (see Chapter 4).

Applying closure for subtraction and division required extending the natural numbers to more "numbers." An early extension was to the rational numbers, since 2 divided by 3 had to be a "number" if division was a permissible operation in arithmetic. Another extension was to zero, since according to closure, 4 – 4, say, has to be some "number." Another extension was to the negative numbers, since according to closure, 3 – 5 has to be some "number."

These extensions all violated the concept of "number" as given by subitizing and the metaphor that Numbers Are Object Collections. You don't subitize negative numbers or fractions. Nor do you have negative or fractional collections of physical objects. Even the idea of a zero collection is strange, as we saw in Chapter 3, since it requires conceptualizing no collection at all as a "collection" of size zero. Throughout the history of mathematics, the technical idea that closure is part of the essence of arithmetic operations on "numbers" has been in conflict with "number" as characterized by subitizing and the grounding metaphors. But within technical mathematics, the imperative of essence implied the imperative of closure.

At each stage of the extensions due to closure, we get a progressively larger collection of entities we call "numbers" that are closed under the operations of arithmetic and obey the laws of arithmetic. By adding zero and the negative numbers to the natural numbers, we get the integers. Since the integers constitute an extension of the natural numbers, all natural numbers are integers. Similarly, by adding fractions to the integers, we get the rationals, and since the rationals are an extension of the integers, all the integers are rationals.

The irrational numbers were brought into being as an attempt to answer the question "Which number, when squared, equals 2?" or, equivalently, "Solve for x in the equation $x^2 = 2$." This equation again takes for granted, via the principle of closure, that x is indeed a "number." Once it was proved that x could not be a rational number, the closure principle entailed that the irrational numbers should exist, with $x = \sqrt{2}$ as the solution to the equation $x^2 = 2$. When the rationals are extended to include the irrationals, we get the reals.

The reals are also not numbers in the sense required by subitizing and by the grounding metaphor that Numbers Are Object Collections. You can neither subitize $\sqrt{2}$ objects nor form a collection of physical objects whose size is $\sqrt{2}$. However, the reals *are* numbers in that they obey closure for the operations and laws of arithmetic.

The reals also have an important property of numbers that comes from subitizing and grouping—namely, all the real numbers can be linearly ordered. Because they obey closure and the laws of arithmetic and can be ordered, such "numbers" as $\sqrt{2}$ are called "real." The ordering of the reals corresponds to the concept of relative magnitude or "size," which is the same concept that fits the subitizing and grouping of physical objects. It is hard to think of something as a "real" number if it does not have a magnitude relative to other numbers.

Where Does *i* Come From, and What Is "Imaginary" About It?

Suppose you are asked, "Find a number that, when multiplied by itself and added to 1, gives you zero." Or, equivalently, solve the equation $x^2 + 1 = 0$. This is a very simple equation, with numbers 1 and zero and the operations of multiplication and addition. As before, the implicit assumption behind such equations is that the x is a number, under all the operations and laws of arithmetic.

Subtracting 1 from both sides of the equation, we get $x^2 = -1$. If x is number, as closure requires, it has to be either a natural number, an integer, a rational, a real, or some other kind of "number." Let us call this number "*i*" for short. *i* has the basic property that $i^2 = -1$ and therefore that $i^4 = 1$.

Is *i* a real number? That is, is it ordered relative to all numbers we call "real?" It is easy to see that it is not. If it were a real number ordered relative to the other reals, then it would have to be either (a) positive, (b) negative, or (c) zero. If *i* were positive, its square would be positive. But its square is -1. So *i* cannot be positive. If *i* were negative, its square would be positive. But its square is -1. So *i* cannot be negative. If *i* were zero, its square would be zero. But its square is -1. So *i* cannot be zero.

That means that *i* is not a real number—not ordered *anywhere* relative to the real numbers! In other words, it does not even have the central property of "numbers," indicating a magnitude that can be linearly compared to all other magnitudes. You can see why *i* has been called imaginary. It has almost none of the properties of the small natural numbers—not subitizability, not groupability, and not even relative magnitude. If i is to be a number, it is a number only by virtue of closure and the laws of arithmetic.

The Conceptualization of Negative Numbers

Since $i^2 = -1$, let us begin our thinking about i with the conceptualization of -1 and other negative numbers. Given the Numbers Are Points on a Line metaphor, we form the Number-Line blend, in which all the real numbers—including the positives, the negatives, and zero—are conceptualized as spread out along a line, with zero at a point called the origin. The positive numbers are conceptualized in this blend as being on one side of the origin (zero) and the negative numbers on the other side of the origin, with $-n$ exactly as far from zero as $+n$, for all nonzero real numbers n. That is, $-n$ and $+n$ are symmetrical points relative to the origin (zero).

This symmetry is conceptualized in terms of mental rotation of the sort that cognitive scientists have studied intensively (see, for example, Shepard & Metzler, 1971; Shepard & Cooper, 1982; Núñez, Corti, & Retschitzki, 1998). Cognitively, we visualize the relationship between the positive and negative numbers using a rotational transformation from the positive to the negative numbers—a "flipping over" of the positive part of the line onto the negative part of the line around the origin, preserving distance—with each $+n$ rotating onto the location of the corresponding $-n$, while each of the $-n$ rotates over onto the location of the corresponding $+n$.

We will call this "rotating" transformation R_{-1}, since it maps $+1$ onto -1. That is, $R_{-1}(+1) = -1$ and $R_{-1}(-1) = +1$. In general, $R_{-1}(+n) = -n$ and $R_{-1}(-n) = +n$. That is, R_{-1} maps positives onto corresponding negatives and vice versa. And if you perform R_{-1} twice, you get back to where you started. In other words, $R_{-1}(R_{-1}(n)) = n$, for all n, positive and negative.

R_{-1} is a conceptual, spatial operation on the Number-Line blend, applying to number-points taken as single complex entities. As a spatial operation, it operates on the line and point-locations on it, but the basic constraint on the operation is defined by the numbers associated with the points.

The mental rotation R_{-1} is correlated one-to-one with multiplication by -1. The number that any positive or negative number n "rotates onto" correlates exactly with the number you would get by multiplying n by -1. This correlation gives rise to the metaphor discussed in Chapter 4: Multiplication of n by -1 Is Rotation to the symmetry point of n. The metaphorical blend of the source and target domains defined by this metaphor is as follows:

The Multiplication-Rotation Blend	
Source Domain	*Target Domain*
Space	Arithmetic

Rotation to the symmetry point of n $\leftrightarrow$	$-1 \cdot n$

This metaphorical blend has the effect of arithmetizing the spatial rotation transformation R_{-1} as the arithmetic operation of multiplication by -1.

Using the blend, we get metaphorical equivalences between applying the spatial operation R_{-1} and multiplication by -1. Here are some examples.

$$R_{-1}(+1) = (-1) \cdot (+1) = (-1)$$
$$R_{-1}(-1) = (-1) \cdot (-1) = (+1) = R_{-1}(R_{-1}(1))$$
$$R_{-1}(-5) = (-1) \cdot (-5) = (+5)$$
$$R_{-1}(R_{-1}(+3)) = (-1) \cdot ((-1) \cdot (+3)) = ((-1) \cdot (-3)) = (+3)$$

Note, incidentally, that applying the rotation twice is metaphorically equivalent to multiplying by $-1 \cdot -1$, which equals 1.

What is important about this is that the spatial operation R_{-1} is part of the *cognitive* apparatus we use to conceptualize the relationship between positive and negative numbers. It is *not* part of the formal mathematics. Yet, as we shall see, this conceptualization of the negative numbers is what unifies our understanding of the square roots of negative numbers.

But we are not quite finished with our conceptualization of negative numbers. Let us locate this conceptualization in the Cartesian plane.

The Rotation-Plane Blend

We saw in Chapter 4 that negative numbers are conceptualized by the cognitive, spatial rotation operation R_{-1}, via the metaphor Multiplication by -1 Is Rotation. We can take the metaphorical blend formed by this metaphor, the Multiplication-Rotation blend, and combine it with the Number-Line blend to form the Rotation-Number-Line blend, in which multiplication by -1 in the number domain of the blend correlates with rotation by $180°$. In the Rotation-Number-Line blend, applying R_{-1} corresponds to multiplying by -1. But this applies to the number line alone. Let us now ask what it implies when this conceptualization of negative numbers is applied to the Cartesian plane. To do so, we must form a new conceptual blend, the Rotation-Plane blend, which combines the Multiplication-Rotation blend with each Number-Line blend in the Cartesian plane.

This conceptual blend has two domains, each of which is a blend: (1) the Rotation-Number-Line blend and (2) the Cartesian Plane blend.

THE ROTATION-PLANE BLEND

Domain 1 THE ROTATION-NUMBER-LINE BLEND		Domain 2 THE CARTESIAN PLANE BLEND
Rotation by 180°	$\leftrightarrow$	Multiplication by –1
The zero point	$\leftrightarrow$	The origin
The number line	$\leftrightarrow$	Every line through the origin

This blend creates a new conceptual entity—the rotation plane, which is the Cartesian plane together with the metaphor characterizing multiplication by –1 as a rotation by 180°.

What is most interesting about this blend is the fact that inferences about rotations R_{-1} on the number line become quite different inferences in the Rotation-Plane blend. This new blend defines a form of metaphorical "multiplication" in the rotation plane—multiplication of one ordered pair of numbers by another, a kind of "multiplication" that does not exist on the real-number line. Thus, given a point (a, b) in the Cartesian plane, rotation by 180° takes you to the point $(-a, -b)$. In the Rotation-Plane blend, this means $(-1, 0) \cdot (a, b) = (-a, -b)$. This is a metaphorical inference in the rotation plane. It is a consequence of cognitive apparatus by which the blend was constructed.

In the Rotation-Plane blend, multiplication of (a, b) by a scalar factor c is, conceptually, "stretching" the x- and y-coordinates by a factor of c. Thus, $(c, 0) \cdot (a, b) = (ac, bc)$.

As a result, the Rotation-Plane blend has a certain number of cases of "multiplication" of ordered pairs by ordered pairs, which are entailed by the formation of the blend.

Addition for ordered pairs is completely determined in the Rotation-Plane blend. Addition on the x- and y-axes is inherited from addition on the number line. Thus, we have $(a, 0) + (b, 0) = (a + b, 0)$ on the x-axis and $(0, a) + (0, b) = (0, a + b)$ on the y-axis.

In the Rotation-Plane blend, rotation and multiplication both exist and are correlated. That means that the rotation R_{-1} applies in this blend. For example,

$R_{-1}(1, 0) = (-1, 0) \cdot (1, 0) = (-1, 0)$
$R_{-1}(-1, 0) = (-1, 0) \cdot (-1, 0) = (1, 0)$
$R_{-1}(5, 0) = (-1, 0) \cdot (5, 0) = (-5, 0)$
$R_{-1}(R_{-1}(3, 0)) = (-1, 0) \cdot ((-1, 0) \cdot (3, 0)) = (-1, 0) \cdot (-3, 0) = (3, 0)$

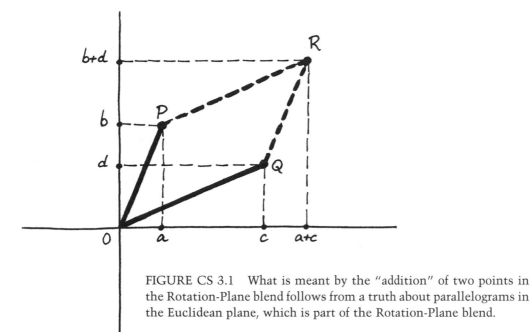

FIGURE CS 3.1　What is meant by the "addition" of two points in the Rotation-Plane blend follows from a truth about parallelograms in the Euclidean plane, which is part of the Rotation-Plane blend.

The Rotation-Plane blend is simple from a conceptual standpoint. It is no more than adding to the Cartesian plane our understanding of negative numbers in terms of a "rotation" operation. What is interesting from a cognitive point of view is how much follows from the simple formation of this conceptual blend. One thing that follows is a principle of addition for all ordered pairs (a, b) and (c, d) in the rotation plane.

What does it mean to "add" two arbitrary points $P = (a, b)$ and $Q = (c, d)$ in the rotation plane? Since the Rotation-Plane blend has the Euclidean plane within it, a relevant theorem of the Euclidean plane holds:

Each pair of intersecting line segments uniquely determines a parallelogram.

Suppose we place the point O at the origin of the rotation plane and let the line segments be OP and OQ, which intersect at O. We get the parallelogram $OPRQ$ in the rotation plane, which includes Cartesian coordinates (see Figure CS 3.1).

The point R is uniquely determined by lines OP and OQ, which are uniquely determined in the rotation plane by points $P = (a, b)$ and $Q = (c, d)$. As Figure CS 3.1 shows, R has an interesting property. $R = (a + c, b + d)$. That is, its coordinates (e, f) are the sums of the coordinates of P and Q. That is, $(e, f) =$

$(a + c, b + d)$, with $e = a + c$ and $f = b + d$. All this simply follows from the conceptual structure of the rotation plane.

What the rotation plane provides is a new concept of a "sum" for points $P = (a, b)$ and $Q = (c, d)$. Does the extension of the concept of "sum" fit the laws of arithmetic and the principle of closure? It should be obvious that associativity and commutativity hold. Moreover, $(0, 0)$ will function as an additive identity, since $(a, b) + (0, 0) = (a + 0, b + 0) = (a, b)$. In addition, each point (a, b) has an inverse $(-a, -b)$, since $(a, b) + (-a, -b) = (0, 0)$.

Notice that the additive inverse $(-a, -b)$ is (a, b) rotated by 180°—that is, (a, b) multiplied by $(-1, 0)$. This is just a special case of scalar multiplication, $(c, 0) \cdot (a, b) = (ca, cb)$, where $c = -1$.

The rotation plane thus has the following arithmetic properties.

ARITHMETIC PROPERTIES OF THE ROTATION PLANE

$(a, b) + (c, d) = (a + c, b + d)$
$(c, 0) \cdot (a, b) = (ca, cb)$
$(-1, 0) \cdot (a, b) = (-a, -b)$
$(0, 0)$ is the additive identity.
$(1, 0)$ is the multiplicative identity.
The additive inverse of (a, b) is $(-a, -b)$.
Addition and scalar multiplication are associative and commutative.
Scalar multiplication is distributive over addition.

The 90° Rotation Plane

Recall that the rotation plane is essentially just the Cartesian plane to which we have added the metaphor characterizing multiplication by -1 as a rotation by 180°. Suppose we extend the rotation plane by extending that metaphor a bit. In the rotation plane, rotation by 180° is multiplication by a number. But rotation by 90° has no arithmetic correlate. Suppose we add such a correlate, a metaphorical submapping in which there is some number n such that rotation by 90° is multiplication by n. It will follow that two rotations by 90° is multiplication by n^2.

THE 90° ROTATION METAPHOR

| Source Domain | | Target Domain |
SPACE		ARITHMETIC
Rotation by 90°	$\rightarrow$	Multiplication by n
Two rotations by 90°	$\rightarrow$	Multiplication by n^2

Let us call the result of adding this metaphor to the rotation plane the *90° rotation plane*.

In the 90° rotation plane, all the arithmetic properties of the Rotation-Plane blend will be preserved. But the additional possibilities added by the 90° Rotation metaphor will yield new properties.

What is the product of the number n and $(1, 0)$? A rotation by 90° applied to $(1, 0)$ is $(0, 1)$. Thus, $n \cdot (1, 0) = (0, 1)$.

What is the product of n and (a, b)? A rotation of (a, b) by 90° yields $(-b, a)$. That is, $n \cdot (a, b) = (-b, a)$.

What is the product of $n^2 \cdot (a, b)$? A rotation of 90° applied to $(-b, a) = (-a, -b)$. Thus, $n^2 \cdot (a, b) = (-a, -b)$

What is the product of $n^2 \cdot (1, 0)$? A rotation by 90° applied to $(1, 0)$ is $(0, 1)$. A second rotation of 90° applied to $(0, 1) = (-1, 0)$. This is equivalent to multiplying $(1, 0)$ by $(-1, 0)$. Since $n^2 \cdot (1, 0) = (-1, 0) \cdot (1, 0)$, $n^2 = (-1, 0)$.

Since $(-1, 0)$ in the rotation plane is just -1 on the Number Line, embedding the number line in the 90° rotation plane yields the result that $n^2 = -1$, and $n = \sqrt{-1}$, which is what has been called "i." Renaming n as $i = \sqrt{-1}$ gives us the following version of the 90° Rotation metaphor, when it occurs as part of the rotation plane.

THE 90° ROTATION METAPHOR (AS PART OF THE ROTATION PLANE)

Source Domain		Target Domain
SPACE		ARITHMETIC
Rotation by 90°	→	Multiplication by i ($= \sqrt{-1}$)
Two rotations by 90°	→	Multiplication by i^2 ($= -1$)

What Is the Complex Plane?

As noted, the rotation plane is just the Cartesian plane with the structure imposed by our normal metaphor for conceptualizing multiplication by -1—namely, Multiplication by -1 Is Rotation to the Symmetric Point on the Line.

The complex plane is just the 90° rotation plane—the rotation plane with the structure imposed by the 90° Rotation metaphor added to it. Multiplication by i is "just" rotation by 90°.

This is not arbitrary; it makes sense. Multiplication by -1 is rotation by 180°. A rotation of 180° is the result of two 90° rotations. Since i times i is -1, it makes sense that multiplication by i should be a rotation by 90°, since two of them yield a rotation by 180°, which is multiplication by -1.

The Properties of the 90° Rotation Plane

The 90° rotation plane has the following properties.

- What is i? $i = (0, 1)$
- What are i^2, i^3, and i^4 ? $i^2 = -1$, [i.e., $(-1, 0)$], $i^3 = -i$ [i.e., $(0, -1)$], and $i^4 = 1$ [i.e., $(1, 0)$].
- What is $b \cdot i$? $b \cdot i = (b, 0) \cdot (0, 1) = (0, b)$.
- What is $a + bi$? $a + bi = (a, 0) + (0, b) = (a, b)$
- What is $(a + bi) + (c + di)$? $(a + bi) + (c + di) = (a + c) + (b + d)i$
- What is $(a + bi) \cdot (c + di)$? $(a + bi) \cdot (c + di) = (ac - bd) + (bc + ad) \cdot i$

The last line isn't obvious; here's the reason why the equality holds. Recall that since we are in a special case of the Cartesian Plane blend, the basic laws of arithmetic—distributivity, associativity, and commutativity—hold.

$$
\begin{aligned}
(a + bi) \cdot (c + di) &= (a + bi) \cdot c + (a + bi) \cdot di && \text{[by the distributive law]} \\
&= ac + bci + adi + bdi^2 && \text{[by the distributive law]} \\
&= ac + bci + adi - bd && \text{[by } i^2 = -1 \text{]} \\
&= (ac - bd) + (bc + ad) \cdot i && \text{[by the associative law]}
\end{aligned}
$$

These are exactly the properties of the complex plane. The 90° rotation plane *is* the complex plane. i is inherently linked to 90° rotations in the complex plane for a simple reason. Because $i^2 = -1$, $i^4 = 1$. In the complex plane, a rotation of 360° gets you back to where you started from. Do four rotations of 90° each, which is equivalent to multiplying by $i^4 = 1$, and you are back where you started.

$a + bi$ is, of course, the general form of all complex numbers. When $b = 0$, you get the real numbers as a special case. The simple fact that addition is vector addition, that $(a + bi) + (c + di) = (a + c) + (b + d)i$, follows from the fact that Euclid's parallelogram theorem is true of the rotation plane, which incorporates the Euclidean plane (Figure CS 3.1).

Finally, in the 90° Rotation-Plane blend, the arithmetic domain contains numbers in a number system, each uniquely paired with a point in the spatial domain of the blend. A complex number, then, is an entity, z, of the form $a + bi$. It functions like any number in a number system. If you want to think of it as isolated from the spatial domain, you can. But then you would be losing all the conceptual structure that motivates the arithmetic properties of this number system.

The word "motivate" is important here. It is part of the paradigm of mathematical idea analysis, part of looking for cognitive explanations in mathematics.

Ignoring the spatial aspects of the conceptual structure of the complex numbers gives the misleading impression that they are "just another formal system"—a disembodied set of elements with operations obeying some constraints or other that happen to meet a particular set of axioms. Ignoring the spatial aspects would be to miss the heart of much of modern mathematics—the conceptual connections across mathematical domains, the great ideas that give meaning to the numbers and motivate the mathematics.

The crux of the issue here is the difference between proof and understanding. As our quotation from Benjamin Peirce (in Case Study 1) indicates, it is possible to prove a deep mathematical result without understanding it at all—for example, when the proof involves a lot of computation and putting together equations that have previously been proved technically. For us, mathematics is not just about proof and computation. It is about ideas.

One can compute with, and prove theorems about, $\sqrt{-1}$ without the idea of the complex plane or rotation. Does that mean that those ideas are extraneous to what $\sqrt{-1}$ is? Do they stand outside the mathematics of complex arithmetic per se? Are they just ways of thinking about $\sqrt{-1}$—mere representations, interpretations, or ways of visualizing $\sqrt{-1}$—useful for pedagogy but not part of the mathematics itself?

From the perspective of cognitive science and mathematical idea analysis, the answer is no. Here's why.

Why Characterize the Complex Plane As the 90° Rotation Plane?

First, because the characterization is true—true given the conceptual metaphors and blends commonplace in classical mathematics.

But mainly because it reveals the conceptual structure inherent in an understanding of what the complex plane is. In particular, it reveals the metaphorical and blend structure of the complex plane.

THE CONCEPTUAL STRUCTURE OF THE COMPLEX PLANE

The Multiplication by –1 Is Rotation by 180° metaphor
The Numbers Are Points on a Line metaphor and its blend,
    the Number-Line blend
The Cartesian Plane blend
The Rotation-Plane blend
The 90° Rotation metaphor in the rotation plane, which is
    the 90° Rotation blend

It is, of course, not the only understanding of the complex plane or the complex number system that one could have. For example, there is a common understanding of the complex plane as just the Cartesian plane together with addition and multiplication, derived not as entailments of conceptual structure but as arbitrary-sounding rules for the addition and multiplication of ordered pairs:

Addition: $(a, b) + (c, d) = (a + c, b + d)$
Multiplication: $(a, b) \cdot (c, d) = (ac - bd, bc + ad)$

Another way to think about the system of complex numbers is to jettison the plane altogether (in the Dedekind-Weierstrass spirit). One can then see the complex numbers as just a set of ordered pairs, with operations + and ·, defined by the rules for addition and multiplication just given and, hence, meeting the usual laws of arithmetic. Here we find no Cartesian plane, complex plane, or vectors, or lines, or rotations, or any geometry at all. It is pure arithmetic.

The last two are the ways the complex numbers are usually taught. But they give very little insight into the conceptual structure of the complex numbers—into *why* they are the way they are. The first method does not explain *why* the i-axis should be 90° degrees from the real axis. In the second, neither an i-axis nor real axis even exist. And neither method explains *why* those laws are there as consequences of the central *ideas* that structure the complex numbers.

We will now turn to $e^{\pi i} + 1 = 0$. Our approach will be there as it was here. $e^{\pi i} + 1 = 0$ uses the conceptual structure of all the cases we have discussed so far—trigonometry, the exponentials, and the complex numbers. Moreover, it puts together all that conceptual structure. In other words, all those metaphors and blends are simultaneously activated and jointly give rise to inferences that they would not give rise to separately. Our job is to see how $e^{\pi i} + 1 = 0$ is a precise consequence that arises when the conceptual structure of these three domains is combined to form a single conceptual blend.

Case Study 4

$e^{\pi i} + 1 = 0$

How the Fundamental Ideas of Classical Mathematics Fit Together

Piecing the Puzzle Together

We have almost all the pieces in place to see just what $e^{\pi i}$ means and why it equals −1. Here are the pieces to the puzzle that we have.

1. There is a distinction between self-multiplication and the exponential function. The exponential function e^x has the following properties:

> e^x maps sums onto products.
> e^x maps 1 onto e.
> e^x is a function whose rate of change is identical to itself; that is, e^x is its
> own derivative.

2. By the Trigonometry metaphor, the cosine of π = −1. This is a rotation of π radians from 1 on the x-axis in the Cartesian plane to −1 on the x-axis. This is a crucial link between π and −1.

3. In the complex plane, rotation by 90° (which is $\pi/2$ radians) is multiplication by i. Rotation by 180° (which is π radians) is multiplication by −1. Here again, in the complex plane, is the same all-important relationship between π and −1.

4. A variable over complex numbers, z, is of the form $x + yi$, where x and y are real variables. πi is an instance of $x + yi$, where $x = 0$ and $y = \pi$. $e^{\pi i}$ is an instance of the complex function $w = e^z$, where $z = \pi i$. $e^{\pi i}$ = −1 indicates that the function $w = e^z$ maps πi onto −1.

From these pieces we can immediately solve an important part of the puzzle: What can it mean to multiply e by itself a πi number of times? The answer, as we have seen, is that it can't mean anything at all. $e^{\pi i}$ is not an instance of the self-multiplication function. It is, rather, a special case of the exponential function $w = e^z$, where $z = \pi i$. This is not the same as the real-valued exponential function $y = e^x$, which maps the real line onto the real line.

$w = e^z$, of course, makes sense only when the exponential function is extended to complex variables, which we have not yet done. But looking ahead, the idea is this: When extended, the complex exponential function $w = e^z$ will map points in one complex plane, z, onto points in another complex plane, w. In the input complex plane, z, πi is the point $(0, \pi)$—that is, the point on the i-axis at a distance π from the origin.

The exponential function $w = e^z$ will map $(1, 0)$ in the input z-plane onto $(e, 0)$ in the output w-plane. That is, $w = e^z$ will map 1 on the x-axis in the z-plane onto e on the x-axis in the w-plane.

The equation $e^{\pi i} = -1$ says that the function $w = e^z$, when applied to the complex number πi as input, yields the real number -1 as the output, the value of w. In the complex plane, πi is the point $(0, \pi)$—π on the i-axis. The function $w = e^z$ maps that point, which is in the z-plane, onto the point $(-1, 0)$—that is, -1 on the x-axis—in the w-plane (see Figure CS 4.1).

This ought to leave you cold. We have not said *why* the mapping works in exactly this way, *why* $w = e^z$ maps πi onto -1 rather than onto some other number. We will get to that shortly.

Even so, this may seem a bit anticlimactic. What's the big deal? So the complex exponential function $w = e^z$ maps π on the i-axis in one complex plane onto -1 in another. So what? Why should anyone care?

Numerically, all that is happening is that 3.141592654358979 . . . on the y-axis in one plane is being mapped by some function onto -1 on the x-axis in another plane, when the same function maps 1 on the x-axis in the first plane onto 2.718281828459045 . . . on the x-axis in the second plane. From a purely numerical point of view, this has no interest at all!

If all that is at issue is calculating values of functions and seeing which number maps onto which other number, the greatest equation of all time is a big bore!

Euler was too smart to care about something that boring. Euler's great equation $e^{\pi i} = -1$ has a lot more meaning than this. But its meaning is not given by the values computed for the function $w = e^z$. Its meaning is conceptual, not numerical. The importance of $e^{\pi i} = -1$ lies in what it tells us about how various branches of mathematics are related to one another—how algebra is related to geometry, geometry to trigonometry, calculus to trigonometry, and how the arithmetic of complex numbers relates to all of them.

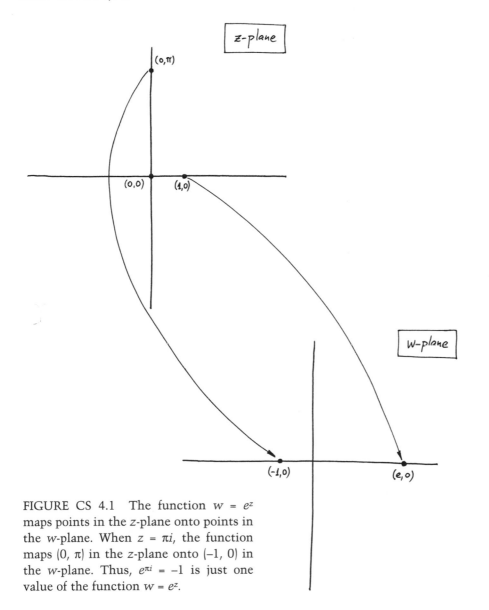

FIGURE CS 4.1 The function $w = e^z$ maps points in the z-plane onto points in the w-plane. When $z = \pi i$, the function maps $(0, \pi)$ in the z-plane onto $(-1, 0)$ in the w-plane. Thus, $e^{\pi i} = -1$ is just one value of the function $w = e^z$.

Mathematical Idea Analysis

As we shall see shortly, answers to the deepest questions we have been asking lie in the conceptual metaphors and blends we have been analyzing. It is through the metaphors and blends that the cognitive linkages across subfields of mathematics occur. When we conceptualize numbers as points, we link arithmetic and algebra with geometry. When we form the unit circle, we link algebra, geometry, and trigonometry. When we form the 90° Rotation-Plane blend, we link complex numbers with geometry and rotations. Each conceptual blend states detailed conceptual relationships among important mathematical concepts—for example, rotation and multiplication. And it states these relationships in sufficient detail to account for how inferences and methods of calculation from one domain are applied to another domain.

The equation $e^{\pi i} + 1 = 0$ is true only by virtue of a large number of profound connections across many fields. *It is true because of what it means!* And it means what it means because of all those metaphors and blends in the conceptual system of a mathematician who understands what it means. To show *why* such an equation is true for conceptual reasons is to give what we have called an idea analysis of the equation. Here are the minimal conditions for an adequate idea analysis:

REQUIREMENTS FOR AN ADEQUATE MATHEMATICAL IDEA ANALYSIS OF AN EQUATION, DEFINITION, AXIOM, OR THEOREM

- The metaphor-and-blend structure that makes the equation, definition, axiom, or theorem true relative to that structure.
- The conceptual relationships among the elements that appear in, or are presupposed by, the equation, definition, axiom, or theorem.
- The ideas expressed by the symbols in the equation, definition, axiom, or theorem.
- The grounding of the concepts used in the idea analysis.

One thing that an idea analysis should make clear is this: *The "truth" of an equation, axiom, or theorem depends on what it means.* That is, truth is relative to a conceptual structure, which at the very least includes an extensive network of metaphors and blends that we call an *idea network.* Thus, $e^{\pi i} + 1 = 0$ is not true of Euclidean geometry or the arithmetic of real numbers for an obvious reason: Those mathematical domains do not have the right idea structure to give meaning to the equation, much less to make it true.

The Idea Network for $e^{\pi i} + 1 = 0$

The idea analysis of $e^{\pi i} + 1 = 0$ requires one conceptual blend in addition to the metaphors and blends we have already given.

THE TRIGONOMETRIC COMPLEX PLANE BLEND

Domain 1 THE POLAR-TRIGONOMETRY BLEND		*Domain 2* THE 90° ROTATION-PLANE BLEND (THE COMPLEX PLANE)
$(r \cos \theta, r \sin \theta)$	$\leftrightarrow$	(a, b); that is, $a + bi$
r	$\leftrightarrow$	The length of the line from the origin to (a, b): $\sqrt{(a^2 + b^2)}$
θ	$\leftrightarrow$	The angle formed by the x-axis and the line from the origin to (a, b)
$(1 \cos \pi/2, 1 \sin \pi/2)$, which equals $(0, 1)$ in the Cartesian Plane blend	$\leftrightarrow$	$(0, 1)$; that is, i
$(1 \cos \pi, 1 \sin \pi)$, which equals $(-1, 0)$ in the Cartesian Plane blend	$\leftrightarrow$	$(-1, 0)$; that is, i^2

Entailments in the Blend

$$r\,(i \sin \theta) = bi$$

$$r \cos \theta = a$$

$$\text{``}r \text{ cis } \theta\text{''}: r\,(\cos \theta + i \sin \theta) = a + bi$$

$$(r_1 \text{ cis } \alpha) \cdot (r_2 \text{ cis } \beta) = (r_1 \cdot r_2) \text{ cis } (\alpha + \beta)$$
$$\text{if and only if}$$
$$(a + bi) \cdot (c + di) = (\,(ac - bd) + (ad + bc)\,i\,)$$

Trigonometry in the complex plane arises from the conceptual blend of the 90° Rotation-Plane blend of Case Study 3 with the Trigonometry blend of Case Study 1. Within this blend, we get entailments not present in previous conceptual blends: characterizations of complex addition and multiplication in trigonometric terms. For example, here is the characterization of complex trigonometric multiplication in terms of complex nontrigonometric multiplication.

$$(r_1 \text{ cis } \alpha) \cdot (r_2 \text{ cis } \beta) = (r_1 \cdot r_2) \text{ cis } (\alpha + \beta)$$

if and only if

$$(a + bi) \cdot (c + di) = (\,(ac - bd) + (ad + bc)\,i\,)$$

This characterization of complex multiplication in trigonometric terms is, of course, metaphorical. The complex plane, with complex addition and multipli-

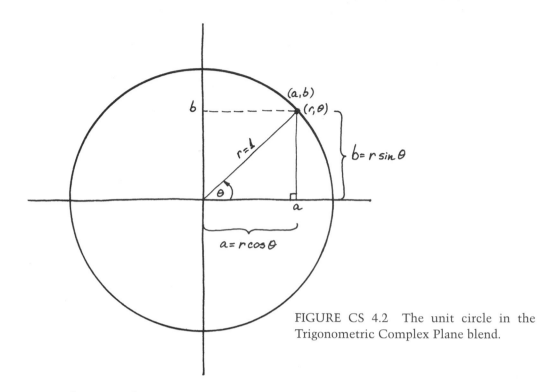

FIGURE CS 4.2   The unit circle in the Trigonometric Complex Plane blend.

cation, has no inherent trigonometric structure. Complex arithmetic does not need trigonometry. Trigonometry is in another mathematical domain altogether. When we understand complex arithmetic in terms of trigonometry, we are conceptualizing one mathematical domain in terms of another, which is what conceptual metaphor does.

A Piece of the Function e^z

As we have seen, πi is an imaginary number of the form yi, where y is a real number. $e^{\pi i}$ is an instance of the complex exponential function e^z, where $z = x + yi$. We are working toward an understanding of e^z. So far, our journey has taken us to the Trigonometric Complex Plane blend. Let us turn now to what the unit circle looks like in that blend (see Figure CS 4.2).

Every point (a, b) on the unit circle can be conceptualized in terms of (r, θ), where $a = r \cos \theta$ and $b = r \sin \theta$, where $r = 1$. The complex number, $a + bi$, on the unit circle can thus be conceptualized via the Trigonometric Complex Plane blend as $1 \cdot (\cos \theta + i \sin \theta)$, or simply as $\cos \theta + i \sin \theta$.

To get some idea of how complex functions work, let us consider a very simple mapping from a segment of the real line onto the unit circle in the complex plane.

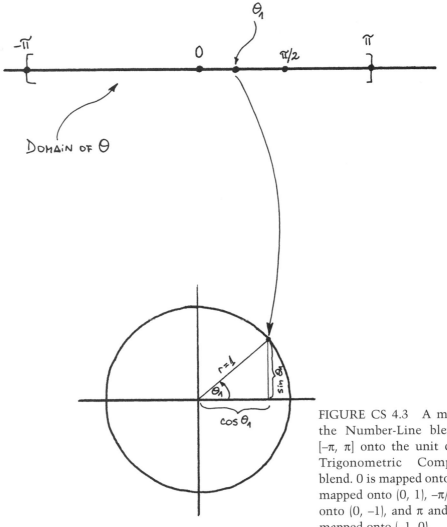

FIGURE CS 4.3   A mapping from the Number-Line blend between $[-\pi, \pi]$ onto the unit circle in the Trigonometric Complex Plane blend. 0 is mapped onto $(1, 0)$, $\pi/2$ is mapped onto $(0, 1)$, $-\pi/2$ is mapped onto $(0, -1)$, and π and $-\pi$ are both mapped onto $(-1, 0)$.

The circumference of the unit circle is 2π. Let us take a segment of length 2π on the real line and map it onto the unit circle. For convenience, we will take the portion of the real line between $-\pi$ and π. Since the points on the unit circle are of the form $\cos\theta + i\sin\theta$, we will let that be our function. That is, we will use θ as a variable over the points on the real line between $-\pi$ and π. Our function will be $f(\theta) = \cos\theta + i\sin\theta$. This function will map every point on the line segment $[-\pi, +\pi]$ onto a point on the unit circle in the complex plane $(\cos\theta, \sin\theta)$, which corresponds to the complex number $\cos\theta + i\sin\theta$ (see Figure CS 4.3).

For example, the number 0 in $[-\pi, +\pi]$ is mapped onto the complex number $\cos 0 + i \sin 0$, which corresponds to the point $(1, 0)$ on the x-axis in the complex plane, since $\cos 0 = 1$ and $\sin 0 = 0$. Here are some other values of the function:

$f(\pi/2) = \cos \pi/2 + i \sin \pi/2 = 0 + 1i = i$, which corresponds to the point $(0, 1)$.

$f(-\pi/2) = \cos -\pi/2 + i \sin -\pi/2 = 0 + -1i = -i$, which corresponds to the point $(0, -1)$.

$f(\pi) = \cos \pi + i \sin \pi = -1 + 0i = -1$, which corresponds to the point $(-1, 0)$.

$f(-\pi) = \cos -\pi + i \sin -\pi = -1 + 0i = -1$, which corresponds to the point $(-1, 0)$.

Note that π and $-\pi$ map onto the same point $(-1, 0)$. This corresponds to the fact that sine and cosine are both periodic functions with a period of 2π. Since $-\pi$ and π are 2π apart, they map onto the same point.

Moreover, each two points on the line segment that are separated by a distance π are mapped onto points that are related by a rotation of 180°, or π radians, on the unit circle. That is, if they are a distance π apart on the line segment, they are π radians apart on the unit circle and one can be imposed on the other by a rotation of π radians. Since Rotation By π Radians Is Multiplication By -1, the numerical value of one point is -1 times the numerical value of the other. Take, for example, the points on the line segment $-\pi/2$ and $\pi/2$, which are separated by a distance π. They map onto $-i$ and i, where each is -1 times the other.

Finally, consider two points on the line segment that are separated by a distance $\pi/2$. They map onto two points on the unit circle that are $\pi/2$ radians apart, so that a rotation of $\pi/2$ radians will map one onto the other. Since a rotation of $\pi/2$ radians is equivalent to multiplication by i, the numerical value of one is i times the numerical value of the other. For example, take the points $-\pi/2$ and 0 on the line segment. They will map onto $-i$ and 1 on the unit circle, where $-i \cdot i = 1$.

What this function does is to wrap the line segment around the unit circle, starting at point $(-1, 0)$ and ending at the same point. In so doing, it maps real numbers onto complex numbers, which are subject to all the constraints of the Complex Plane (90° Rotation-Plane blend), including the metaphors linking rotation to multiplication.

Another Piece of the Function e^z

Since the complex variable $z = x + yi$, the corresponding exponential function is of the form $e^z = e^{x+yi}$. The function $w = e^z$ maps points of a complex plane z onto points of a complex plane w. If e^z is an exponential function, then e^z maps sums onto products, and so $e^{x+yi} = e^x \cdot e^{yi}$. When $y = 0$, $e^{yi} = e^0 = 1$, and so e^z reduces to e^x. We can now see exactly what e^z does when $y = 0$. In the complex plane z, all there is when $y = 0$ is the x-axis. The function $w = e^z$ is now reduced to $w = e^x$, which will map the points on the x-axis of the z-plane onto points in the w-plane. What does e^x map the x-axis onto in the w-plane?

As an exponential function applying to a real line, e^x maps the origin $(0, 0)$ in the z-plane onto $(1, 0)$ in the w-plane, since $e^0 = 1$. Next, e^x maps $(1, 0)$ in the z-plane onto $(e, 0)$ in the w-plane, since e is the base of the exponential function. Correspondingly, e^x will map $(2, 0)$ onto $(e^2, 0)$, $(3, 0)$ onto $(e^3, 0)$, and so on. It will also map $(-3, 0)$ onto $(e^{-3}, 0)$, which equals $(1/e^3, 0)$. e^x will thus map all the negative numbers of the x-axis onto the real numbers between 0 and 1 and all the positive numbers onto numbers greater than 1 (see Figure CS 4.4).

- The entire negative half of the real line from $-\infty$ up to 0 is mapped onto the interval from 0 to 1.
- The entire positive half is mapped onto the portion of the positive half from 1 to ∞.
- The entire x-axis in the z-plane is mapped onto the positive half of the horizontal axis in the w-plane.

One of the basic properties of e^x is that it maps sums onto products. For example, the sum $2 + 3$ will be mapped onto the product $e^2 \cdot e^3$. Thus, arithmetic progressions are mapped onto geometric progressions. Intervals on the positive side of the x-axis are expanded exponentially, while intervals on the negative side of the x-axis are contracted exponentially. Suppose one were to take the positive half of the horizontal axis in the w-plane. And suppose one were to mark each point on it with the number in the z-plane that e^x maps onto it. For example, one would mark the point $(1, 0)$ with the number 0, the point $(e, 0)$ with the number 1, the point $(e^2, 0)$ with the number 2, and so on. The resulting "marked up" positive half of the line would look like a logarithmic scale of a slide rule (see Figure CS 4.5).

Why $e^{yi} = \cos y + i \sin y$

We have seen that $e^z = e^{x+yi} = e^x \cdot e^{yi}$. We have seen the effect of e^x when $y = 0$. And we have seen the effect of the function $f(y) = \cos y + i \sin y$. With one more

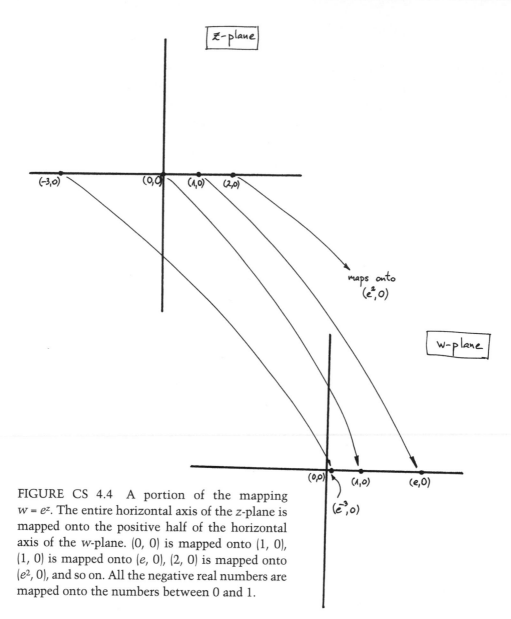

FIGURE CS 4.4 A portion of the mapping $w = e^z$. The entire horizontal axis of the z-plane is mapped onto the positive half of the horizontal axis of the w-plane. $(0, 0)$ is mapped onto $(1, 0)$, $(1, 0)$ is mapped onto $(e, 0)$, $(2, 0)$ is mapped onto $(e^2, 0)$, and so on. All the negative real numbers are mapped onto the numbers between 0 and 1.

step, we can see the effect of the whole function e^z by showing that $e^{yi} = \cos y + i \sin y$. To do so, we need to know something about Taylor series.

We saw in Case Study 2 that there is a set of metaphors that jointly allow us to conceptualize any continuous function $f(x)$ as an infinite sum of functions of the form

$$f(x) = \sum_{n=0}^{\infty} (a_n \cdot x^n) = a_0 + a_1 x + a_2 x^2 + \dots$$

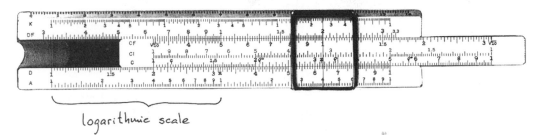

FIGURE CS 4.5   A slide rule with its logarithmic scales.

under certain conditions for convergence. This sum of functions can be differentiated term by term. And each derivative of the function is another infinite series of the same form, which can in turn be itself differentiated to give an infinite series of that form, ad infinitum. Let $f'(x)$ be the first derivative of $f(x)$, $f''(x)$ be the second, and so on.

$$f'(x) = a_1 + 2a_2x + 3a_3x^2 + \ldots$$
$$f''(x) = 1 \cdot 2a_2 + 2 \cdot 3a_3x + 3 \cdot 4a_4x^2 + \ldots$$
$$f'''(x) = 1 \cdot 2 \cdot 3a_3 + 2 \cdot 3 \cdot 4a_4x + 3 \cdot 4 \cdot 5a_5x^2 + \ldots$$

In general, the nth derivative

$$f^{(n)}(x) = n! \cdot a_n + \text{terms containing } x \text{ as a factor.}$$

Suppose we set $x = 0$ for each function. Then,

$$f(0) = a_0 \qquad\qquad a_0 = f(0)$$
$$f'(0) = a_1 \qquad\qquad a_1 = f'(0)$$
$$f''(0) = 2! \cdot a_2 \qquad\qquad a_2 = f''(0)/2!$$
$$f'''(0) = 3! \cdot a_3 \qquad\qquad a_3 = f'''(0)/3!$$
$$\ldots \qquad\qquad\qquad \ldots$$
$$f^{(n)}(0) = n! \cdot a_n \qquad\qquad a_n = f^{(n)}(0)/n!$$

Given that $f(x) = \sum_{n=0}^{\infty}(a_n \cdot x^n) = a_0 + a_1x + a_2x^2 + \ldots$, we now have values for all the a_n's in terms of the nth derivatives of the function itself and n factorial. Thus, the values of the function are determined by a *unique* power series determined by the derivatives of the function itself at $x = 0$. This is the Taylor series for the function.

Let us stop for a moment to consider what this means. If you take the function evaluated at one point, 0, and you take all its derivatives at that point, you can determine the entire function! That is, there is enough information in an arbitrarily small neighborhood around that point to determine every value of the function everywhere.

When we think about this from a conceptual perspective, we can see why this is so. Every neighborhood around zero contains an infinity of points—as many as there are on the whole real line (in Cantor's sense). The derivative is a limit of an infinite sequence—a sequence of values of the function at points in that neighborhood. In other words, a derivative presupposes the assignment of an infinity of values of the function in an arbitrarily small neighborhood.

Moreover, each successive derivative is the limit of another such infinite sequence. In short, there is a huge amount of information about a continuous function in every neighborhood around every point where the function is defined. In the infinite number of sequences given by the infinity of derivatives, there is enough information to determine the entire function. When you think of it from this perspective, the existence of Taylor series is not all that surprising.

Let us now turn to the relevance of the Taylor series for our story. We want to see why $e^{yi} = \cos y + i \sin y$. We know that the Taylor series for each function is unique, because it is determined by the function's own derivatives evaluated at zero. That means that if two allegedly different functions have the same Taylor series, they must be the same functions; they must be the same sets of ordered pairs of numbers (given the metaphor that A Function Is a Set of Ordered Pairs).

Moreover, since the Taylor series of a function is determined by the function's own derivatives, something very important follows. One entailment of the metaphor A Function Is a Set of Ordered Pairs is that the same set of ordered pairs may be characterized by different concepts, symbolized in different ways. Traditionally a function was a rule that gave a unique output, given a fixed input. The rule was characterized in conceptual terms. When functions became reconceptualized in terms of set theory—via this metaphor—the "function" became the set of ordered pairs, not the rule.

It can be a substantive discovery in mathematics that two conceptually quite different functions (in the rule sense) can actually determine the same set of ordered pairs, and thus *be* the *same* function (in the metaphorical set-of-ordered pairs sense typical of contemporary mathematics).

Because of this, two conceptually different rules that have all the same derivatives must therefore have the same Taylor series. As such, they must determine the same set of ordered pairs and must "be" the same function.

For example, e^{yi} and cos $y + i$ sin y are two conceptually different rules for mapping inputs to outputs. It is a matter to be discovered whether they happen to determine the same set of ordered pairs, and hence are the "same function."

One way to determine this is to look at their Taylor series. If the Taylor series are the same, then they do constitute the same function (under the metaphor).

Thus, to see why e^{yi} and cos $y + i$ sin y are the same function, we need to know only why these two functions have all the same derivatives at $y = 0$. That is easy to show. Let us start with cos $y + i$ sin y. Recall that the derivative of the sine is the cosine and that the derivative of the cosine is the negative of the sine.

DERIVATIVES	DERIVATIVES AT $y = 0$
$f(y) = \cos y + i \sin y$	$f(0) = 1$
$f'(y) = -\sin y + i \cos y$	$f'(0) = i$
$f''(y) = -\cos y - i \sin y$	$f''(0) = -1$
$f'''(y) = \sin y - i \cos y$	$f'''(0) = -i$
$f''''(y) = \cos y + i \sin y$	$f''''(0) = 1$
. . .	. . .

Notice that the fourth derivative is the same as the function itself. After that, the sequence of derivatives recurs. The fifth derivative will be the same as the first, the sixth the same as the second, and so on.

Now let us compare this to the derivatives of e^{yi}. Consider the value of a derivative of $f(y) = e^{yc}$, for a constant $c \cdot f'(y) = c \cdot e^{yc}$. When the constant $c = i$, we get the following derivatives:

DERIVATIVES	DERIVATIVES AT $y = 0$
$f(y) = e^{yi}$	$f(0) = 1$
$f'(y) = i \cdot e^{yi}$	$f'(0) = i$
$f''(y) = -e^{yi}$	$f''(0) = -1$
$f'''(y) = -i \cdot e^{yi}$	$f'''(0) = -i$
$f''''(y) = e^{yi}$	$f''''(0) = 1$
. . .	. . .

Again we see that the fourth derivative is the same as the function itself, and that the derivatives recur after that. Note that the values of the derivative in the right-hand column are the same as for $f(y) = \cos y + i \sin y$. Therefore, the infinite number of derivatives at $y = 0$ for these two functions are the same. Therefore, they have the same Taylor series. Therefore, they perform the same mappings—that is, determine the same set of ordered pairs. By the metaphor that Functions Are Sets of Ordered Pairs, they are the "same function."

However, they are conceptualized, and hence symbolized, in very different ways. Recall from Chapter 16 that the symbol "=" can be used in mathematics to relate ideas that are cognitively very different. In this case, when we write e^{yi} = cos y + i sin y, the "=" has a special meaning, based on the metaphor that functions are sets of ordered pairs. The "=" means that the two conceptually different mapping rules on the two sides of the equal sign have the same *referent*, the same set of ordered pairs.

Finally!

We can now evaluate the function e^{yi} by evaluating cos y + i sin y. Setting $y = \pi$, $e^{\pi i}$ = cos π + i sin π.

Since sin π = 0, i sin π = 0. Since cos π = –1, the entire expression cos π + i sin π = –1. Thus, $e^{\pi i}$ = –1!

But What Does It Mean?

We know what it is, but what does it mean? At first, all this sounds like mumbo-jumbo. Functions with the same power series compute the same mappings and are linked by "=." Two functions with all the same derivatives at 0 have the same Taylor series. If you want to find out what the value of e^{yi} is when $y = \pi$, plug π in for y in cos y + i sin y. When $y = \pi$, cos y + i sin y = –1. Therefore, when $y = \pi$, e^{yi} = –1. Okay. But again, why does it matter that e^{yi} = –1 when $y = \pi$? And what does it mean?

What Are the Ideas?

What we have described in our discussion of Taylor series are the usual mathematical reasons given ("informally" from a cognitive perspective) for why $e^{\pi i}$ + 1 = 0. Now we need a mathematical idea analysis to complete the cognitive account we have been building up of why the equation is true.

Let us begin with the blend implicit in the mathematical account. We refer to it as Euler's blend, partly in honor of Euler and partly because we suspect that he must have had, at least intuitively and unconsciously, an implicit understanding of the relationships between the branches of classical mathematics that the blend represents.

EULER'S BLEND				
Domain 1 THE POWER SERIES BLEND	*Domain 2* THE TAYLOR SERIES BLEND	*Domain 3* THE EXPONENTIAL COMPLEX PLANE BLEND	*Domain 4* THE TRIG. COMPLEX PLANE BLEND	*Domain 5* SETS OF ORDERED PAIRS
The power series $$\sum_{n=0}^{\infty} a_n z^n$$ (via the BMI)	The Taylor series $$\sum_{n=0}^{\infty} \frac{f^{(n)}(0)}{n!} z^n$$ (via the BMI)	$f(\theta) = e^{i\theta}$	$f(\theta) = \cos\theta + i\sin\theta$	$\{(\theta, f(\theta))\}$

Here is the structure of this conceptual blend:

- Domain 2 is a special case of domain 1, since a Taylor series is a type of power series. This brings the entire Power Series blend into this blend.
- The Taylor series in domain 2 is the Taylor series of $e^{i\theta}$ in domain 3.
- The Taylor series in domain 2 is the Taylor series of $f(\theta) = \cos\theta + i\sin\theta$ in domain 4.
- The set of pairs $\{(\theta, f(\theta))\}$ in domain 5 is determined by the Taylor Series in domain 2, and the functions in domains 3 and 4.

This blend incorporates other metaphors and blends that we have discussed both in this case study and throughout the book. If one traces back through all of them, one gets a picture of the idea network used in Euler's equation. It is a significant part of the idea network of classical mathematics.

An inspection of this idea network and of the idea analyses given throughout the book provides answers to the questions we have raised in this case study. Here are some of the answers to those questions.

1. What is the conceptual structure of e^{yi}?

As an exponential function, e^{yi} can also be written in the notation given in Case Study 2 as $\text{Exp}_e(yi)$. The mathematical ideas are as follows:

a. An exponential function metaphorically conceptualizes products in terms of sums. It does this via a mapping from sums to products, 0 to 1, and 1 to e.
b. The derivative $f'(y) = i \cdot e^{yi}$ of the function $f(y) = e^{yi}$ is proportional to the function itself, with i being the factor expressing that proportionality via multiplication.

c. Conceptually, derivatives are mathematizations of the concept of rate of change, and a function with a derivative proportional to its size changes at a rate proportional to its size. (Here the "proportion" is metaphorical: It is $\sqrt{-1} = i$. We will discuss the meaning of this metaphorical value below.)

d. The two ideas (a) and (b) are intimately linked via two metaphors: the derivative metaphor (Instantaneous Change Is Average Change over an Infinitely Small Interval) and the special case of limits in the Basic Metaphor of Infinity that characterizes what an "infinitely small interval" is.

The derivative metaphor introduces the expression $b^{t+\Delta t}$. The metaphorical exponential function that maps sums onto products will map $b^{t+\Delta t}$ onto $b^t \cdot b^{\Delta t}$. That, as we saw, allows b^t to be factored out. This b^t is the exponential function that expresses both of the following ideas: (i) change in proportion to itself and (ii) the mapping of sums to products. These two ideas are mutually entailing, given the above metaphors.

e. i is cyclic with a cycle of four. That is, $i^4 = 1$, which means that any number n multiplied by i four times $= n$, itself. Geometrically, multiplication by i is equivalent to 1/4 of the rotation of a circle. Four such rotations get you back to where you began. This idea derives ultimately from the metaphor for multiplication by negative numbers, Multiplication by -1 is Rotation by 180°. This is extended naturally to Multiplication by $\sqrt{-1}$ is Rotation by 90°.

f. Conceptually, circles and cyclicity are about recurrence: either about change that brings you back to the same place or about the regular recurrence of events or actions. The metaphor here is Recurrence Is Circularity.

g. The second derivative mathematizes the concept of acceleration. e^{yi} has the derivative $i \cdot e^{yi}$ and the second derivative $-e^{yi}$. This expresses the idea that its acceleration is the negative of its size. The fourth derivative—the acceleration of its acceleration—is e^{yi} itself. e^{yi} is thus a mathematization of the concept *a periodically recurrent process that changes in proportion to its size but is self-regulated*. The self-regulating character comes from the fact that the acceleration is the negative of the size.

Thus, when the function is increasing in value, its deceleration is also increasing—until the deceleration stops the increase. Then the function starts decreasing in value. As it progressively decreases, there is a progressive increase in deceleration in the negative direction—until its deceleration stops the decrease and the function starts increasing again. This is a self-regulating cycle.

h. e^{yi} is therefore the mathematization of a very rich and important idea with a great deal of conceptual structure.

2. What is the conceptual structure of $\cos y + i \sin y$?

    a. It has a very different conceptual structure and seems on the surface to express a very different idea: the sum of the cosine of an angle plus the product of i times the sine of that angle. The question is why this should have anything to do with "a periodically recurrent process that changes in proportion to its size but is self-regulated."

    b. Since there are a sine and a cosine, it makes sense that there should be periodicity involved, since they are defined in terms of the unit circle and the metaphor Recurrence Is Circularity.

But why should multiplication of the $\sin y$ by i added to $\cos y$ entail change in proportion to its size and be self-regulating? And why should that entail the mapping of sums to products by this complex trigonometric function? Here is the reason:

    a. Each point on the unit circle in the complex plane is of the form (a, b), where $a = \cos y$ and $b = \sin y$ and y is the angle formed by the x-axis and the line segment connecting the origin to point (a, b) (via the Trigonometry metaphor applying in the complex plane).

    b. The point (a, b) therefore corresponds to the complex number $\cos y + i \sin y$ (via the metaphor that Complex Numbers Are Points In a Plane).

    c. In polar coordinates, the product of two complex numbers on the unit circle is $\cos (\alpha + \beta) + i \sin (\alpha + \beta) = ((\cos \alpha + i \sin \alpha) \cdot (\cos \beta + i \sin \beta))$. In cis-notation, this is: cis $(\alpha + \beta)$ = (cis α) · (cis β)

    d. In other words, the function cis $y = \cos y + i \sin y$ maps sums onto products on the unit circle in the complex plane.

Now, why does the function $\cos y + i \sin y$ change in proportion to its size, just as an exponential function does?

    a. The limit metaphor says that Instantaneous Change is Average Change over an Infinitely Small Interval.

    b. A function that maps sums onto products maps $f(t+\Delta t)$ onto $f(t) \cdot f(\Delta t)$ and allows $f(t)$ to be factored out of the average change expression in the limit metaphor (see Case Study 2).

    c. This entails that the change in such a function will be in proportion to its size (again, see Case Study 2).

But, conceptually, what does the function cos y + i sin y have to do with the number e?

    a. Any function $f(y)$ that maps sums onto products must map 0 onto 1 and 1 onto some base.
    b. The rate of change of such a function $f(y)$ is proportional to its size.
    c. When the rate is exactly equal to its size, the function maps 1 onto e.
    d. Therefore, cos y + i sin y maps 1 onto e.

Why is the function cos y + i sin y periodic and self-regulating?

    a. Sine and cosine are both periodic functions, since they are conceptualized via the Trigonometry metaphor in terms of the unit circle and the Recurrence Is Circularity metaphor (see Case Study 1).
    b. The rate of change of sin y is cos y, and the rate of change of cos y is –sin y.
    c. As a result, the second derivative of cos y + i sin y is itself multiplied by –1 and the fourth derivative is itself multiplied by 1.
    d. That means that the acceleration (the second derivative) of the function is the negative of the function itself, which means that the function cannot get out of control and get progressively greater. Moreover, the acceleration of that negative acceleration (the fourth derivative) is the function itself. Thus, if the function is controlled, its negative acceleration is controlled.

Thus, conceptually, both e^{yi} and cos y + i sin y are functions that map sums onto products, change in exact proportion to their size, are periodic, and are self-regulating.

Though they are characterized by different combinations of concepts, they entail, *by virtue of what they mean*, exactly the same properties. From the perspective of mathematical idea analysis, this is why they are equal. As a consequence, these functions perform exactly the same mappings of complex numbers onto complex numbers.

The significance of $e^{\pi i}$ +1 = 0 is thus a conceptual significance. What is important is not just the numerical values of e, π, i, 1, and 0 but their *conceptual* meaning. After all, e, π, i, 1, and 0 are not just numbers like any other numbers. Unlike, say, 192,563,947.9853294867, these numbers have conceptual meanings in a system of common, important nonmathematical concepts, like change, acceleration, recurrence, and self-regulation.

They are not mere numbers; they are the *arithmetizations of concepts*. When they are placed in a formula, the formula incorporates the ideas the function expresses as well as the set of pairs of complex numbers it mathematically determines by virtue of those ideas.

Because arithmetization works via conceptual metaphor, the conceptual inferences are thereby expressed in the arithmetic—if you understand the arithmetization metaphors.

Why do e, π, i, 1, and 0 come up all the time when we do mathematics, while most numbers (e.g., 192,563,947.9853294867) do not?

The reason is that these numbers express common and important concepts via arithmetization metaphors. Those concepts, like recurrence, rotation, change, and self-regulation are important in our everyday life.

From the perspective of a mathematics where concepts have no place— in which numbers are just numbers—e, π, i, 1, and 0 have no conceptual significance at all and each has the same status that a number like 192,563,947.9853294867 has. It is the ideas that these numbers express that make them significant. And it is the combination of these common, important ideas taken together that makes the truth of the equation $e^{\pi i} + 1 = 0$ significant.

It is no accident that our branches of mathematics are linked in the way they are. Those conceptual connections—via conceptual metaphors and blends—express ideas that matter to us. The way the branches of mathematics are interrelated is a consequence of what is important to us in our everyday lives and how we conceptualize those concerns. The various branches of mathematics mathematicize our concerns. The mechanism for doing so is the all too human conceptual metaphor and blending.

References

Mathematical Cognition and Mathematics Education

Anderson, S., A. Damasio, & H. Damasio (1990). Troubled letters but not numbers: Domain specific cognitive impairments following focal damage in frontal cortex. *Brain*, 113: 749–766.

Antell, S. E., & D. P. Keating (1983). Perception of numerical invariance in neonates. *Child Development*, 54: 695–701.

Baillargeon, R. (1994). Physical reasoning in young infants: Seeking explanations for impossible events. *British Journal of Developmental Psychology*, 12: 9–33.

Beth, E. W., & J. Piaget (1961). *Epistémologie mathématique et psychologie. Essai sur les relations entre la logique formelle et la pensée réelle.* Paris: P.U.F.

Bideaud, J. (1996). La construction du nombre chez le jeune enfant: Une bonne raison d'affûter le rasoir d'Occam. *Bulletin de Psychologie*, 50 (427): 19–28.

Bijeljac-Babic, R., J. Bertoncini, & J. Mehler (1991). How do four-day-old infants categorize multisyllabic utterances? *Developmental Psychology*, 29: 711–721.

Boysen, S. T., & G. G. Berntson (1996). Quantity-based interference and symbolic representations in chimpanzees *(Pan troglodytes)*. *Journal of Experimental Psychology: Animal Behavior Processes*, 22: 76–86.

Boysen, S. T., & E. J. Capaldi (eds.) (1993). *The Development of Numerical Competence: Animal and Human Models.* Hillsdale, N.J.: Erlbaum.

Brainerd, C. (1982). *Children's Logical and Mathematical Cognition: Progress in Cognitive Development Research.* New York: Springer-Verlag.

Butterworth, B. (1999). *What Counts: How Every Brain is Hardwired for Math.* New York: Free Press.

Changeux, J.-P., & A. Connes (1995). *Conversations on Mind, Matter, and Mathematics.* Princeton: Princeton University Press.

Chi, M. T., & D. Klahr (1975). Span and rate of apprehension in children and adults. *Journal of Experimental Child Psychology*, 19: 434–439.

Chiu, M. (1996). *Building Mathematical Understanding During Collaboration.* Ph.D. dissertation, School of Education, University of California at Berkeley.

Church, R. M., & W. H. Meck (1984). The numerical attribute of stimuli. In H. L. Roitblat, T. G. Bever, & H. S. Terrace (eds.), *Animal Cognition.* Hillsdale, N.J.: Erlbaum.

Cipolotti, L., E. K. Warrington, & B. Butterworth (1995). Selective impairment in manipulating Arabic numerals. *Cortex*, 31: 73–86.

Davis, H., & R. Pérusse (1988). Numerical competence in animals: Definitional issues, current evidence, and new research agenda. *Behavioral and Brain Sciences*, 11: 561–615.

Dehaene, S. (1996). The organization of brain activations in number comparison: Event-related potentials and the additive-factors method. *Journal of Cognitive Neuroscience*, 8: 47–68.

———. (1997). *The Number Sense: How the Mind Creates Mathematics*. New York: Oxford University Press.

Dehaene, S., & L. Cohen (1991). Two mental calculation systems: A case study of severe acalculia with preserved approximation. *Neuropsychologia*, 29: 1045–1074.

———. (1994). Dissociable mechanisms of subitizing and counting: Neuropsychological evidence from simultanagnosic patients. *Journal of Experimental Psychology: Human Perception and Performance*, 20: 958–975.

Dehaene, S., G. Dehaene-Lambertz, & L. Cohen (1998). Abstract representations of numbers in the animal and human brain. *Trends in Neuroscience*, 21: 355–361.

Dehaene, S., E. Spelke, P. Pinel, R. Stanescu, & S. Tsivkin (1999). Sources of mathematical thinking: Behavioral and brain-imaging evidence. *Science*, 284: 970–974.

Deloche, G., & X. Serone (eds.) (1987). *Mathematical Disabilities: A Cognitive Neuropsychological Perspective*. Hillsdale, N.J.: Erlbaum.

Desanti, J. T. (1967). Une crise de développement exemplaire: La "découverte" des nombres irrationels. In J. Piaget (ed.), *Logique et connaissance scientifique*. Encyclopédie de la Pléiade: 439–464. Paris: Gallimard.

Duval, R. (1983). L'obstacle du dedoublement des objets mathématiques. *Educational Studies in Mathematics*, 14: 385–414.

Edwards, L., & R. Núñez (1995). Cognitive science and mathematics education: A non-objectivist view. *Proceedings of the 19th Conference of the International Group for the Psychology of Mathematics Education*, 2: 240–247.

Evans, D. (1984). *Understanding Zero and Infinity in the Early School Years*. Ph.D. dissertation, University of Pennsylvania. *Dissertation Abstracts International*, 44, 2265B.

Fischbein, E., D. Tirosh, & P. Hess (1979). The intuition of infinity. *Educational Studies in Mathematics*, 10: 3–40.

Freudenthal, H. (1983). *Didactical Phenomenology of Mathematical Structures*. Dordrecht: D. Reidel.

Gallistel, C. R. (1988). Counting versus subitizing versus the sense of number. *Behavioral and Brain Sciences*, 11 (4): 565–586.

Gelman, R., & C. R. Gallistel (1978). *The Child's Understanding of Number*. Cambridge: Harvard University Press.

Giaquinto, M. (1995). Concepts and calculation. *Mathematical Cognition*, 1 (1): 61–81.

Ginsburg, H. P. (ed.) (1983). *The Development of Mathematical Thinking*. New York: Academic Press.

Hauser, M. D., P. MacNeilage, & M. Ware (1996). Numerical representations in primates. *Proceedings of the National Academy of Sciences USA*, 93: 1514–1517.

Kaufmann, E. L., M. W. Lord, T. W. Reese, & J. Volkmann (1949). The discrimination of visual number. *American Journal of Psychology*, 62: 498–525.

Koechlin, E., S. Dehaene, & J. Mehler (1997). Numerical transformations in five-month-old human infants. *Mathematical Cognition*, 3 (2): 89–104.

Lancy, D. (1983). *Cross-Cultural Studies in Cognition and Mathematics*. New York: Academic Press.

Langford, P. E. (1974). Development of concepts of infinity and limit in mathematics. *Archives de Psychologie*, 42: 311–322.

Lave, J. (1988). *Cognition in Practice: Mind, Mathematics and Culture in Everyday Life*. New York: Cambridge University Press.

Lesh, R., & M. Landau (eds.) (1983). *Acquisition of Mathematics Concepts and Processes*. New York: Academic Press.

Longo, G. (1997). Géométrie, mouvement, espace: Cognition et mathématiques. *Intellectica*, 25: 195–218.

———. (1998). The mathematical continuum: From intuition to logic. In J. Petitot et al. (eds.), *Naturalizing Phenomenology: Issues in Contemporary Phenomenology and Cognitive Science*. Stanford: Stanford University Press.

———. (1999). Mathematical intelligence, infinity, and machines. In R. Núñez & W. J. Freeman (eds.), *Reclaiming Cognition: The Primacy of Action, Intention, and Emotion*. Thorverton, U.K.: Imprint Academic.

Mandler, G., & B. J. Shebo (1982). Subitizing: An analysis of its component processes. *Journal of Experimental Psychology: General*, 111 (1): 1–22.

Mangan, F., & G. E. Reyes (1994). Category theory as a conceptual tool in the study of cognition. In J. MacNamara & G. E. Reyes (eds.), *The Logical Foundations of Cognition*. New York: Oxford University Press.

Matsuzawa, T. (1985). Use of numbers by a chimpanzee. *Nature*, 315: 57–59.

———. (1997). Nineteen years with Ai. Videotape. Japan: NHK.

Mechner, F. (1958). Probability relations within response sequences under ratio reinforcement. *Journal of the Experimental Analysis of Behavior*, 1: 109–121.

Mechner, F., & L. Guevrekian (1962). Effects of deprivation upon counting and timing in rats. *Journal of the Experimental Analysis of Behavior*, 5: 463–466.

Moreno, L. E., & G. Waldegg (1991). The conceptual evolution of actual mathematical infinity. *Educational Studies in Mathematics*, 22: 211–231.

Nesher, P., & J. Kilpatrick (eds.) (1990). *Mathematics and Cognition: A Research Synthesis by the International Group for the Psychology of Mathematics Education*. New York: Cambridge University Press.

Nunes, T., A. D. Schliemann, & D. W. Carraher (1993). *Street Mathematics and School Mathematics*. New York: Cambridge University Press.

Núñez, R. (1990). Infinity in mathematics as a scientific subject matter for cognitive psychology. *Proceedings of the 14th Conference of the International Group for the Psychology of Mathematics Education*, 1: 77–84.

———. (1993a). Approaching infinity: A view from cognitive psychology. *Proceedings of the 15th Annual Meeting of the International Group for the Psychology of Mathematics Education*, North American chapter, 1: 105–111. Asilomar, California.

———. (1993b). Big and small infinities: Psycho-cognitive aspects. *Proceedings of the 17th International Conference for the Psychology of Mathematics Education*, 2: 121–128. Tsukuba, Japan.

_____. (1993c). *En deçà du transfini: Aspects psychocognitifs sous-jacents au concept d'infini en mathématiques.* Fribourg, Switzerland: Éditions Universitaires.

_____. (1994). Cognitive development and infinity in the small: Paradoxes and consensus. In A. Ram & K. Eiselt (eds.), *Proceedings of the 16th Annual Conference of the Cognitive Science Society.* Hillsdale, N.J.: Erlbaum.

_____. (1997). Infinito en lo pequeño y desarrollo cognitivo: Paradojas y espacios consensuales. *Educación Matemática,* 9 (1): 20–32.

_____. (2000). Mathematical idea analysis: What embodied cognitive science can say about the human nature of mathematics. Opening plenary address in *Proceedings of the 24th International Conference for the Psychology of Mathematics Education,* 1:3–22. Hiroshima, Japan.

_____. (in press). Conceptual metaphor and the embodied mind: What makes mathematics possible? In F. Hallyn (ed.), *Metaphor and Analogy in the History and Philosophy of Science.* Dordrecht: Kluwer.

Núñez, R., L. Edwards, & J. F. Matos (1999). Embodied cognition as grounding for situatedness and context in mathematics education. *Educational Studies in Mathematics,* 39 (1–3): 45–65.

Núñez, R., & G. Lakoff (1998). What did Weierstrass really define? The cognitive structure of natural and ε–δ continuity. *Mathematical Cognition,* 4 (2): 85–101.

Piaget, J. (1952). *The Child's Conception of Number.* New York: W. W. Norton.

_____. (1954). *The Construction of Reality in the Child.* New York: Basic Books.

Piaget, J., & B. Inhelder (1948/1956). *The Child's Conception of Space.* London: Routledge.

Powel, A. B., & M. Frankenstein (eds.) (1997). *Ethnomathematics: Challenging Eurocentrism in Mathematics Education.* Albany: State University of New York.

Radford, L. (1997). On psychology, historical epistemology, and the teaching of mathematics: Towards a socio-cultural history of mathematics. *For the Learning of Mathematics,* 17 (1): 26–33.

Rey, A. (1944). Le problème des "quantités limites" chez l'enfant. *Revue Suisse de Psychologie,* 2: 238–249.

Robert, A. (1982). L'acquisition de la notion de convergence de suites numériques dans l'enseignement supérieur. *Recherches en Didactique des Mathématiques,* 3: 307–341.

Sierpinska, A. (1985). Obstacles épistémologiques relatifs à la notion de limite. *Recherches en Didactique des Mathématiques,* 6: 5–67.

Simon, T. J., S. J. Hespos, & P. Rochat (1995). Do infants understand simple arithmetic? A replication of Wynn (1992). *Cognitive Development,* 10: 253–269.

Skemp, R. R. (1971). *The Psychology of Learning Mathematics.* London: Penguin.

Starkey, P., & R. G. Cooper (1980). Perception of numbers by human infants. *Science,* 210: 1033–1035.

Starkey, P., E. S. Spelke, & R. Gelman (1983). Detection of intermodal numerical correspondences by human infants. *Science,* 222: 179–181.

_____. (1990). Numerical abstraction by human infants. *Cognition,* 36: 97–127.

Strauss, M. S., & L. E. Curtis (1981). Infant perception of numerosity. *Child Development,* 52: 1146–1152.

Taback, S. (1975). The child's concept of limit. In M. F. Rosskopf (ed.), *Children's Mathematical Concepts: Six Piagetian Studies in Mathematics Education.* New York: Teachers College Press.

Tall, D. (1980). The notion of infinite measuring number and its relevance in the intuition of infinity. *Educational Studies in Mathematics*, 11: 271–284.

Tall, D., & S. Vinner (1981). Concept image and concept definition in mathematics with particular reference to limits and continuity. *Educational Studies in Mathematics*, 12: 151–169.

Trick, L. M., & Z. W. Pylyshyn (1993). What enumeration studies can show us about spatial attention: Evidence for limited capacity preattentive processing. *Journal of Experimental Psychology: Human Perception and Performance*, 19: 331–351.

_____. (1994). Why are small and large numbers enumerated differently? A limited capacity preattentive stage in vision. *Psychological Review*, 100: 80–102.

van Loosbroek, E., & A. W. Smitsman (1990). Visual perception of numerosity in infancy. *Developmental Psychology*, 26: 916–922.

Wells, C. (1999). *A Handbook of Mathematical Discourse*. Electronic version available at: www.cwru.edu/artsci/math/wells/pub/abouthbk.htm.

Woodruff, G., & D. Premack (1981). Primative (sic) mathematical concepts in the chimpanzee: Proportionality and numerosity. *Nature*, 293: 568–570.

Wynn, K. (1990). Children's understanding of counting. *Cognition*, 36: 155–193.

_____. (1992a). Addition and subtraction by human infants. *Nature*, 358: 749–750.

_____. (1992b). Evidence against empiricist accounts of numerical knowledge. *Mind and Language*, 7 (4): 315–332.

_____. (1995). Origins of numerical knowledge. *Mathematical Cognition*, 1 (1): 35–60.

Xu, F., & S. Carey (1996). Infants' metaphysics: The case of numerical identity. *Cognitive Psychology*, 30 (2): 111–153.

Embodied Mind

Bateson, G. (1979). *Mind and Nature: A Necessary Unity*. New York: Bantam.

Berthoz, A. (1997). *Le sens du mouvement*. Paris: Odile Jacob.

Csordas, T. J. (1994). *Embodiment and Experience: The Existential Ground of Culture and Self*. New York: Cambridge University Press.

Damasio, A. R. (1994). *Descartes' Error: Emotion, Reason, and the Human Brain*. New York: Putnam.

Edelman, G. (1992). *Bright Air, Brilliant Fire: On the Matter of the Mind*. New York: Basic Books.

_____. (1987). *Neural Darwinism: The Theory of Neuronal Group Selection*. New York: Basic Books.

Freeman, W. J. (1995). *Societies of Brains: A Study in the Neuroscience of Love and Hate*. Hillsdale, N.J.: Erlbaum.

_____. (1999). *How Brains Make Up Their Minds*. London: Weidenfeld & Nicolson.

Freeman, W. J., & R. Núñez (1999). Restoring to cognition the forgotten primacy of action, intention, and emotion. In R. Núñez & W. J. Freeman (eds.), *Reclaiming Cognition: The Primacy of Action, Intention, and Emotion*. Thorverton, U.K.: Imprint Academic.

Husserl, E. (1928/1964). *Leçons pour une phénoménologie de la conscience intime du temps*. [Vorlesungen zur Phänomenologie des inneren Zeitbewusstseins]. Paris: P.U.F.

Iverson, J. M., & E. Thelen (1999). Hand, mouth, and brain: The dynamic emergence of speech and gesture. In R. Núñez & W. J. Freeman (eds.), *Reclaiming Cognition: The Primacy of Action, Intention, and Emotion*. Thorverton, U.K.: Imprint Academic.

Johnson, M. (1993). Conceptual metaphor and embodied structures of meaning. *Philosophical Psychology*, 6 (4): 413–422.

_____. (1987). *The Body in the Mind: The Bodily Basis of Meaning, Imagination, and Reason.* Chicago: University of Chicago Press.

Lakoff, G. (1987). *Women, Fire, and Dangerous Things: What Categories Reveal about the Mind.* Chicago: University of Chicago Press.

Lakoff, G., & M. Johnson (1999). *Philosophy in the Flesh.* New York: Basic Books.

Lock, M. (1993). Cultivating the body: Anthropology and epistemologies of bodily practice and knowledge. *Annual Review of Anthropology*, 22: 133–155.

Maturana, H. (1969). The neurophysiology of cognition. In P. Garvin (ed.), *Cognition: A Multiple View.* New York: Spartan Books.

_____. (1975). The organization of the living: A theory of the living organization. *International Journal of Man-Machine Studies*, 7: 313–332.

Maturana, H., & F. Varela (1980). *Autopoiesis and Cognition: The Realization of the Living.* Boston: Reidel.

_____. (1987). *The Tree of Knowledge: The Biological Roots of Human Understanding.* Boston: New Science Library.

McNeill, D. (1992). *Hand and Mind: What Gestures Reveal About Thought.* Chicago: Chicago University Press.

Merleau-Ponty, M. (1945/1994). *Phenomenology of Perception.* London: Routledge.

_____. (1964). *L'oeil et l'esprit.* Paris: Gallimard.

Narayanan, S. (1997). *Embodiment in Language Understanding: Sensory-Motor Representations for Metaphoric Reasoning about Event Descriptions.* Ph.D. dissertation, Department of Computer Science, University of California at Berkeley.

Núñez, R. (1995). What brain for God's-eye? Biological naturalism, ontological objectivism, and Searle. *Journal of Consciousness Studies*, 2 (2): 149–166.

_____. (1997). Eating soup with chopsticks: Dogmas, difficulties, and alternatives in the study of conscious experience. *Journal of Consciousness Studies*, 4 (2): 143–166.

_____. (1999). Could the future taste purple? Reclaiming mind, body, and cognition. In R. Núñez & W. J. Freeman (eds.), *Reclaiming Cognition: The Primacy of Action, Intention, and Emotion.* Thorverton, U.K.: Imprint Academic.

Núñez, R., L. Edwards, & J. F. Matos (1999). Embodied cognition as grounding for situatedness and context in mathematics education. *Educational Studies in Mathematics*, 39 (1–3): 45–65.

Núñez, R., & W. J. Freeman (eds.) (1999). *Reclaiming Cognition: The Primacy of Action, Intention, and Emotion.* Thorverton, U.K.: Imprint Academic.

Pfeifer, R., & C. Scheier (1999). *Understanding Intelligence.* Cambridge: MIT Press.

Regier, T. (1996). *The Human Semantic Potential.* Cambridge: MIT Press.

Thelen, E. (1995a). Motor development: A new synthesis. *American Psychologist*, 50 (2): 79–95.

_____. (1995b). Time-scale dynamics and the development of an embodied cognition. In R. F. Port & T. van Gelder (eds.), *Mind as Motion: Explorations in the Dynamics of Cognition.* Cambridge: MIT Press.

Thelen, E., & L. Smith (1994). *A Dynamic Systems Approach to the Development of Cognition and Action.* Cambridge: MIT Press.

Thompson, E. (1995). *Colour Vision: A Study in Cognitive Science and the Philosophy of Perception.* London: Routledge.

Varela, F. (1979). *Principles of Biological Autonomy.* New York: Elsevier North-Holland.

_____. (1989). *Connaître les sciences cognitives: Tendances et perspectives.* Paris: Seuil.

Varela, F., E. Thompson, & E. Rosch (1991). *The Embodied Mind: Cognitive Science and Human Experience.* Cambridge: MIT Press.

von Foerster, H. (1974). Notes for an epistemology of living beings. In E. Morin & M. Piatelli (eds.), *L'unité de l'homme.* Paris: Seuil.

Winograd, T., & F. Flores (1986). *Understanding Computers and Cognition.* Norwood, N.J.: Ablex-Wesley.

Cognitive Linguistics and Conceptual Metaphor Theory

Fauconnier, G. (1985). *Mental Spaces: Aspects of Meaning Construction in Natural Language.* Cambridge: MIT Press.

_____. (1997). *Mappings in Thought and Language.* New York: Cambridge University Press.

Fauconnier, G., & E. Sweetser (1996). *Spaces, Worlds, and Grammar.* Chicago: University of Chicago Press.

Fauconnier, G., & M. Turner (1998). Conceptual integration networks. *Cognitive Science,* 22 (2): 133–187.

Fillmore, C. (1982). Frame semantics. In Linguistic Society of Korea (eds.), *Linguistics in the Morning Calm.* Seoul: Hanshin.

_____. (1985). Frames and the semantics of understanding. *Quaderni di Semantica,* 6: 222–253.

Gentner, D., & D. R. Gentner (1982). Flowing waters or teeming crowds: Mental models of electricity. In D. Gentner & A. L. Stevens (eds.), *Mental Models.* Hillsdale, N.J.: Erlbaum.

Gibbs, R. (1994). *The Poetics of Mind: Figurative Thought, Language, and Understanding.* New York: Cambridge University Press.

Gibbs, R., & J. O'Brien (1990). Idioms and mental imagery: The metaphorical motivation for idiomatic meaning. *Cognition,* 36: 35–68.

Gibbs, R., & G. Steen (eds.) (1999). *Metaphor in Cognitive Linguistics.* Amsterdam: John Benjamins.

Grady, J. (1997). *Foundations of Meaning: Primary Metaphors and Primary Scenes.* Ph.D. dissertation, Linguistics Department, University of California at Berkeley.

_____. (1998). The conduit metaphor revisited: A reassessment of metaphors for communication. In J. P. Koenig (ed.), *Discourse and Cognition: Bridging the Gap.* Stanford: CSLI/Cambridge.

Johnson, C. (1997). Metaphor vs. conflation in the acquisition of polysemy: The case of SEE. In M. K. Hiraga, C. Sinha, & S. Wilcox (eds.), *Cultural, Typological and Psychological Issues in Cognitive Linguistics.* Current Issues in Linguistic Theory. Amsterdam: John Benjamins.

Johnson, M. (1987). *The Body in the Mind: The Bodily Basis of Meaning, Imagination, and Reason.* Chicago: University of Chicago Press.

Johnson, M. (ed.) (1981). *Philosophical Perspectives on Metaphor.* Minneapolis: University of Minnesota Press.

Kövecses, Z. (1988). *The Language of Love: The Semantics of Passion in Conversational English.* Lewisburg, Pa.: Bucknell University Press.

Lakoff, G. (1993). The contemporary theory of metaphor. In A. Ortony (ed.), *Metaphor and Thought*, 2nd ed. New York: Cambridge University Press.

Lakoff, G., & M. Johnson (1980). *Metaphors We Live By.* Chicago: University of Chicago Press.

Lakoff, G., & R. Núñez (1997). The metaphorical structure of mathematics: Sketching out cognitive foundations for a mind-based mathematics. In L. English (ed.), *Mathematical Reasoning: Analogies, Metaphors, and Images.* Mahwah, N.J.: Erlbaum.

———. (1998). Conceptual metaphor in mathematics. In J. P. Koenig (ed.), *Discourse and Cognition: Bridging the Gap.* Stanford: CSLI/Cambridge.

Lakoff, G., & M. Turner (1989). *More than Cool Reason: A Field Guide to Poetic Metaphor.* Chicago: University of Chicago Press.

Núñez, R., & G. Lakoff (1998). What did Weierstrass really define? The cognitive structure of natural and ε–δ continuity. *Mathematical Cognition,* 4 (2): 85–101.

Núñez, R., V. Neumann, & M. Mamani (1997). Los mapeos conceptuales de la concepción del tiempo en la lengua Aymara del norte de Chile. [Conceptual mappings in the understanding of time in northern Chile's Aymara]. *Boletín de Educación de la Universidad Católica del Norte,* 28: 47–55.

Reddy, M. (1979). The conduit metaphor. In A. Ortony (ed.), *Metaphor and Thought.* New York: Cambridge University Press.

Sweetser, E. (1990). *From Etymology to Pragmatics: Metaphorical and Cultural Aspects of Semantic Structure.* New York: Cambridge University Press.

Talmy, L. (1988). Force dynamics in language and cognition. *Cognitive Science,* 12: 49–100.

———. (1996). Fictive motion in language and "ception." In P. Bloom, M. Peterson, L. Nadel, & M. Garrett (eds.), *Language and Space.* Cambridge: MIT Press.

———. (2000). *Toward a Cognitive Linguistics.* Cambridge: MIT Press.

Taub, S. (1997). *Language in the Body: Iconicity and Metaphor in American Sign Language.* Ph.D. dissertation, Linguistics Department, University of California at Berkeley.

Turner, M. (1991). *Reading Minds: The Study of English in the Age of Cognitive Science.* Princeton: Princeton University Press.

———. (1996). *The Literary Mind.* New York: Oxford University Press.

Turner, M., & G. Fauconnier (1995). Conceptual integration and formal expression. *Journal of Metaphor and Symbolic Activity,* 10 (3): 183–204.

Winter, S. (1989). Transcendental nonsense, metaphoric reasoning and the cognitive stakes for law. *University of Pennsylvania Law Review,* 137.

Yu, N. (1998). *The Contemporary Theory of Metaphor: A Perspective from Chinese.* Amsterdam: John Benjamins.

Mathematics

Aczel, P. (1988). *Non-Well-Founded Sets.* Stanford: CSLI.

Adler, I. (1958). *The New Mathematics.* New York: New American Library.

Ahlfors, L. V. (1979). *Complex Analysis: An Introduction to the Theory of Analytic Functions of One Complex Variable.* New York: McGraw-Hill.

Alexandroff, P. (1961). *Elementary Concepts of Topology.* New York: Dover.

Apostol, T. M. (1957). *Mathematical Analysis: A Modern Approach to Advanced Calculus.* Reading, Mass.: Addison-Wesley.

Aubin, J.-P. (1993). *Optima and Equilibria: An Introduction to Nonlinear Analysis.* Berlin: Springer-Verlag.

Barr, S. (1964). *Experiments in Topology.* New York: Thomas Crowell.

Barwise, J., & L. Moss (1991). Hypersets. *The Mathematical Intelligencer,* 13 (4): 31–41.

Bell, J. L., & A. B. Slomson (1969). *Models and Ultraproducts: An Introduction.* Amsterdam: North-Holland.

Berlekamp, E. R. (1988). Blockbusting and domineering. *Journal of Combinatorial Theory, Series A,* 49: 67–116.

Berlekamp, E. R., J. H. Conway, & R. K. Guy (1982). *Winning Ways for Your Mathematical Plays.* London: Academic Press.

Berlekamp, E. R., & D. Wolfe (1994). *Mathematical Go: Chilling Gets the Last Point.* Wellesley, Mass.: A. K. Peters.

Blackwell, D., & M. A. Girshick (1954). *Theory of Games and Statistical Decisions.* New York: Dover.

Bolzano, B. (1847/1950). *Paradoxes of the Infinite* [Paradoxien des Unendlichen]. London: Routledge & Kegan Paul.

Boole, G. (1854/1958). *An Investigation of the Laws of Thought on Which Are Founded the Mathematical Theories of Logic and Probabilities.* New York: Dover.

Bronshtein, I. N., & K. A. Semendyayev (1985). *Handbook of Mathematics.* New York: Van Nostrand Reinhold.

Buck, R. C. (1956). *Advanced Calculus.* New York: McGraw-Hill.

Carnap, R. (1958). *Introduction to Symbolic Logic and Its Applications.* New York: Dover.

Chang, C. C., & H. J. Keisler (1990). *Model Theory.* Amsterdam: Elsevier.

Clapham, C. (1990). *A Concise Oxford Dictionary of Mathematics.* New York: Oxford University Press.

Cohen, P. J. (1966). *Set Theory and the Continuum Hypothesis.* New York: W. A. Benjamin.

Combès, M. (1971). *Fondements des mathématiques.* Paris: P.U.F.

Conway, J. H. (1976). *On Numbers and Games.* London: Academic Press.

Copi, I. M. (1954). *Symbolic Logic.* New York: Macmillan.

Courant, R., & H. Robbins (1941). *What Is Mathematics? An Elementary Approach to Ideas and Methods.* London: Oxford University Press.

Davis, M. (1958). *Computability and Unsolvability.* New York: McGraw-Hill.

———. (1977). *Applied Nonstandard Analysis.* New York: John Wiley & Sons.

Davis, M., & R. Hersh (1972). Nonstandard analysis. *Scientific American,* June: 78–86.

Dedekind, R. (1888/1976). Was sind und was sollen die Zahlen. In P. Dugac, *Richard Dedekind et les fondements des mathématiques.* Paris: J. Vrin.

———. (1872/1901). *Essays on the Theory of Numbers.* La Salle, Ill.: Open Court.

Der kleine Duden (1986). *Mathematik.* Mannheim: Bibliographisches Institut.

Descartes, R. (1637/1954). *The Geometry of René Descartes with a Facsimile of the First Edition.* New York: Dover.

Dilke, O. A. W. (1987). *Mathematics and Measurement.* Berkeley: University of California Press.

Euler, L. (1748/1988). *Introduction to Analysis of the Infinite.* New York: Springer-Verlag.

———. (1770/1984). *Elements of Algebra.* New York: Springer-Verlag.

Feferman, S. (1987). Infinity in mathematics: Is Cantor necessary? In *L'infinito nella scienza,* Istituto della Enciclopedia Italiana, pp. 151–210.

_____. (1993). Working foundations—'91. In G. Corsi et al. (eds.), *Bridging the Gap: Philosophy, Mathematics, and Physics.* Dordrecht: Kluwer.

Frege, G. (1893/1980). *The Foundations of Arithmetic.* Evanston, Ill.: Northwestern University Press.

_____. (1964). *The Basic Laws of Arithmetic.* Berkeley: University of California Press.

Galuzzi, M. (1979). Funzioni. *Enciclopedia,* 6: 432–497. Torino: Giulio Einaudi.

Godement, R. (1984). Calcul infinitésimal. B: Calcul à une variable. *Encyclopedia Universalis,* 754–762. Paris: Encyclopedia Universalis.

Guy, R. K. (1989). *Fair Game: How to Play Impartial Combinatorial Games.* Arlington, Mass.: COMAP.

Guy, R. K. (ed.) (1991). Combinatorial games. *Proceedings of the Symposia in Applied Mathematics.* Providence: American Mathematical Society.

Hahn, H. (1968). Geometry and intuition. In *Readings from* Scientific American: *Mathematics in the Modern World.* San Francisco: W. H. Freeman.

Halmos, P. R. (1960). *Naive Set Theory.* New York: Van Nostrand Reinhold.

Hardy, G. H. (1955). *A Course of Pure Mathematics.* New York: Cambridge University Press.

Heath, T. L. (1908/1956). *The Thirteen Books of Euclid's Elements,* 3 vols. New York: Dover.

_____. (1931/1963). *A Manual of Greek Mathematics.* New York: Dover.

Hempel, C. G. (1945). On the nature of mathematical truth. *The American Mathematical Monthly,* 52: 543–556.

Hersh, R. (1978). Introducing Imre Lakatos. *The Mathematical Intelligencer,* 1 (3): 148–151.

Heyting, A. (1968). *Logic and Foundations of Mathematics.* New York: Gordon & Breach Science.

Hight, D. W. (1977). *A Concept of Limits.* New York: Dover.

Hilbert, D. (1925). Über das Unendliche. *Mathematische Annalen,* 95: 161–190.

_____. (1971). *Foundations of Geometry* (Grundlagen der Geometrie). La Salle, Ill.: Open Court.

Hogg, R. V., & A. T. Craig (1978). *Introduction to Mathematical Statistics.* New York: Macmillan.

Hughes, P., & G. Brecht (1979). *Vicious Circles and Infinity: An Anthology of Paradoxes.* New York: Penguin.

Kamke, E. (1950). *Theory of Sets.* New York: Dover.

Kasner, E., & J. Newman (1940). *Mathematics and the Imagination.* New York: Simon & Schuster.

Keisler, H. J. (1976a). *Elementary Calculus.* Boston: Prindle, Weber, & Schmidt.

_____. (1976b). *Foundations of Infinitesimal Calculus.* Boston: Prindle, Weber, & Schmidt.

Keisler, J. (1994). The hyperreal number. In P. Ehrlich (ed.), *Real Numbers, Generalizations of the Reals, and the Continua.* Dordrecht: Kluwer.

Kleene, S. C. (1967). *Mathematical Logic.* New York: John Wiley & Sons.

Kleiner, I., & N. Movshovitz-Hadar (1997). Proof: A many-splendored thing. *The Mathematical Intelligencer,* 19 (3): 16–26.

Knopp, K. (1945). *Theory of Functions.* New York: Dover.

Lakatos, I. (1978). Cauchy and the Continuum: The significance of non-standard analysis for the history and philosophy of mathematics. *The Mathematical Intelligencer,* 1 (3): 151–161.

Landau, E. (1957). *Foundations of Analysis.* New York: Chelsea.

Lauwerier, H. (1991). *Fractals: Endlessly Repeated Geometrical Figures*. Princeton: Princeton University Press.

Lavine, S. (1994). *Understanding the Infinite*. Cambridge: Harvard University Press.

Lawvere, F. W. (1969). Adjointness in foundations. *Dialectica*, 23 (3/4): 282–296.

_____. (1992). Categories of space and of quantity. In J. Echeverria, A. Ibarra, & T. Mormann (eds.), *The Space of Mathematics: Philosophical, Epistemological, and Historical Explorations*. Berlin: Walter de Gruyter.

Lawvere, F. W., & S. H. Schanuel (1997). *Conceptual Mathematics: A First Introduction to Categories*. New York: Cambridge University Press.

MacLane, S. (1986). *Mathematics, Form and Function*. New York: Springer-Verlag.

Mandelbrot, B. (1983). *The Fractal Geometry of Nature*. New York: W. H. Freeman.

Mansfield, M. J. (1963). *Introduction to Topology*. Princeton: Van Nostrand.

Marsden, J. E. (1974). *Elementary Classical Analysis*. San Francisco: W. H. Freeman.

Massey, G. J. (1970). *Understanding Symbolic Logic*. New York: Harper & Row.

Meder, A. E. (1967). *Topics from Inversive Geometry*. New York: Houghton Mifflin.

Mendelson, B. (1990). *Introduction to Topology*. New York: Dover.

Mendelson, E. (1964). *Introduction to Mathematical Logic*. Princeton: Van Nostrand.

Monk, J. D. (1969). *Introduction to Set Theory*. New York: McGraw-Hill.

Mueller, F. J. (1964). *Arithmetic: Its Structure and Concepts*. Englewood Cliffs, N.J.: Prentice-Hall.

Nagel, E., & J. Newman (1958). *Gödel's Proof*. New York: New York University Press.

Nelson, E. (1977). Internal set theory: A new approach to nonstandard analysis. *Bulletin of the Mathematical Society*, 83 (6): 1165–1198.

Newman, M. H. A. (1992). *Elements of the Topology of Plane Sets of Points*. New York: Dover.

Northrop, E. P. (1944). *Riddles in Mathematics: A Book of Paradoxes*. New York: Van Nostrand.

Ore, O. (1948). *Number Theory and Its History*. New York: McGraw-Hill.

Peitgen, H.-O., H. Jürgens, & D. Saupe (1992). *Fractals for the Classroom: Part 1 and 2. Introduction to Fractals and Chaos*. New York: Springer-Verlag.

Pierpont, J. (1899). On the arithmetization of mathematics. *Bulletin of the American Mathematical Society*, 5: 394–406.

Pinter, C. (1971). *Set Theory*. Reading, Mass.: Addison-Wesley.

Quine, W. V. O. (1965). *Elementary Logic*. New York: Harper & Row.

Reichenbach, H. (1947). *Elements of Symbolic Logic*. New York: Macmillan.

Robert, A. (1989). L'analyse non-standard. In *Encyclopédie philosophique universelle, Vol. I, L'Univers philosophique*. Paris: P.U.F.

Robinson, A. (1966). *Non-Standard Analysis*. Amsterdam: North-Holland.

_____. (1979). *Selected Papers of Abraham Robinson. Vol. 2: Nonstandard Analysis and Philosophy*. New Haven: Yale University Press.

Rosenbloom, P. (1950). *The Elements of Mathematical Logic*. New York: Dover.

Sagan, H. (1993). A geometrization of Lebesgue's space-filling curve. *The Mathematical Intelligencer*, 15 (4): 37–43.

_____. (1994). *Space-Filling Curves*. New York: Springer-Verlag.

Scientific American (1968). *Mathematics in the Modern World: Readings from* Scientific American. San Francisco: W. H. Freeman.

Sierpinski, W. (1975). Sur une courbe dont tout point est un point de ramification. In W. Sierpinski, *Oeuvres choisies*, Vol. 2. Warsaw: Éditions Scientifiques de Pologne (PWN).

_____. (1975). Sur une courbe cantorienne qui contient une image biunivoque et continue de toute courbe donnée. In W. Sierpinski, *Oeuvres choisies*, Vol. 2. Warsaw: Éditions Scientifiques de Pologne (PWN).

Simmons, G. F. (1983). *Introduction to Topology and Modern Analysis*. Malabar, Fla.: Krieger.

_____. (1985). *Calculus with Analytic Geometry*. New York: McGraw-Hill.

Sinaceur, H. (1994). L'infini. *La Recherche*, 25: 904–910.

Steen, L. A., & J. A. Seebach (1970). *Counterexamples in Topology*. New York: Holt, Rinehart & Winston.

Stein, S. K. (1963). *Mathematics, the Man-Made Universe: An Introduction to the Spirit of Mathematics*. San Francisco: W. H. Freeman.

_____. (1982). *Calculus and Analytic Geometry*. New York: McGraw-Hill.

Stewart, B. M. (1952). *Theory of Numbers*. New York: Macmillan.

Stewart, I. (1995). *Concepts of Modern Mathematics*. New York: Dover.

Stewart, I., & D. Tall (1977). *The Foundations of Mathematics*. Oxford: Oxford University Press.

_____. (1983). *Complex Analysis*. New York: Cambridge University Press.

Stoll, R. R. (1961). *Sets, Logic, and Axiomatic Theories*. San Francisco: W. H. Freeman.

Stroyan, K. D., & W. A. J. Luxemburg (1976). *Introduction to the Theory of Infinitesimals*. New York: Academic Press.

Swokowski, E. W. (1988). *Calculus with Analytic Geometry*. Boston: PWS-Kent.

Tall, D. (1980). Looking at graphs through infinitesimal microscopes, windows and telescopes. *The Mathematical Gazette*, 64 (427): 22–49.

Thomas, G. B. (1953). *Calculus and Analytic Geometry*. Reading, Mass.: Addison-Wesley.

Tiles, M. (1989). *The Philosophy of Set Theory: An Introduction to Cantor's Paradise*. New York: Basil Blackwell.

White, A. J. (1973). *Real analysis: An introduction*. Reading, Mass.: Addison-Wesley.

Whitehead, A. N., & B. Russell (1910). *Principia Mathematica*. New York: Cambridge University Press.

Wilder, R. (1952). *Introduction to Foundations of Mathematics*. New York: John Wiley & Sons.

_____. (1967). The role of the axiomatic method. *The American Mathematical Monthly*, 74: 115–127.

Zippin, L. (1962). *Uses of Infinity*. New York: Random House.

Philosophy and History of Mathematics

Aczel, A. D. (1996). *Fermat's Last Theorem: Unlocking the Secret of an Ancient Mathematical Problem*. New York: Dell.

Allard, A. (1995). La révolution arithmétique du Moyen Age. *La Recherche*, 26 (278): 742–748.

Aspray, W., & P. Kitcher (eds.) (1988). *History and Philosophy of Modern Mathematics*. Minneapolis: University of Minnesota Press.

Ball, W.W.R. (1893/1960). *A Short Account of the History of Mathematics*. New York: Dover.

Barbaras, R. (ed.) (1998). *Merleau-Ponty: Notes de cours sur l'origine de la géometrie de Husserl.* Paris. P.U.F.

Barker, S. F. (1964). *Philosophy of Mathematics.* Englewood Cliffs, N.J.: Prentice-Hall.

Baron, M. E. (1987). *The Origins of the Infinitesimal Calculus.* New York: Dover.

Barrow, J. H. (1992). *Pi in the Sky: Counting, Thinking, and Being.* Boston: Little, Brown.

Beckmann, P. (1974). *A History of Pi.* New York: Dorset.

Bell, E. T. (1931). *The Queen of Sciences.* Baltimore: Williams & Wilkins.

_____. (1937). *Men of Mathematics.* New York: Simon & Schuster.

Benacerraf, P., & Putnam, H. (eds.) (1964). *Philosophy of Mathematics: Selected Readings.* Englewood Cliffs, N.J.: Prentice-Hall.

Beth, E. W., & J. Piaget (1961). *Epistémologie mathématique et psychologie: Essai sur les relations entre la logique formelle et la pensée réelle.* Paris: P.U.F.

Blay, M. (1998). *Reasoning with the Infinite: From the Closed World to the Mathematical Universe.* Chicago: University of Chicago Press.

Bloor, D. (1991). *Knowledge and Social Imagery.* Chicago: University of Chicago Press.

Bonola, R. (1955). *Non-Euclidean Geometry: A Critical and Historical Study of Its Developments.* New York: Dover.

Bos, H. J. M. (1981). On the representation of curves in Descartes' *Géométrie. Archive for History of Exact Science,* 24: 295–338.

Bourbaki, N. (1969). *Éléments d'histoire de mathématiques.* Paris: Hermann.

Boyer, C. (1949). *The History of the Calculus and Its Conceptual Development.* New York: Dover.

_____. (1968). *A History of Mathematics.* New York: John Wiley & Sons.

Bruter, C. P. (1987). *De l'intuition à la controverse: Essais sur quelques controverses entre mathématiciens.* Paris: Albert Blanchard.

Cajori, F. (1894). *A History of Mathematics.* New York: Macmillan.

_____. (1928/1993). *A History of Mathematical Notations.* New York: Dover.

Caraça, B.-J. (1998). *Conceitos fundamentais da matemática.* Lisboa: Gradiva.

Cartier, P. (1995). Kepler et la musique du monde. *La Recherche,* 26 (278): 750–755.

Cavaillès, J. (1962). *Philosophie mathématique.* Paris: Hermann.

Changeux, J.-P., & A. Connes (1995). *Conversations on Mind, Matter, and Mathematics.* Princeton: Princeton University Press.

Couturat, L. (1973). *De l'infini mathématique.* Paris: Blanchard.

Dahan-Dalmedico, A., and J. Peiffer (1986). *Une histoire des mathématiques: Routes et dédales.* Paris: Seuil.

Dauben, J. W. (1983). Georg Cantor and the origins of transfinite set theory. *Scientific American,* June: 122–154.

_____. (1979). *Georg Cantor: His Mathematics and Philosophy of the Infinite.* Princeton: Princeton University Press.

Davis, P. J., & R. Hersh (1981). *The Mathematical Experience.* Cambridge: Birkhäuser Boston.

Delahaye, J.-P. (1996). Des jeux infinis et des grands ensembles. *Pour la Science,* June (224): 60–66.

Deledicq, A., & F. Casiro (1997). *Apprivoiser l'infini.* Paris: ACL-Éditions.

Desanti, J. T. (1967). Une crise de développement exemplaire: La "découverte" des nombres irrationels. In J. Piaget (ed.), *Logique et connaissance scientifique.* Encyclopédie de la Pléiade: 439–464. Paris: Gallimard.

_____. (1968). *Les idéalités mathématiques.* Paris: Seuil.

_____. (1984). Infini mathématique. *Encyclopedia Universalis.* Corpus 9: 283–289. Paris: Encyclopedia Universalis.

Dugac, P. (1976). *Richard Dedekind et les fondements des mathématiques.* Paris: Vrin.

Echeverria, J., A. Ibarra, & T. Mormann (eds.) (1992). *The Space of Mathematics: Philosophical, Epistemological, and Historical Explorations.* Berlin: Walter de Gruyter.

Eves, H. (1981). *Great Moments in Mathematics, Vol. 2: After 1650.* New York: Mathematical Association of America.

Falletta, N. (1983). *The Paradoxicon.* Garden City, N.Y.: Doubleday.

Garavaso, P. (1998). *Filosofia della matematica: Numeri e strutture.* Milano: Guerini.

Gardner, M. (1959). *The Scientific American Book of Mathematical Puzzles and Diversions.* New York: Simon & Schuster.

George, A. (ed.) (1994). *Mathematics and Mind.* New York: Oxford University Press.

Gillies, D. (ed.) (1992). *Revolutions in Mathematics.* New York: Oxford University Press.

Glaser, A. (1981). *History of Binary and Other Nondecimal Numeration.* Los Angeles: Tomash.

Gleick, J. (1987). *Chaos: Making a New Science.* New York: Penguin.

Gödel, K. (1964). What is Cantor's continuum problem? In P. Benacerraf & H. Putnam (eds.), *Philosophy of Mathematics: Selected Readings.* Englewood Cliffs, N.J.: Prentice-Hall.

Guillen, M. (1983). *Bridges to Infinity: The Human Side of Mathematics.* Los Angeles, Calif.: Tarcher.

Grattan-Guinness, I., & G. Bornet (1997). *George Boole: Selected Manuscripts on Logic and Its Philosophy.* Cambridge: Birkhäuser Boston.

Hadamard, J. (1945). *An Essay on the Psychology of Invention in the Mathematical Field.* New York: Dover.

Hardy, G. H. (1940/1993). *A Mathematician's Apology.* New York: Cambridge University Press.

Hersh, R. (1986). Some proposals for reviving the philosophy of mathematics. In T. Tymoczko (ed.), *New Directions in the Philosophy of Mathematics.* Cambridge: Birkhäuser Boston.

_____. (1997). *What Is Mathematics, Really?* New York: Oxford University Press.

Hintikka, J. (1969). *The Philosophy of Mathematics.* London: Oxford University Press.

Ifrah, G. (1994). *Histoire universelle des chiffres.* Paris: Robert Laffont.

_____. (1985). *From One to Zero: A Universal Theory of Numbers.* New York: Viking Penguin.

Kitcher, P. (1976). Hilbert's epistemology. *Philosophy of Science,* 43: 99–115.

_____. (1983). *The Nature of Mathematical Knowledge.* New York: Oxford University Press.

_____. (1988). Mathematical naturalism. In W. Aspray & P. Kitcher (eds.), *History and Philosophy of Modern Mathematics.* Minneapolis: University of Minnesota Press.

Kline, J. (1968). *Greek Mathematical Thought and the Origin of Algebra.* New York: Dover.

Kline, M. (1959). *Mathematics and the Physical World.* New York: Dover.

_____. (1962). *Mathematics: A Cultural Approach.* Reading, Mass.: Addison-Wesley.

_____. (1965). *Mathematics in Western Culture.* New York: Oxford University Press.

_____. (1972). *Mathematical Thought from Ancient to Modern Times.* New York: Oxford University Press.

_____. (1980). *Mathematics: The Loss of Certainty.* New York: Oxford University Press.

Knuth, D. E. (1974). *Surreal Numbers.* Reading, Mass.: Addison-Wesley.

Kolgomorov, A. N., & A. P. Yushkevich (eds.) (1992). *Mathematics of the 19th Century: Mathematical Logic, Algebra, Number Theory, Probability Theory.* Basel: Birkhäuser.

Körner, S. (1968). *The Philosophy of Mathematics: An Introductory Essay*. London: Hutchinson.

Kramer, E. (1970). *The Nature and Growth of Modern Mathematics*. New York: Hawthorn.

Kreisel, G. (1964). Hilbert's programme. In P. Benacerraf & H. Putnam (eds.), *Philosophy of Mathematics: Selected Readings*. Englewood Cliffs, N.J.: Prentice-Hall.

Lakatos, I. (1976). *Proofs and Refutations: The Logic of Mathematical Discoveries*. New York: Cambridge University Press.

_____. (1978). *Mathematics, Science and Epistemology*. New York: Cambridge University Press.

Lawvere, F. W. (1990). Some thoughts on the future of category theory. Unpublished manuscript. Department of Mathematics, S.U.N.Y. at Buffalo.

Le Lionnais, F. (ed.) (1948). *Les grands courants de la pensée mathématique*. Paris: Cahiers du Sud.

Le Méhauté, A. (1991). *Fractal Geometries: Theory and Applications*. London: Prenton Press.

Macintosh, W. A. (1995). *The Infinite in the Finite*. New York: Oxford University Press.

MacLane, S. (1981). Mathematical models: A sketch for the philosophy of mathematics. *The American Mathematical Monthly*, August–September, pp. 462–472.

Maddy, P. (1990). *Realism in Mathematics*. New York: Oxford University Press.

Mancosu, P. (1998). *From Brouwer to Hilbert: The Debate on the Foundations of Mathematics in the 1920s*. New York: Oxford University Press.

Maor, E. (1991). *To Infinity and Beyond: A Cultural History of the Infinite*. Princeton: Princeton University Press.

_____. (1994). *e: The Story of a Number*. Princeton: Princeton University Press.

McLaughlin, W. (1994). Resolving Zeno's paradoxes. *Scientific American*, November: 66–71.

Monnoyeur, F. (ed.) (1992). *Infini des mathématiciens, infini des philosophes*. Paris: Éditions Belin.

Moore, A. W. (1990). *The Infinite*. London: Routledge.

_____. (1995). A brief history of infinity. *Scientific American*, April: 112–116.

Motz, L., & J. H. Weaver (1993). *The Story of Mathematics*. New York: Avon Books.

Nahin, P. (1998). *An Imaginary Tale: The Story of $\sqrt{-1}$*. Princeton: Princeton University Press.

Newman, J. R. (1969). *The World of Mathematics*. New York: Simon & Schuster.

Noël, E. (1985). *Le matin des mathématiciens*. Paris: Belin.

Odifreddi, P. (2000). *La matematica del novecento: Dagli insiemi alla complessità*. Torino: Einaudi.

Osserman, R. (1995). *Poetry of the Universe: A Mathematical Exploration of the Cosmos*. New York: Doubleday.

Pascal, B. (1897/1976). *Pensées*. (Text by Léon Brunschvieg). Paris: Garnier, Flammarion.

Pedoe, D. (1958). *The Gentle Art of Mathematics*. New York: Dover.

Péter, R. (1962). *Playing with Infinity: Mathematics for Everyman*. New York: Simon & Schuster.

Piaget, J. (1950). *Introduction à l'épistémologie génétique, Vol. I: La pensée mathématique*. Paris: P.U.F.

Poincaré, H. (1902/1968). *La science et l'hypothèse*. Paris: Flammarion.

_____. (1905/1970). *La valeur de la science*. Paris: Flammarion.

_____. (1913/1963). *Mathematics and Science: Last Essays* [Dernières pensées]. New York: Dover.

Polya, G. (1945). *How to Solve It*. Princeton: Princeton University Press.

Ribet, K. A., & B. Hayes (1994). Fermat's last theorem and modern arithmetic. *American Scientist,* 82: 144–156.

Rotman, B. (1987). *Signifying Nothing: The Semiotics of Zero.* New York: St. Martin's Press.

_____. (1993). *Ad Infinitum: The Ghost in Turing's Machine: Taking God Out of Mathematics and Putting the Body Back In.* Stanford: Stanford University Press.

Rucker, R. (1982). *Infinity and the Mind: The Science and Philosophy of the Infinite.* Cambridge: Birkhäuser Boston.

Russell, B. (1937/1903). *The Principles of Mathematics.* London: Allen & Unwin.

_____. (1971). *Introduction to Mathematical Philosophy.* New York: Simon & Schuster.

Saaty, T., & F. J. Weyl (eds.) (1969). *The Spirit and the Uses of the Mathematical Sciences.* New York: McGraw-Hill.

Salanskis, J.-M. (1984). Continu et discret. In *Encyclopedia Universalis,* pp. 458–462. Paris: Encyclopedia Universalis.

Schaaf, W. L. (ed.) (1963). *Our Mathematical Heritage.* New York: Collier Books.

Scientific American (1957). *Lives in Science: A* Scientific American *Book.* New York: Simon & Schuster.

Singh, J. (1959). *Great Ideas of Modern Mathematics: Their Nature and Use.* New York: Dover.

Singh, S., & K. A. Ribet (1997). Fermat's last stand. *Scientific American,* November: 36–41.

Sondheimer, E., & A. Rogerson (1981). *Numbers and Infinity: A Historical Account of Mathematical Concepts.* New York: Cambridge University Press.

Stewart, I. (1989). *Does God Play Dice? The Mathematics of Chaos.* Cambridge, Mass.: Blackwell.

_____. (1995). *Nature's Numbers: The Unreal Reality of Mathematics.* New York: Basic Books.

_____. (1996). *From Here to Infinity: A Guide to Today's Mathematics.* New York: Oxford University Press.

Struick, D. J. (1967). *A Concise History of Mathematics.* New York: Dover.

Taton, R. (1984). Calcul infinitésimal. A: Histoire. In *Encyclopedia Universalis:* 748–754. Paris: Encyclopedia Universalis.

Thom, R. (1982). Mathématique et théorisation scientifique. In J. Dieudonné, M. Loi, & R. Thom (eds.), *Penser les mathématiques.* Paris: Seuil.

Tieszen, R. L. (1989). *Mathematical Intuition: Phenomenology and Mathematical Knowledge.* Dordrecht: Kluwer.

Tymoczko, T. (ed.) (1986). *New Directions in the Philosophy of Mathematics.* Cambridge: Birkhäuser Boston.

Van Fraassen, B. C. (1971). *Formal Semantics and Logic.* New York: Macmillan.

Vilenkin, N. Y. (1995). *In Search of Infinity.* Cambridge: Birkhäuser Boston.

Waldrop, M. M. (1992). *Complexity: The Emerging Science at the Edge of Order and Chaos.* New York: Simon & Schuster.

Weyl, H. (1918/1994). *The Continuum: A Critical Examination of the Foundation of Analysis.* New York: Dover.

_____. (1949/1927). *Philosophy of Mathematics and Natural Science.* Princeton: Princeton University Press.

_____. (1951). A half-century of mathematics. *The American Mathematical Monthly,* 58: 523–553.

_____. (1952). *Symmetry*. Princeton: Princeton University Press.

Wigner, E. (1960). The unreasonable effectiveness of mathematics in the natural sciences. *Communications on Pure and Applied Mathematics*, 13: 1–14.

Wilder, R. (1944). The nature of mathematical proof. *The American Mathematical Monthly*, 51: 309–323.

_____. (1986). The Cultural Basis of Mathematics. In T. Tymoczco (ed.), *New Directions in the Philosophy of Mathematics*. Cambridge: Birkhäuser Boston.

Wittgenstein, L. (1967/1956). *Remarks on the Foundations of Mathematics*. Cambridge: MIT Press.

Zellini, P. (1996). *Breve storia dell'infinito*. Milano: Adelphi.

More Cognitive Science and the Study of the Mind in General

Andler, D. (ed.) (1992). *Introduction aux sciences cognitives*. Paris: Gallimard.

Bailey, D. (1997). *A Computational Model of Embodiment in the Acquisition of Action Verbs*. Ph.D. dissertation, Computer Science Division, EECS Department, University of California at Berkeley.

Barsalou, L. W. (1983). Ad-hoc categories. *Memory and Cognition*, 11: 211–227.

_____. (1987). The instability of graded structure: Implications for the nature of concepts. In U. Neisser (ed.), *Concepts and Conceptual Development: Ecological and Intellectual Factors in Categorization*. Cambridge, U.K.: Cambridge University Press.

Bloom, F. E., & A. Lazerson (1988). *Brain, Mind, and Behavior*. New York: W. H. Freeman.

Carpenter, P. A., & P. Eisenberg (1978). Mental rotation and the frame of reference in blind and sighted individuals. *Perception and Psychophysics*, 23: 117–124.

Churchland, P. S. (1986). *Neurophilosophy: Toward a Unified Science of the Mind/Brain*. Cambridge: MIT Press.

_____. (1995). *The Engine of Reason, the Seat of the Soul: A Philosophical Journey into the Brain*. Cambridge: MIT Press.

Churchland, P. S., & T. J. Sejnowski (1992). *The Computational Brain*. Cambridge: MIT Press.

Clark, A. (1997). *Being There: Putting Brain, Body, and World Together Again*. Cambridge: MIT Press.

Comrie, B. (1976). *Aspect: An Introduction to the Study of Verbal Aspect and Related Problems*. Cambridge Textbooks in Linguistics. New York: Cambridge University Press.

Cooper, L. A., & R. N. Shepard (1973). Chronometric studies of the rotation of mental images. In W. G. Chase (ed.), *Visual Information Processing*. New York: Academic Press.

Crick, F., & C. Koch (1995). Are we aware of neural activity in primary visual cortex? *Nature*, 375: 121–123.

Damasio, A. R., & H. Damasio (1992). Brain and language. *Scientific American*, 267 (3): 88–95.

Deacon, T. W. (1997). *The Symbolic Species: The Co-Evolution of Language and the Brain*. New York: W. W. Norton.

Dreyfus, H. L. (1979). *What Computers Can't Do: The Limits of Artificial Intelligence*. New York: Harper Colophon.

Dupuy, J.-P. (1994). *Aux origines des sciences cognitives*. Paris: La Découverte.

Gallese, V., L. Fadiga, L. Fogassi, and G. Rizzolatti (1996). Action recognition in the premotor cortex. *Brain,* 119: 593–609.

Ganascia, J.-G. (1996). *Les sciences cognitives.* Paris: Flammarion.

Gardner, H. (1985). *The Mind's New Science: A History of the Cognitive Revolution.* New York: Basic Books.

Gazzaniga, M. S. (ed.) (1996). *Conversations in the Cognitive Neurosciences.* Cambridge: MIT Press.

_____. (ed.) (1999). *The New Cognitive Neurosciences,* 2nd ed. Cambridge: MIT Press.

Hobson, J. A. (1988). *The Dreaming Brain.* New York: Basic Books.

_____. (1994). *The Chemistry of Conscious States.* Boston: Little, Brown.

Hutchins, E. (1980). *Culture and Inference: A Trobriand Case Study.* Cambridge: Harvard University Press.

Kay, P. (1983). Linguistic competence and folk theories of language: Two English hedges. In *Proceedings of the 9th Annual Meeting of the Berkeley Linguistics Society.* Berkeley: Berkeley Linguistics Society.

Kay, P., & C. McDaniel (1978). The linguistic significance of the meanings of basic color terms. *Language,* 54: 610–646.

Kerr, N. H. (1983). The role of vision in "visual imagery" experiments: Evidence from the congenitally blind. *Journal of Experimental Psychology: General,* 112 (2): 265–277.

Lakoff, G. (1972). Hedges: A study in meaning criteria and the logic of fuzzy concepts. In *Papers from the 8th Regional Meeting, Chicago Linguistic Society.* Chicago: Chicago Linguistic Society. Reprinted in the *Journal of Philosophical Logic,* 2 (1973): 458–508.

Langacker, R. (1986/1991). *Foundations of Cognitive Grammar.* 2 vols. Stanford: Stanford University Press.

Marmor, G. S., & L. A. Zaback (1976). Mental rotation by the blind: Does mental rotation depend on visual imagery? *Journal of Experimental Psychology: Human Perception and Performance,* 2 (4), 515–521.

Mayr, E. (1984). Biological classification: Toward a synthesis of opposing methodologies. In E. Sober (ed.), *Conceptual Issues in Evolutionary Biology.* Cambridge: MIT Press.

Minsky, M. (1985). *The Society of Mind.* New York: Simon & Schuster.

Núñez, R. (1996). Ecological naturalism: Conscious experience as a supra-individual biological (SIB) phenomenon. *Consciousness Research Abstracts, Toward a Science of Consciousness 1996,* p. 178.

Núñez, R., D. Corti, & J. Retschitzki (1998). Mental rotation in children from Ivory Coast and Switzerland. *Journal of Cross-Cultural Psychology,* 29 (3): 493–505.

Piaget, J. (1967). *Biologie et connaissance.* Paris: Gallimard.

Port, R. F., & T. Van Gelder (eds.) (1995). *Mind as Motion: Explorations in the Dynamics of Cognition.* Cambridge: MIT Press.

Posner, M. I. (ed.) (1990). *Foundations of Cognitive Science.* Cambridge: MIT Press.

Putnam, H. (1988). *Representation and Reality.* Cambridge: MIT Press.

Ramachandran, V. S. (1992). Blind spots. *Scientific American,* 266: 85–91.

Ramachandran, V. S., & S. Blakeslee (1998). *Phantoms in the Brain: Probing the Mysteries of the Human Mind.* New York: William Morrow.

Ramachandran, V. S., & R. L. Gregory (1991). Perceptual filling-in of artificially induced scotomas in human vision. *Nature,* 350: 699–702.

Rizzolatti, G., L. Fadiga, V. Gallese, and L. Fogassi (1996). Premotor cortex and the recognition of motor actions. *Cognitive Brain Research*, 3: 131–141.

Rogoff, B. (1990). *Apprenticeship in Thinking: Cognitive Development in Social Context.* New York: Oxford University Press.

Rogoff, B., & J. Lave (eds.) (1984). *Everyday Cognition: Its Development in Social Context.* Cambridge: Harvard University Press.

Rosch, E. (1973). Natural categories. *Cognitive Psychology*, 4: 328–350.

_____. (1975a). Cognitive reference points. *Cognitive Psychology*, 7: 532–547.

_____. (1975b). Cognitive representations of semantic categories. *Journal of Experimental Psychology: General*, 104: 192–233.

_____. (1977). Human categorization. In N. Warren (ed.), *Studies in Cross-Cultural Psychology.* London: Academic.

_____. (1978). Principles of categorization. In E. Rosch and B. B. Lloyd (eds.), *Cognition and Categorization.* Hillsdale, N.J.: Erlbaum.

_____. (1981). Prototype classification and logical classification: The two systems. In E. Scholnick (ed.), *New Trends in Cognitive Representation: Challenges to Piaget's Theory.* Hillsdale, N.J.: Erlbaum.

_____. (1994). Categorization. In V. S. Ramachandran (ed.), *The Encyclopedia of Human Behavior.* San Diego: Academic Press.

_____. (1999). Reclaiming Concepts. In R. Núñez & W. J. Freeman (eds.), *Reclaiming Cognition: The Primacy of Action, Intention, and Emotion.* Thorverton, U.K.: Imprint Academic.

Rose, S. (1992). *The Making of Memory: From Molecules to Mind.* New York: Anchor Books, Doubleday.

Schacter, D. I. (1996). *Searching for Memory: The Brain, the Mind, and the Past.* New York: Basic Books.

Searle, J. R. (1992). *The Rediscovery of the Mind.* Cambridge: MIT Press.

_____. (1995). *The Construction of Social Reality.* New York: Free Press.

Shepard, R. N., & L. A. Cooper (1982). *Mental Images and Their Transformations.* Cambridge: MIT Press.

Shepard, R. N., & J. Metzler (1971). Mental rotation of three-dimensional objects. *Science*, 171: 701–703.

Shore, B. (1996). *Culture in Mind: Cognition, Culture, and the Problem of Meaning.* New York: Oxford University Press.

Slobin, D. (1985). Cross-linguistic evidence for the language-making capacity. In D. Slobin (ed.), *A Cross-Linguistic Study of Language Acquisition: Vol. 2, Theoretical Issues.* Hillsdale, N.J.: Erlbaum.

Smith, E. E., & D. L. Medin (1981). *Categories and Concepts.* Cambridge: Harvard University Press.

Snyder, S. H. (1986). *Drugs and the Brain.* New York: Scientific American Library.

Taylor, J. (1989). *Linguistic Categorization: Prototypes in Linguistic Theory.* Oxford: Clarendon Press.

Thompson, D. W. (1942/1992). *On Growth and Form: The Complete Revised Edition.* New York: Dover.

Thompson, R. F. (1985). *The Brain: A Neuroscience Primer.* New York: W. H. Freeman.

Tomasello, M. (2000). *The Cultural Origins of Human Cognition*. New York: Harvard University Press.

Tomasello, M., & J. Call (1997). *Primate Cognition*. New York: Oxford University Press.

Vygotsky, L. (1980). *Mind in Society: The Development of Higher Psychological Processes*. New York: Harvard University Press.

_____. (1986). *Thought and Language*. Cambridge: MIT Press.

Wittgenstein, L. (1953). *Philosophical Investigations*. New York: Macmillan.

Zadeh, L. (1965). Fuzzy sets. *Information and Control*, 8: 338–353.

Zeki, Z. (1993). *A Vision of the Brain*. Oxford: Blackwell Science.

_____. (1992). The visual image in mind and brain. *Scientific American*, 267 (3): 68–76.

Zimler, J., & J. M. Keenan (1983). Imagery in the congenitally blind: How visual are visual images? *Journal of Experimental Psychology: Learning, Memory, and Cognition*, 9 (2): 269–282.

Index

Abstract elements, 112
  of Boolean algebra, 127
Abstract essences, 117
Abstraction, 354
Abstract operations, 112
Abstract reasoning, 43
Acalculia, 26
Acceleration, 448, 450
Accessible Pointed Graphs (APGs), 147
Accumulation points, 275
Actual infinity, 202, 207
  of Cantor's proof, 212
  characterized by BMI, 180, 203
  final resultant state of, 167, 169, 172, 176,
    178, 196
  as metaphorical process, 158
  as unique, 160
Addition, 61, 65
  Boole's new, 127
  object collection roles in, 57
  for transfinite cardinals, 214, 219
Addition modulo 3, 112
  concept, 117
  literal interpretation of, 114
  metaphor, 113
  and rotations of triangle, 117
AE metaphors. *See* Algebraic Essence
    metaphors
Ai, 22
Algebra, 110, 128
  abstract, 119
  and arithmetic, 25–26, 111, 114
  Boolean. *See* Boolean algebra
  cognitive structure of, 119
  fundamental theorem of, 82
Algebra groups, 113, 116

Algebraic Essence metaphors, 113
  form of, 117–118
Algebraic laws, to fit classes, 129
Algebraic structures, 113, 117
Algorithms, 137, 261
  cognitive activity with, 86
  structure, 98
  symbolic, 98, 99
All, 178
  meaning with infinite sets, 175, 179
All parallel lines meet at infinity axiom,
    167
American Mathematical Society, 307
American Sign Language, 47
Analyst, The (Berkeley), 251
Analytic geometry, 397
  and calculus, 261
  and Cartesian plane, 384
Angles as numbers, 387
Animals
  numerical abilities of, 21, 22
Anti-Foundation axiom, 147, 148
APGs. *See* Accessible Pointed Graphs
Approach, 188
Approaching a limit, 199, 200
Arbitrary distance condition, 168
Archimedean principle, 298
  and Number-Line blend, 301
  and real numbers, 225, 305
Aristotelian logic, 123
Aristotle
  potential infinity, 158
  predication metaphor, 123, 129, 134
  theory of categories, 109, 122
Arithmetic, 24, 29, 125. *See also* Innate
    arithmetic

and algebra, 25–26, 113–114
axiomatic, 175
capabilities of infants, 17(fig). *See also* Babies, numerical abilities of
elements for conceptual system, 99
and ERFs, 87, 98
inferior parietal cortex, 25
of infinitesimals, 234
laws of, 92, 95, 96, 98, 421
and link with world, 96
of natural numbers, 95, 294
and rotation plane, 428
as target domain, 55, 72
Arithmetic Cut metaphor, 300
Arithmetic Is Motion Along a Path metaphor, 72, 75, 92
extension of, 89
and zero, 367
Arithmetic Is Object Collection metaphor, 55, 56, 59, 63, 67. *See also* Object-collection metaphor
extended to multiplication and division, 61–62
and innate arithmetic, 60
and mapping of ERF, 87
and zero, 64, 366
Arithmetic Is Object Construction metaphor, 65, 66, 67, 404. *See also* Object-construction metaphor
and prime numbers, 82
and Roman notation, 83
Arithmetization, 450
of calculus by Dedekind, 305
of calculus by Weierstrass, 293
of granular numbers, 246
metaphors, 100, 224, 451
program, 261
Ascending categories, 162
As Many As concept, 144
Aspect, 35, 180
and infinity, 170
in motor-control system, 156
schema, 35
Aspectual systems, 36, 155, 156
Associative laws, 89, 132
for addition and multiplication, 126, 200
for arithmetic, 88, 96
holding for classes, 125
Associativity, 63
Axiomatic method, 110, 118, 145

Axiomatic set theory, 174
Axiom of all parallel lines meet at infinity, 167
Axiom of Choice, 145
Axiom of Extension, 145
Axiom of Foundation, 146, 148
Axiom of Infinity, 145, 232
in set theory, 174
Axiom of Least Upper Bounds, 200
Axiom of Mathematical Induction, 175–176
Axiom of Pairing, 145
Axiom of Powers, 145
Axiom of Projective Geometry, 169–170
Axiom of Specification, 145
Axiom of Union, 145, 232
Axioms, 89, 119, 205, 355
expressed as constraints, 370
first nine, 200, 227, 228, 235
and idea of essence, 358
for ordered fields, 205
for set theory, 145, 174, 215, 355

Babies' numerical abilities, 15–19. *See also* Children
Bailey, David, 34
Basal ganglia, 25
Basic Metaphor of Infinity (BMI), 8, 158, 159, 161, 180, 189, 192, 373
and actual infinity, 179, 207
applied for limits, 189(fig), 194, 200, 243
applied to projective geometry, 169(fig)
applied recursively, 249
and Cantor's proof, 212
characterizing infinite sequences, 186, 220
and class of all infinite decimals, 184
and point at infinity, 170, 173
creating least upper bound, 202
and creation of infinitesimals, 248
to define new entities, 229
and equal symbol, 377
and forming numbers process, 234, 247
implicitly used by Newton, 228
for infinite class with all numerals, 182
for infinite nesting property, 275, 277, 311
and infinitesimal numbers, 233, 255
and infinite sum existence, 367
and infinity as a number, 165

linking discrete and continuous
      mathematics, 259
and naturally continuous space, 293
and natural numbers set, 174, 221, 216
neural, cognitive plausibility of, 163
parameters, 167
and point concept, 266, 278
and real numbers, 203, 206
and transfinite cardinals and ordinals, 222
Berkeley, George, 251, 252
Bideaud, J.,19
Binary operations, 177
   of addition, 84, 176
   of addition modulo 3, 112
   of Boolean algebra, 127
Binary system, 82, 83
Blends, 131, 282, 267, 372. *See also under
      specific names*; Conceptual blending;
      Metaphorical blends
   of geometry, arithmetic, algebra, 385
   metaphorical, 48, 94
   of positive and negative numbers, 91–92
   of source and target domains, 70, 260
   of space, sets, numbers, 283, 295
Blindness and visual imagery, 34
BMI. *See* Basic Metaphor of Infinity
Bombelli, Rafael, 73
Boolean algebra, 133
   classes in, 140
   properties of, 127
   source domain, 128
Boolean classes, 124, 133
   inferential laws of, 134
   Venn diagrams of, 122
Boolean logic, 45, 122
Boole, George, xiii, 123
   algebra of classes, 124, 131
   algebraic notation, 129, 131
   and Aristotle's predication metaphor, 129
   concept of class, 130
   empty class, 127, 130
   new addition table, 127
   and symbolic logic, 357
   universal class, 124–125, 127, 130
Boole's metaphor, 124, 128–129, 138, 140,
      373
   first-stage, 124, 125, 126, 127
   as linking metaphor, 150
   mapping of, 134, 367
Bounded regions in space, 43, 275, 353

Bourbaki program, 349
Boysen, Sarah T., 22–23
Brain, 33–34, 138. *See also* Inferior parietal
      cortex
   damage of Mr. M, 383
   neural structure of, 134, 347
Bumpy-curve-sets, 328
Butterworth, B., 19, 26

Calculation, 22, 85, 98, 99
   and embodied understanding, 89
   and notation, 86
Calculus, 29, 233, 251, 309
   arithmetization of, 320
   of change, 223, 260, 314
   for classes by Boole, 129
   Dedekind's arithmetization of, 305
   and granulars, 250–251
   with hyperreals, 249
   with infinitesimals, 253
   Keisler's development of, 250
   with notion of limit, 242
   propositional, 133, 134, 137
   from Robinson's perspective, 242
   Weierstrass's arithmetization of, 293,
      311
Cantor, Georg, xiii, 144, 162, 322
   and BMI, 219
   diagonalization proof, 210, 211(fig)
   infinite numbers, 254
   and infinite sequences, 218
   numbers of infinite size, 208
   Ordinal metaphor, 218, 221
   pairability concept, 143
   set theory, 215, 216
Cantor's Continuum hypothesis, 215, 289,
      290
Cantor size, 208
   increasing, 215
Cantor's metaphor, 143, 144, 150, 210, 213,
      255, 373
   and BMI, 368
   and Cantor size, 221
   and cardinality of real numbers, 216, 290
   and continuum, 289
   and creation of infinite sets, 214
   and two sets having same size, 208
Cantor's proof, 211, 212–213
Cardinality, 208, 215. *See also* Cantor size
   and Cantor's metaphor, 216

same, 210
  of set of real numbers, 290
Cardinal numbers, 214, 216, 217
  assignment, 51
Cartesian plane, 38, 170, 305, 306, 315, 325,
    326, 328, 330, 333, 334, 344, 384–385
  arithmetic functions in, 296
  blend, 331, 333, 385, 386, 396, 426
  coordinates, 170, 171
  ordered pairs of points in, 386
  and trigonometry, 396
  and unit circle, 48, 388
Categories, 109, 110, 354
  as entities in world, 161
  formation of ascending, 162
  hierarchy of, 161
  as kinds, 108
Categories Are Containers metaphor, 43, 44,
    45. See also, Classes Are Containers
    metaphor
Category of Being, 162
Cauchy, Augustin-Louis, 305, 308
Causality of essences, 108, 119
Change
  average, 449
  mathematicization of, 407–408, 409, 411
  and motion, 314, 386
  pattern of, 108
  rate of, 448, 450
Change Is Motion metaphor, 408
Change of Function Is Coordinated Motion
    of Two Trajectors, 198
Chieu, Ming Ming, xii, 101
Children
  and concept of a point, 270
  and innate arithmetic, 83
  and language development, 47
  subitizing by, 54
Chimpanzees' use of numerical symbols, 22
Christianity and infinity, 162
Circle, 264, 388
Circle-set, 271, 272
Classes, 28, 123, 130
  Boole's law for, 124
  as collections, 122
  logic of, 125, 130
  operations on, 124, 129
  similarities to arithmetic, 123, 124, 127,
    129
  as target domain, 123, 125, 128

Classes and propositions blend, 131
Classes Are Containers metaphor, 122, 123,
    124, 134, 138, 367, 373
  and inferential laws, 135
  and model theory, 146
Class-membership statement, 123
Closed interval, 205, 284
Closeness, 310, 311
  preservation of. See Preservation of
    closeness
Closure, 82, 112, 253
  and arithmetical operations, 422
  and metaphors, 81, 89
  principles, 290, 421
  requirements, 93, 94
  as special case of BMI, 177
Cognition, 33, 180, 420
Cognitive linguistics, 138
  and image schema, 31
  methodology, 100
Cognitive mechanisms, 161
  as linking metaphors, 99
  for multiplication, 60
Cognitive psychology, 100, 138
Cognitive reality, 372
Cognitive science, 2, 4, 337, 339
  and embodied mathematics, 347
  and infinity as a number, 165
  of mathematics, 273, 348, 366, 397
Cognitive unconscious, 5, 28, 41
Cohen, Laurent, 23, 383
Cohen, Paul, 215
Collection metaphor. See Object-collection
    metaphor
Collections, 54, 56
  fixed, 217
  and items in sequence, 216
Combinatorial game theory, 343
Combinatorial-grouping capacity, 52
Commutative groups, 115
  with three elements, 112, 114, 117
Commutative laws, 125, 133
  for addition and multiplication, 126,
    200
Commutativity, 58, 62
Compactness theorem, 230, 231, 232
Completion in mathematics, 36, 37, 159
Complex numbers, 421, 432
Complex plane, 431, 437
  constraints, 440

as 90° rotation plane, 429
properties of, 430
Computational neural modeling, 138
Computers, 360, 361
Conceptual blending, 48, 52. *See also under specific names*; Blends; Metaphorical blends
as element of embodied cognition, 347
as linking mechanisms, 102
Conceptual mapping, 74, 84. *See also* Mapping
of BMI, 160
from algebraic structures, 113
Conceptual metaphors, 5, 41, 47, 55, 100–101, 116, 123, 138, 412, 438. *See also* Metaphors; Metaphors cited in text
for abstract concepts, 39
as cognitive mechanisms, 6, 52, 351
as definitions, 151
as elements of embodied cognition, 347
and generalizability, 98
as inference preserving, 137, 353
as linking mechanisms, 102
and mathematical thought, 6
Conflation, 96, 157
in embodied cognition, 42
of subitizing plus, 95
Container schemas, 33, 39, 122, 123, 133, 138
cognitive, 32(fig)
embodied, 135
logic of, 43, 136(fig)
in mathematics, 30
and sets, 141, 145, 146, 148
structure of, 31
Containment, 33, 147
Continuity, 8, 314
for numbers, 299, 304
for prototypical curve, 317
for space, 304
of Weierstrass, 20, 309, 310, 318, 407, 408
Continuity and Irrational Numbers (Dedekind), 294, 308
Continuity for Function Is Preservation of Closeness metaphor, 322
Continuity for Line Is Numerical Gaplessness metaphor, 322
Continuity Is Gaplessness metaphor, 299
Continuity Is Numerical Completeness metaphor, 299

Continuous, 157
conceptualized as discrete, 262, 323, 324
as repeated actions, 157
Continuum, 8, 323. *See also* Cantor's Continuum hypothesis
in Cantor's proof, 212
granular, 239
meaning of, 288–289
Continuum, The (Weyl), 323
Convergence, 186, 192
Convergent sequences, 193, 205
and BMI, 192
Cosine, 253, 449
functions, 395(fig)
metaphorical definition of, 394
Counting, 54
cognitive mechanisms for, 51
independent-order capacity, 52
morpheme, 68
pairing capacity, 51
Course of Pure Mathematics, A (Hardy), 164
Cultural relativism, 362
Culture, 356
as dimension of mathematics, 355, 358, 359
Curvature, 264
Curves, 306, 315, 325–334, 387
in Cartesian plane, 330
as continuous, 260, 317
metaphorical, 287
as motions of points, 385
Pierpont's eight properties of, 321
replaced by sets, 327, 328, 329
Curves (and Lines) Are Ordered Pairs of Numbers metaphor, 328, 329

Dantzig, Tobias, 29
Davis, M., 251
Deceleration, 448
Dedekind, Richard, xiii, xiv, 295, 298, 315, 322
Arithmetic Cut metaphor, 302
continuity in arithmetic terms, 293, 294
Continuity and Irrational Numbers, 308
Continuity Is Numerical Completeness metaphor, 299
Continuity metaphor, 305
correspondence between numbers and points, 297
cut frame, 299, 300

and discretization program, 289, 292
Geometric Cut metaphor, 300, 301
and infinitesimal numbers, 304–305
link between real numbers and
    continuity, 296
Number-Line blend, 301
and Weierstrass's metaphorical world
    view, 322
Definitions, 151
    Aristotle's definition of, 110
    as necessary, sufficient properties, 109
    in symbols, 273
Dehaene, Stanislas, 21, 23, 24, 26, 29, 49,
    383
De Morgan, Augustus, 124, 129
Denseness property, 301
Derivative metaphor, 448
Derivatives, 445
    cumulative, 253
    defined by metaphor, 252
    idea of, 225(fig), 226(fig)
    properties, 412, 414
    and rate of change, 448
Descartes, René, xiii, 315, 386
    analytic geometry, 260, 384
    metaphors, 261, 306, 394
    numbers as points on a line, 261, 292
    thought as mathematical calculation,
        357
Diagonalization argument, 210
Differences, infinitesimal, 252–253
Differential calculus, 408
Dimension, 264
Dimensionality, 260, 264
Disc, 266–267
    of zero diameter, 278
Disc/Line-Segment Blend, 267
Discrete numbers, 261
Discrete space, 287
Discretization, 369
Discretization program, 261, 262, 291
    rigor, 293
    and Dedekind, 289
    meaning of continuum, 288
    and Pierpont, 320
    points as different entities, 273
    of space, 272, 288
Discretized geometry, 272
Discretized mathematics, 271, 275, 282
    and logicism, 261

Discretized models, 274
Discretized Number-Line blend, 288, 295,
    296
Discretized plane, 287
Discretized real-number line, 283, 284, 288
Disembodied mind, 5
Disjunction, 133
Distance, 188, 274, 275
    as Cartesian Plane blend, 407
    between point-locations, 310
    -time ratio, 315
Distributive laws, 125, 129, 200
    of arithmetic, 127
    for conjunction and disjunction, 133
    for intersection and union, 132
    for multiplication, 126
Division, 61, 62
    as iteration of subtraction, 72
    by negative numbers, 93
Dreams, 34

e , 441
    and mathematicization of change, 409
    meaning of, 411, 416
    numerical value of, 414, 415
Electromagnetism, 345
Elemente (Heine), 308
Elements, 112, 213, 214
    in arithmetic structure, 111
    and sets, 177, 212, 221, 263, 369
Embodied cognition, 347
    and mathematics, 49, 349, 365
    and metaphor, 97
Embodied mathematics, 33, 45, 348, 362,
    365
    historical dimension, 359
    properties, 351–352
    theory of, 9, 346, 347
Embodied realism, 9
Embodiment of mind, 5, 121, 348, 349,
    362
Emptiness, 75, 76
Empty class, 129, 367
    of Boole, 127, 130
    as unique object, 368
Empty collection, 366
Empty set, 121, 191
    as number one, 151
    as unique entity, 141
    zero as, 343

Entailments, 59, 162, 167, 302–303
  of BMI, 171, 198
  of 4Gs, 93, 96, 98
  and human cognition, 180
  of metaphors, 56, 64, 68, 92, 97, 367
Entity-creating metaphor, 64
Epilepsia arithmetices, 23
Epsilon-delta, 198, 199, 200
Epsilon disc, 275
  frame, 276, 311
Epsilons, 194, 199
Equality, 376, 377
Equal symbol, 376–377
Equivalent Result Frame (ERF), 87, 88, 98, 99
ERF. See Equivalent Result Frame
Essence Is Form, 110
Essence of all Being, 357
Essence of Mathematical System metaphor, 110, 119
Essences, 109, 110, 112, 308, 358
  of arithmetic, 111, 295
  as causal source of natural behavior, 109, 119, 356
  and pattern of change, 108
  defined, 108
  as defining numbers, 422
  folk theory of, 107–108, 109
  of mathematical system, 118, 119
Essences Are Forms metaphor, 119
Euclid, 109
  axiomatic method, 118
  definition of surface, 265
  geometry, 70, 145, 356
  Properties Are Functions metaphor, 266
  properties of point, 269
Euclidean plane, 385, 388, 427
Eudoxus, 71
Euler, Leonard, 315, 317, 381, 411
Euler's blend, 446, 447
Euler's equation, 7, 49, 381, 436
  conceptual meaning of, 434
Euler's proof, 383, 384
Even numbers, 144(fig)
Event-concepts, 156
Event structure metaphor, 157
Everyday classes, 133, 135
Evolution of mathematics, 354, 359
Excluded middle, 44, 135, 137
Exhaustion-detection capacity, 51

Exponential blend, 406
Exponential function, 381, 434, 447
  as conceptual metaphor, 405
  distinct from self-multiplication, 404, 433
  extended to granular numbers, 238
  rate of change, 410
Exponential mapping, 403, 404, 405
Extensions, 66, 422

Fibonacci sequences, 163, 357
Fictive motion, 38–39, 163
  schema of Talmy, 191, 198
Fillmore, Charles, 87
Final resultant state, 160, 162
  of actual infinity, 167, 169, 172, 176, 178, 196
  in BMI, 159
Fitting together extension, 66, 69
Fixed collection, 217
Floating-point arithmetic, 360, 361
Floating-point numbers, 181
Folk theory, 107
  of essences, 107–108, 109, 118, 119
  of measurement and magnitude, 298, 305
Formal Foundations metaphor, 142
Formal Foundations program, 261, 372–373, 376
Formal mathematics, 372, 374
Formal Reduction metaphor, 369, 370, 373, 375
  cognitive interpretation of, 372
  as metaphorical schema, 371
Foundational metaphors, 100
Foundations of Mathematics movement, 6, 130, 254, 305, 358
4G metaphors. See Four grounding metaphors; Grounding metaphors
4Gs. See Four grounding metaphors; Grounding metaphors
Four grounding metaphors, 77, 79, 80, 88
  mapping of ERFs, 89, 98
  and metaphorical extensions, 92, 94
  needed for zero, 373
Fractions, 66–67, 95
  and Arithmetic Is Motion metaphor, 73
  and Measuring Stick metaphor, 70
  and object-construction metaphor, 97
  as point-locations, 89
Frame for Disc, 266
Frame semantics, 87

Functions, 198, 386
    as curves in Cartesian plane, 394
    infinite sum of, 413
    limits of, 200
    with same derivatives, 445
    with same power series, 446
    as spaces, 264
Functions Are Numbers metaphor, 386,
    387, 412, 413
Functions Are Ordered Pairs of Numbers
    metaphor, 322
Functions Are Sets of Ordered Pairs
    metaphor, 444, 446
Fundamental Metonymy of Algebra
    metaphor, 75, 119, 421
    and cognitive frames, 98
    and generalizations over numbers, 111

Gallistel, C. Randy, 21
Gaps, 296, 298, 299, 303
Gauss, Carl Friedrich, 303
Gelman, Rochel, 21
Generative closure, 176, 177, 179
Geometric curves, 330
Geometric Cut metaphor, 300
Geometric cuts, 302
Geometric figures
    discretized notion of, 278
    as part of space, 271, 272, 263
Geometric Figures Are Objects in Space
    metaphor, 272
Geometric paradigm, 306, 307, 322
    changes in, 308
    and monster functions, 318
Geometry, 293
    blended with arithmetic, algebra,
        385
    elimination of, 308
    rotations, 116
Gödel, Kurt, 215, 262
Granular arithmetic, 235
Granular calculus, 250
Granularity, 233, 239
Granular numbers, 233, 235, 246, 255
    advantages for calculus, 250–251
    for infinitesimals, 234, 242
    and infinite sum, 240
    integral in, 243, 244(fig)
    layers of, 248
    line, 237

model theory, 254
    and points on a line, 237
Granulars. See Granular numbers
Greek mathematics, 118
Greek philosophy, 107, 161
    and essence, 110, 357
Grounding domains, 102
Grounding metaphors, 99, 102, 150, 367,
        404. See also Four grounding
    metaphors
    for arithmetic, 65, 421
    for classes, 123, 140
    defined, 52, 53
    experiential basis of, 96
    for natural numbers, 98
    and zero and one, 75, 76

H. See Huge number
Hardy, G. H., 164–165, 166
Heine, H. E., 308
Henkin, Leon, 230, 232
    model of all hyperreals, 254
Hersh, R., 251, 356
Hilbert curve, 285(fig), 319
Hilbert, David, 285, 319
Hobbes, Thomas, 357
Huge number (H), 234
    as infinitely beyond natural numbers, 240
    as metaphorical integer, 238
Hyperinteger, 250
Hyperreal numbers. See Hyperreals
Hyperreal Numbers Are Points on a Line
    metaphor, 249
Hyperreals, 251, 290
    calculus in, 249
    and infinitesimals, 247, 254
    and linearly ordered sets, 284
    naturally continuous, 283
    variables used over, 250
Hyperreal unit square, 288
Hypersets, 152
    and graphs, 148, 149(fig)
    theory of, 147, 370
Hypothetical syllogism, 44, 135, 137

i as number, 424
Idea network, 436, 447
Idempotent laws, 127, 129
Image schemas, 30, 31, 33, 39, 54
    and embodied cognition, 347

and metaphorical mappings, 42
and motion, 37
neural circuitry, 33
Imperfective aspect, 36, 156, 180
Imperfective processes, 156, 158
Indefinite Continuous Processes metaphor, 157
Induction. *See* Mathematical induction
Inductive proof, 175–176
Infants. *See* Children; Babies' numerical abilities
Inference, 47, 54
of innate arithmetic, 77
stability of, 353
Inferential laws, 137
of Boolean classes, 134
and embodied container schemas, 135
Inferential structure, 48, 138
of source domain, 42, 44
Inferior parietal cortex, 29
lesions in, 24, 25(fig)
and nonrote arithmetic abilities, 25
and symbolic numerical abilities, 23
Infinite decimals, 184, 186, 201, 205, 206
in BMI, 212
as cases of actual infinity, 183
constructed by Cantor, 211
of Euler's equation, 381
and infinite polynomials, 185
representing real numbers, 210
symbolizing same number, 204
Infinite extremity, 179, 184
Infinite polynomials, 185, 186, 201
Infinite sequences, 219
and BMI, 197, 202
with limit L, 190, 195
meaning of, 189
of nested intervals, 206
and numbers, 186, 217, 218
sums of, 233, 241
Infinite series. *See* Infinite sequences
Infinite sets, 179, 209, 213, 232
Cantor's investigation of, 143
Cantor size of, 221
and cardinal numbers, 214
and infinite numbers, 176
Infinitesimal discs as points, 270, 271, 272
Infinitesimal numbers, 229, 304–305
Infinitesimals, 226, 230, 234, 247, 253
arithmetic, 233

combined with real numbers, 227
creating, 236, 247–248
existence of, 254
inverses of, 255
Leibniz's hypothesis, 224, 359
as mathematical entities, 223
philosophical consequences, 233
ratios, 229
Robinson's treatment, 242, 255, 359
Infinite sum metaphor, 197
Infinite sums, 239, 240, 250
Infinite totality, 176, 179, 183
and BMI, 175, 184
Infinite unions of sets, 232
Infinity, 8, 160, 249, 444
degrees of, 208
embodied, 155
literal concept of, 156
as a number, 164–166
theological importance of, 162
Infinity paradox, 325
Inhelder, Bärbel, 270
Innate arithmetic, 19, 23, 51, 59, 83
conflation with, 102
and embodied mathematics, 347
extension of, 60, 78
and 4Gs, 77, 89
linearity of, 59
and subitizing plus, 95
Innate mathematics, 100
Instantaneous Change Is Average Change metaphor, 223, 252, 448, 449
Instantaneous Speed Is Average Speed metaphor, 408
Integers, 24, 290
Intersections, 126, 132
infinite, 205
Interval of zero length, 278
Intuition, 321, 322
Intuitive sets, 140
Inverses, 200
of infinitesimals, 253
Inversive geometry, 171(fig), 172
frame, 171
and point at infinity, 170, 173
Irrational numbers, 71, 422–423
with holes, 320
and rationals, 205, 294
Isomorphisms, 78
across source domains of 4Gs, 79(fig), 80

Isosceles triangle frame, 168
Iteration extension, 63, 66, 72
    and physical segment metaphor, 69
    and Object Collection Metaphor, 62
Iterative processes, 156, 158, 161
    in BMI, 182
    for infinite closure, 177
    language of, 157
    resultant state after, 167
    as step-by-step condition, 167, 168
    syntactic forms of, 157
    as target domain, 159, 162, 165, 168, 172

Japanese counting morpheme, 68
Judaism and infinity, 162

Kabbalistic concept of God, 162
Kaufmann, E. L., 19
Keisler, H. J., 242, 249, 250

Lack, 265
Lakoff, George, xii, 9, 357
Landmark, 31, 33, 39
    trajector relation, 37
Language, 130, 156, 340
    as conceptual metaphor, 55
    natural expressions, 138
Larger, 215
Larger Than concept, 208, 221
Laws, 344
    of arithmetic, 124
    of Container Schemas, 134
    of thought, 124, 129
Least upper bound, 206, 227
    defined, 202
    and greatest lower bound, 232
Least Upper Bound axiom, 200, 201
    characterizing real numbers, 203, 205
Leibniz, Gottfried Wilhelm von, 224
    calculus of instantaneous change, 223
    idea of derivative, 226(fig)
    and infinitesimals, 228, 229, 359
    monads, 227
    thought as mathematical calculation, 357
Length, 264, 329
Length paradox, 325, 332
Limit metaphor, 224, 449
Limit points, 275, 290
Limits, 187, 229, 243, 314
    and BMI, 243

defined, 187, 242
    of functions, 198, 199, 200, 312, 313
    of infinite sequences, 186, 197
    and infinite sums, 233, 239, 246
    Newton's notion of, 228
    of sequences, 195, 198, 327
Limit-set metaphor, 328, 329
    and paradox, 330, 334
Line, 191, 260
    at infinity, 368
    segments, 68, 70. See also Physical
        segments
    as set, 263, 372
Linear notation, 97
Linear symbol systems, 86
Line-Segment frame, 267, 268(fig)
Linguistics, 74, 103
Linking metaphors, 102, 131, 142, 150
    between arithmetic and classes, 124, 125
    as cognitive mechanisms, 99
    defined, 53
Logarithms, 399
    and exponentials, 406
    mapping, 401, 402(fig), 403
    values, 400
Logic, 32(fig), 356
    of classes, 125
Logical compactness frame, 231
Logical Independence metaphor, 407
Love Is a Journey metaphor, 98
Love Is a Partnership metaphor, 46

Malcev, Anatolii, 230
Mandler, G., 19
Map-comparison structures, 33
Mapping, 78, 80, 84, 284
    log values of, 400
    numeral-number, 98, 99
    and object collections, 55, 57
    onto set-theoretical structure, 369
    point-locations to numbers, 90
    source domain, 143, 218, 282
    and symbolization, 99, 129
Mathematical concepts, 29, 48
Mathematical Function Is Curve in
        Cartesian Plane metaphor, 306
Mathematical idea analysis, 29, 49, 119,
        120, 152, 215, 222, 338, 373
    of Euler's proof, 384
    of infinity concepts, 155

and precise cognitive terms, 375
requirements for, 436
and symbolic logic, 121
Mathematical Idea Genome Project, 338
Mathematical ideas, 371
as metaphorical in nature, 365
and neural motor control, 34
Mathematical induction, 175, 176
Mathematical logic, 138, 357
Mathematical Platonism, 80
Mathematical reasoning, 358
as version of logic, 356
Mathematical truth, 8, 81
and embodied cognition, 366
as universal, 339
Mathematics, 37, 346, 352
Aristotle's definition of, 110
concepts, 2, 9, 354, 376
creation, 379
and culture, 356, 362, 363
and folk theory of essences, 109, 118, 340
formal logic and set theory, 373
inaccessibility, 341
as linear notational system, 86
in 19th-century Europe, 293
objective truth, 339, 343, 361
as part of physical universe, 344
Platonic, 2, 3, 4, 97
as progressive discretization, 261
properties, 364
remetaphorization, 322
self-characterization, 348
social constructivist theory, 365
McGwire, Mark, 345
Measurement, 297, 298
and completeness of real numbers, 305
and magnitude, 298, 305
Measurement Criterion for Completeness, 303
Measuring stick, 78
Measuring Stick metaphor, 68, 70, 71(fig)
role of one in, 75
and zero, 366–367
Members of sets, 146, 212
Mental rotation, 91, 92, 102
Metaphorical analysis, 48
Metaphorical blends, 48. *See also under
specific names;* Conceptual blending;
Blends
and arithmetic, 94

corresponding to AE metaphors, 117
of 4Gs domains, 96
for multiplication, 60, 425
of object collections, 61
of positive and negative numbers, 91–92
Metaphorical conceptions, 62, 162
Metaphorical mapping, 43, 55, 90. *See also*
Mapping
in cognitive science, 41
to conceptualize numbers, 65
from groups to rotations, 119
and image schemas, 42
from object collections to numbers, 59
of triangle rotations, 116
Metaphorizing capacity, 5, 52, 54
Metaphors. *See also* Conceptual metaphors;
Metaphors cited in text
for characterizing essence, 108
defining discretized mathematics, 273
forming, 99
linguistic examples of, 55–56
and meaning of numbers, 398
redefinitional, 150
Metaphors cited in text. *See also under
specific titles*
Addition Modulo 3 Forms Commutative
Group, 113
Aristotle's Predication, 123, 129
Arithmetic Is Motion along a Path, 72, 92
Arithmetic Is Object Collection, 55, 56,
59, 60, 63
Arithmetic Is Object Construction, 65,
66, 82, 404
Basic Metaphor of Infinity, 8, 158, 161,
180, 192, 373
Categories Are Containers, 43, 44, 45
Change Is Motion, 408
Classes Are Containers, 122, 123
Continuity for Function Is Preservation of
Closeness, 322
Continuity Is Gapless, 299
Continuity Is Numerical Completeness,
299
Continuity for Line Is Numerical
Gaplessness, 322
Curves (and Lines) Are Ordered Pairs,
328, 329
Essence of Mathematical System Is
Algebraic Structure, 110
Functions Are Numbers, 386, 387, 412, 413

Functions Are Ordered Pairs of Numbers, 322

Functions Are Sets of Ordered Pairs, 444, 446

Fundamental Metonymy of Algebra, 75

Geometric Figures Are Objects in Space, 272

Hyperreal Numbers Are Points on a Line, 249

Indefinite Continuous Processes Are Iterative Processes, 157

Infinite Sums Are Limits of Infinite Sequences, 197

Instantaneous Change Is Average Change, 252, 448

Instantaneous Speed Is Average Speed, 408

Logical Independence Is Geometrical Orthogonality, 407

Love Is a Journey, 98

Love Is a Partnership, 46

Mathematical Function Is Curve in Cartesian Plane, 306

Measuring Stick, 68, 70, 75

Multiplication Is Addition metaphor, 405

Natural Numbers Are Sets, 142, 150, 370

90° Rotation, 428–429

Numbers Are Object Collections, 56, 96, 422

Numbers Are Physical Segments, 70

Numbers Are Points on a Line, 6, 48, 188, 237, 261, 280, 322, 424

Numbers Are Sets, 6

Numbers Are Things in World, 80, 81

Ordered Pair, 141

Polar Coordinate, 396

Properties Are Functions, 266

Propositional Logic, 131

Recurrence Is Circularity, 395, 448, 449, 450

Rotation Group, 116

Sets are Objects, 141

Spaces Are Sets of Points, 264, 271, 273, 284, 322, 369

States Are Locations, 42

Trigonometry, 388

Weierstrass's Continuity, 311

Zero Collection, 64

Zero Object, 67

Zooming In Is Multiplication by Infinitesimal, 234

Metonymy, 74, 75

Metric property, 274

Mind-based mathematics, 4, 8, 9

Model-building methodology, 101, 102

Models, discretized, 274

Model theory, 146, 254

Modular arithmetic, 113

Modus ponens, 44, 135, 137

Modus tollens, 44, 135, 137

Monads
    as clusters of infinitesimals, 227
    formation, 237
    of real numbers, 236–238

Monkeys. *See* Rhesus monkeys

Monster functions
    and geometric paradigm, 307, 308, 316, 318, 323
    and nongeometric paradigm, 315
    and Pierpont, 321–322
    in Weierstrass's paradigm, 320

Montague, Richard, 357

Moore, E. H., 285

More Than concept, 143, 144

Motion, 29, 92
    arithmetic as, 71. *See also* Arithmetic Is Motion along a Path metaphor
    conceptualized as change, 314–315
    conceptualized as points along axes, 385–386
    as 4Gs source domain, 88
    language, 74
    mathematicization of, 407
    path, 72, 78
    zero as origin of, 367

Motor-control system, 34–36
    as embodiment of aspect, 156

Motor schemas, 34, 35

Moving along a path, 95

Mr. M, 49, 383
    arithmetic disabilities of, 23–24

Multiplication, 63
    commutative law for, 126
    as iteration of addition, 72
    metaphorical conceptions of, 62
    by negative numbers, 91, 92
    as operations on collections, 60
    by pooling, 61

Multiplication Is Addition metaphor, 405

Multiplication-Rotation blend, 425
Multiplicative identity, 63
Multiplicative inverse, 229

Nahin, P., 420
Napier, John, 399, 401
Narayanan, Srini, 34, 35, 156
Natural continuity, 317, 374
Natural continuous space, 280
Natural numbers, 94, 95
  arithmetic, 96
  cardinal and ordinal uses, 216
  and closure, 81
  extension, 82
  infinite class of numerals for, 182
  and infinity, 166
  laws governing, 294
  properties, 57
  and rational numbers, 209(fig)
  successor operation, 173
  in terms of sets, 141, 143, 182, 186
Natural Numbers Are Sets metaphor, 142, 370
  as linking metaphor, 150
  and set theory, 151
Naturally continuous line, 282, 288, 290
Naturally continuous space, 260, 265, 288, 311, 385
  and naturally continuous change, 261
  properties of, 274
  as stasis and discreteness, 293
Nearness, 274
Negative acceleration, 450
Negative numbers, 441
  adding and multiplying, 89–90, 91
  and Arithmetic Is Motion metaphor, 73
  conceptualization of, 425
  as point-locations, 92, 97
Neighborhood, 274, 277
Nested intervals, 206
Nesting properties, 275, 277
  and sequence of discs, 312
Neural computational structure, 156
Neural connections, 54–55
Neural mechanisms, 49, 348
Neural model, 33
Neural motor-control, 34, 35, 36
Neural structures, 35, 99, 347
Neuropsychology, 100
Neuroscience, 100, 138

Newton, Sir Isaac, 223
  arithmetization, 224, 314
  calculus, 292, 306
  fluxions, 252
  geometric curves, 315
  limit metaphor, 224
  and notion of limit, 228, 229
Newtonian calculus, 230
Newtonian derivative, 225(fig), 314
Newtonian mathematics, 226
Newton-Leibniz metaphor, 315
90° Rotation metaphor, 428–429
90° rotation plane, 430
90° Rotation-Plane blend, 430, 437
  constraints on, 440
No distance, 271
Noncontainer metaphor, 147
Non-Euclidean geometries, 355
Nongeometric paradigm, 315
Nonstandard analysis, 232, 249
Nonstandard models, 254
Notation, 83
  for classes, 129
  and mathematics, 120, 378
Number line, 278, 279
  Dedekind's concept of, 305
  point-location of, 283
  two conceptions of, 282
Number-Line blend, 70, 71, 280(fig), 424
  described by Dedekind, 295, 298–299
  with gap in points, 298
  mapping from, 439(fig)
  and Numbers Are Points on a Line metaphor, 296
  and Weierstrass, 310
Number/Physical Segment blend. See Number-Line blend
Number-points, 48
Numbers, 57, 58, 60, 65, 81
  and abstract elements, 114
  as collections, 54
  and average speed, 406–407
  meaning of, 397
  and numerals, 83–85
  as objectively existing entities, 80, 97, 166
  as objects, 110
  and physical segments, 68, 70
  and points, 73, 279
  relationship properties for, 59

symbolic representations of, 84
symmetry between positive and negative, 91
as values of functions, 386
as zero-dimensional geometric objects, 342
Numbers Are Object Collections metaphor, 56
and associative law for arithmetic, 96
and subitizing, 422
Numbers Are Physical Segments metaphor, 70
Numbers Are Points on a Line metaphor, 6, 48, 188, 191, 198, 260, 295, 424
of Descartes, 261
extended to granular numbers, 237
forming Number-Line blend, 296–297
fully discretized version, 281
for naturally continuous space, 279
with Space-Set blend, 280
Numbers Are Sets metaphor, 6
as literal statement, 373
as set theory model, 354
Numbers Are Things in World metaphor, 80
consequences of, 81
as generalization of 4Gs, 97
Numeral-number mapping, 89, 98
Numerals
application of BMI to, 213
as infinite decimals, 184
for natural numbers, 182, 183
new, 213
and numbers, 83–85
Numerical abilities, 15, 21
Numerical distinctions, 16
Numerical estimation, 22
Numerosity, 51
Núñez, Rafael E., xii, xvi

Object collection, 95. See also Object-collection metaphor; Arithmetic Is Object Collection metaphor
associative ERF for, 87
mapping of, 60
metaphoric blend of, 61
as source domain, 55, 56–57, 88
Object-collection metaphor, 67, 93. See also Arithmetic As Object Collection metaphor
individuality symbolized, 75

Object construction. See also Object-construction metaphor; Arithmetic Is Object Construction metaphor
conflation with object collection, 96
primary experiences with, 95
as source domain, 65, 88
Object-construction metaphor, 67, 413. See also Arithmetic As Object Construction metaphor
and destruction, 76
and object-collection metaphor, 78
role of one in, 75
and unity, 75
Object location, 18
Occam's razor, 227, 230
One, 75, 197
Ontology, 365
Open set property, 275
Operations, 81, 97, 111
Ordered fields, 200, 205
Ordered number systems, 290
Ordered Pair metaphor, 141, 150
and set theory, 151
Ordered-pair-of-number set, 328, 329, 331
Ordered pairs
constrained sets of, 370
infinite set of, 186
metaphorical definition of, 141
of real numbers, 313
Ordered Pairs Are Sets metaphor, 370
Ordinal numbers, 218, 220
Orientational schemas, 33

Pairability, 143, 144
Parabolic trajectories, 344
Paradigm shift, 309
Paradox of infinity, 325
Paradox of length, 327, 330
Parallel lines, 167, 170
meeting at infinity, 169
Parallel preattentive processing, 21
Path of motion, 71–72
Path schema, 141
Peano, Giuseppe, 284
Peirce, Benjamin, 383, 431
Perfective aspect, 180
and language of iteration, 157
of verbs, 36
Perpetual motion, 156
Personality, 109

Philosophy in the Flesh (Lakoff and Johnson), 5
Physical laws, 344
Physical regularity, 345
Physical segment metaphor. *See* Measuring Stick metaphor
Physical segments, 68
 as constructed objects, 70, 78
 and path of motion, 71–72
 primary experiences with, 95
 as source domain 4Gs, 88
Physics, 109
 laws in, 344
 mathematics of, 340, 345
Piaget, Jean, 270
Pierpont, James, 307
 arithmetization of calculus, 320
 geometric understanding of functions, 374
 mathematical ideas, 321, 322
 prototypical properties of a curve, 307, 315
Planck length, 181
Plane geometry, 109
Planes, 260, 263
Plato, 110
Platonic mathematics, 2, 3, 4, 97
Platonism, 80
Poincaré, Jules Henri, xiii
Point-location metaphor, 73
Point-locations, 90, 283
 and choice of numbers, 310
 in naturally continuous space, 282
 and negative numbers, 92
 sequence of, 191
Points, 278, 291
 as abstract objects, 265
 child's concept of, 270
 correspondence with numbers, 297
 on a curve, 287
 in discretized mathematics, 273
 as discs, 268, 269(fig)
 as elements of sets, 33, 263, 272, 328, 372
 as infinitesimal discs, 270, 271, 272
 infinite smallness of, 266
 at infinity, 167, 171, 173, 175
 on a line, 186, 283, 372, 271
 in space, 264, 282
Point-set topology, 261
Points on a Line Are Numbers metaphor, 322
Polar Coordinate metaphor, 396

Polar-Trigonometric blend, 437
Pólya, George, 285
Polynomials, 82
 cumulative derivatives of, 253
 infinite, 185, 186
Polysemy, 47, 103
Pooling, 61, 62, 63
Positional notation, 97
Positive numbers, 92
Potential infinity, 158, 159(fig), 202
Power series, 414
Power-Series blend, 413
Power set proof, 213
Power sets, 141, 215
 of Cantor, 212, 213
 members, 255
 number of elements in, 214
Precision, 350, 351
Predication metaphor, 123, 129, 134
Prefrontal cortex, 24, 34
Preservation of closeness, 314, 318
 and space-filling curves, 319
 of Weierstrass, 323
Pre-Socratic philosophy, 161
Primates numerical abilities, 22–23
Prime numbers, 82
Principles of Mathematics (Russell), 408
Processes, 163, 187
 as motion, 157
 as static things, 163
Projective geometry, 167, 168
 basic axiom, 169–170
 infinity of points, 173
Proofs, 75, 372
 concept, 362
 done by computers, 361
 proving, 342
 symbolizing, 369
Properties, 364
 of external objects, 350
 inherent, 265
 of naturally continuous space, 274
Properties Are Functions metaphor, 266
Propositional calculus, 133
 grounded in container schemas, 134
 mechanization of, 137
Propositional logic, 133
 and Boole's algebra of classes, 131
Propositional Logic metaphor, 131, 134, 138
 mapping inferential laws, 135

Prototypical properties of curve, 307
Puppet experiments for babies, 16–18
Pythagorean essence of all Being, 357
Pythagorean theorem, 70, 349, 393

Rational array diagram, 208
Rationality, 340
Rational-Number-Line blend, 300, 301
Rational numbers, 209
Rats, 21, 22
Real-approximation operator, 241, 243
    and granulars, 250
Real line, 282
    misnaming of, 291
Real-Number-Line blend, 301
Real numbers, 181, 184, 185, 283, 440
    adding infinitesimals to, 238
    and Archimedean principle, 225
    axioms of, 200, 202, 227, 228, 235
    closure for, 290, 423
    commensurability, 226
    and continuity link, 296
    defined, 289
    e as, 415
    embedded in granulars, 241
    as exhausting continuum, 288
    within hyperreal interval, 287
    and infinite decimals, 204
    infinite intersections, 205
    and infinite polynomials, 186
    and Least Upper Bound axiom, 205
    lines, 283, 284, 289, 291
    and monads, 237
    on naturally continuous line, 289
    and natural numbers, 212, 214
    and rational numbers, 201, 210, 212, 309
Real-valued points, 288
Reason, 121
Reclaiming Cognition (Núñez and
    Freeman), 5
Recurrence Is Circularity metaphor, 448
    in language, 395
    and unit circle, 449, 450
Recursion, 28
Reduction, 151, 371
    of mathematics to set theory, 370
Regier, Terry, 33
Relationship properties, 59
Relativistic view of science, 362
Religion and mathematics, 162–163

Repeated action, 28
Repetition, 157
Result of an infinite process, 158
Rhesus monkeys, 22
Rigor, 321
    concept, 293
    increased, 307
    myth, 321
    and Pierpont, 322
Robinson, Abraham, 234, 249
    hyperreals concept, 250
    mathematics of infinitesimals, 242, 255,
        359
    and nonstandard analysis, 232
Romance of Mathematics, 339, 340, 341,
    368
    claim of, 361
    and cognitive science, 346
    and culture, 355
    disconfirmation of, 364–365
Roman numerals, 83, 85
Romeo and Juliet, 46
Rotation, 37
    applying twice, 425
    as a commutative group, 114–115
    concept, 29
    mental, 90(fig)
    plane, 428
    to symmetrical point, 91
    of triangle forming a group, 115(fig)
Rotation Group metaphor, 116
Rotation-Number-Line blend, 426
Rotation-Plane blend, 426, 427
Rote memorization of arithmetic, 24, 49
    and cortico-subcortical loops, 25
Russell, Bertrand, 357, 408

Same Number As concept, 142, 143
Same Size As concept, 152, 208
Schemas, 40(fig). *See also* Image schemas
Secant-tangent difference, 224
Self-membership of sets, 146
Self-multiplication and exponentiation, 404,
    433
Semantics of propositional logic, 131
Sensory-motor system, 54–55
    and grounding of abstract concepts, 101
Sequence and Limit Frame
    general version, 195
    prototypical version, 189

Sequence of differences, 202, 204
Sequence of functions, 286
Sequence of length, 220
Sequence of partial sums, 197
Sequence of sums, 246
Sequences, 188, 209, 218
  of all odd numbers, 219
  approaching limit, 194
  convergent, 192, 193, 205
  along diagonal, 212
  elements of, 195, 216
  formation process of, 217
  mapping, 218
  of natural numbers, 210, 217
Set-of-points, 263
Set-of-points metaphor. *See* Spaces Are Sets
    of Points metaphor
Sets, 140, 147, 196, 215, 371
  of all natural numbers, 173, 174, 182,
    210, 212, 214, 221
  of all rational numbers, 297(fig)
  closed, 82
  conceptualized as containers, 141, 145,
    146, 148
  containing all members of sequence, 190
  extensions, 177
  and infinite nesting property, 276
  infinite operations on, 221
  length, 329
  linearly ordered, 284
  linked to natural numbers, 142
  objectively existing, 373
  of ordered pairs of numbers, 328, 329
  pairable, 144
  process of forming, 247
  replacing spaces, curves, points, 328
  self-membership of, 146
  size of, 208
  real numbers, 212
  two infinite sequences of, 195
  union, 219
Sets Are Graphs metaphor, 147
  and hypersets, 152
  as linking metaphor, 150
Sets Are Objects metaphor, 141
Set-theoretical structure
  constraints on, 370
  with infinitesimals, 232
  mapping onto, 369
  models, 232

Set theory, 343
  axiom of infinity in, 174
  axioms for, 145, 215
  defined by axioms, 355
  modern, 140
  and Zermelo-Fraenkel axioms, 148
Sheba, 22–23
Shebo, B. J., 19
Sierpinsky, W., 285
Sine, 253
  functions of, 395(fig)
  metaphorical definition of, 394
  and periodicity, 449
Slide rule, 443
Small concept, 56, 269
Smaller defined by Cantor's metaphor, 215
Source domains, 42
  algebra, 128
  algebra groups, 113, 116
  arithmetic, 125
  classes, 123, 131
  completed iterative processes, 159
  container schemas, 123
  of 4Gs, 77
  inferences of, 102
  inferential structure of, 44, 48, 80
  limits of infinite sequences, 197
  mappings, 143, 218
  measuring stick, 68
  motion, 72, 74, 79, 91
  nonisomorphic characterization of, 81
  numbers (values of functions), 386
  object collection, 55, 79, 56–57
  object construction, 65
  sets, 141, 142, 371, 263
  space, 425
  Space-Set blend, 280
  Unit Circle blend, 393, 396
Source location, 37
Source-Path-Goal schema, 37, 38(fig)
Space, 304, 425
  as axis on Cartesian plane, 407
  bounded region of, 28
  discretized, 271, 279
  naturally continuous, 260, 385
  set-of-points conception of, 263
Space-filling curves, 9, 284, 291
  concept, 265
  of Hilbert, 285, 286
  and naturally continuous unit square, 287

Space Is a Set of Points metaphor, 263, 264, 265
Spaces
  properties of, 264
  replaced by sets, 328
Spaces Are Sets of Points metaphor, 271, 273, 275, 277, 279, 284, 295, 322, 369
  as literal truth, 373
Space-Set blend, 279, 280, 282
Spatial abilities, 24
Spatial logic, 33
  of Container schema, 44, 45
  internal, 37
Spatial metaphor, 186
Spatial relations, 353
  systems of language, 30
  terms of, 33
Special cases
  formation of ascending categories, 162
  granular numbers, 235
  infinite class of numerals, 182
  infinitesimals, 228
  inversive geometry, 172
  mathematical induction, 176
  projective geometry, 168
  real numbers, 184
  set of natural numbers, 174
  unending sequence of integers, 165
Speck concept, 223, 233
Speed, 406
  concept of average, 406–407, 408
  mathematicization of, 407
Splitting up concept, 61
Stability, 350, 352
Standard part, 250
  operator of hyperreal numbers, 250
States Are Locations metaphor, 42
Structured connectionism, 33
Structure of events, 35
Subcortical basal ganglia. See Basal ganglia
Subitizing, 52
  as embodied capacity, 51, 351
  extensions, 54
  as inborn ability, 19–21
  and innate arithmetic, 96
  and neural structure, 99
  and object collections metaphor, 422
  of small numbers, 29
Subitizing plus, 95

Subject matter, 372
  of mathematics, 351, 354
  multiple versions of, 355
  in terms of sets, 369
Subject-matter frame, 167
Substance, 108
Subtraction, 57, 61
Superinfinitesimal number, 248
Support schema, 30
Symbolic calculation, 97
Symbolic domain, 129
  and infinite decimals, 186
  of propositional calculus, 133
Symbolic logic, 8
  development of, 130
  to discretize reason, 261
  implications of, 121
  mapping, 133, 134, 138
  as propositional calculus, 132
  requiring cognitive analysis, 5
Symbolic numerical abilities, 23
Symbolization, 95
  capacity for, 52, 102–103, 351
  mappings, 99, 370
  mathematical, 120
  of operations on classes, 129
Symbols, 129
  manipulation of, 137
  mathematical, 49
  for numbers, 83, 84
  used by chimpanzees, 22
Symmetrical point, 90
  -location, 92
  rotation to, 91
Symmetry, 91
Syntactic forms of iterated action, 157

Talmy, Len, 38, 163, 191, 198
Target domains, 42
  arithmetic, 55, 65, 68, 72
  classes, 123, 125, 128
  functions, 386
  geometry rotations, 116
  and inferential structure, 48
  iterative processes, 159, 162, 165, 168, 172
  mathematical ideas, 371
  modular arithmetic, 113
  naturally continuous space, 311
  natural numbers, 142

new elements in, 46
numeration, 143
ordered pairs, 141
ordinal numbers, 218
predication, 123
propositional logic, 132
trigonometric functions, 393, 396
Tautology, 137
Taylor series, 444
  for functions, 443, 445
Teaser elements, 196
  convergence, 192, 193, 194
  and critical elements, 195
Theodorus of Cyrene, 71(fig)
Theological importance of infinity, 162
Thought, 5
  as mathematical calculation, 357
Time and distance, 315, 407
Topographic maps of visual fields, 134
Totalities involving sets, 175
Total ordering axiom, 200
Touching of points on a line, 271
Trajector, 33
  of image schema, 31
  landmark relation, 37
  in motion, 191
Transcendent mathematics
  existence of, 342
  and objective truth, 363
  philosophical paradigm of, 343
Transfinite addition, 221
Transfinite cardinals
  and Cantor's metaphor, 214
  and elements in infinite set, 221
  hierarchy, 216
  as infinite numbers, 254
  and line at infinity, 368
Transfinite numbers, 290
  defined, 208
  embodied, 222
  formed by Cantor's metaphor, 216
Transfinite ordinal numbers, 217, 220
  arithmetic of, 219
  hierarchy of, 220
Triangle rotations, 114, 116, 117
Trigonometric Complex Plane blend, 437
Trigonometric functions, 393, 396
Trigonometry, 397
Trigonometry blend, 396
Trigonometry metaphor, 393, 433
  and angles as numbers, 387

entailments, 394
and unit circle, 450

Unconscious conceptualization
  of mathematics, 29
  metaphors, 55
  system, 108, 180, 339
Unconscious memory, 27–28
Unending process, 266
  for creating infinitesimals, 248
  literal, 160
Unending sequence of integers, 166
Union
  of a class with itself, 126
  commutative law for, 132
  of set size of natural numbers, 219
Unique mathematics, 374
Unit circle, 170
  on Cartesian plane, 48, 388
  circumference of, 439
  in complex plane, 449
Unit Circle blend, 388
  entailment of, 393
  first stage of, 389(fig)
  and sine/cosine functions, 395
  as source domain, 393, 396
  Stage 3, 390, 392(fig)
Unit square, 287
Unity, 75
Universal class, 131
  of Boole, 127, 130
  as correlated to one, 124–125
  metaphorical creation of, 129
Universality of external objects and
    mathematics, 350, 351
Upper bound, 201, 202

Venn diagrams, 45
  for Boolean classes, 122
  and member concepts, 145
  of propositional calculus, 134
Verbs, 36
Violation-of-expectation paradigm, 16, 22
Visual systems, 33, 34
Von Neumann, John, 141, 150
  axiom of Foundation, 146
Von Neumann metaphor, 148

Weierstrass, Karl Theodor, xiv, 305, 308,
    322, 407
  arithmetization of calculus, 293

characterization of derivative, 314
descretizing continuity, 292, 310, 408
elimination of geometry, 315
implicit conceptual metaphors, 309
linkage between continuity and limit, 313
nongeometric arithmetization, 230
nongeometric paradigm, 315
notation for concept of continuity, 313
notation for concept of limits, 312
Weierstrass's continuity, 320
and preservation of closeness, 318
versus natural continuity, 374
Weierstrass's Continuity metaphor, 311
Weierstrass's paradigm, 316, 318
Weyl, Hermann, xiii, 323, 324
Wholeness, 75
Wigner, Eugene, 3
Wilder, R., 356
Writing systems, 378

Zermelo-Fraenkel axioms, 145, 148
Zero, 366
Boole's arithmetic law for, 124
on Cartesian plane, 170
as empty collection, 366
as empty set, 343
as infinitesimal, 250
and least upper bound, 202
meanings of, 75
as number, 366
and one as natural correlates, 124
as point-location, 73, 97
Zero Collection Metaphor, 64
Zero-diameter discs as points, 270, 271, 272
Zero-dimensional geometric objects, 342, 343
Zeroes (two), 360, 361
Zero Object Metaphor, 67
ZFC axioms. *See* Zermelo-Fraenkel axioms
Zooming In Is Multiplication metaphor, 234

A Note on the Text

Trump Mediaeval was designed by Georg Trump for the Weber foundry and released between 1954 and 1960. A student of F. H. Ernst Schneidler, Trump was a prolific type designer who considered himself first and foremost a teacher of the graphic and lettering arts. With its crisp angularity and wedge-shaped serifs, Trump Mediaeval appears carved in stone. It is a strong text typeface that is highly legible. Mediaeval is the German term for oldstyle; this design is a modern rethinking of the oldstyle theme.